OCR
Design &
Technology
For A Level

Official Publisher Partnership

OCR
Design &
Technology
For A Level

John Grundy
Denis Hallam
Mike Hopkinson
Sharon McCarthy

HODDER
EDUCATION
AN HACHETTE UK COMPANY

Orders: please contact Bookpoint Ltd, 130 Milton Park, Abingdon, Oxon OX14 4SB. Telephone: (44) 01235 827720. Fax: (44) 01235 400454. Lines are open from 9.00 – 5.00, Monday to Saturday, with a 24 hour message answering service. You can also order through our website www.hoddereducation.co.uk

If you have any comments to make about this, or any of our other titles, please send them to educationenquiries@hodder.co.uk

British Library Cataloguing in Publication Data
A catalogue record for this title is available from the British Library

ISBN: 978 0 340 96634 1

First Edition Published 2008
Impression number   10 9 8 7 6 5 4 3 2
Year   2012 2011 2010 2009

Hachette Livre UK's policy is to use papers that are natural, renewable and recyclable products and made from wood grown in sustainable forests. The logging and manufacturing processes are expected to conform to the environmental regulations of the country of origin.

Cover photo © Getty Images/Digital Vision/Tom Brakefield
Typeset by Fakenham Photosetting Ltd, Fakenham, Norfolk
Printed in Italy for Hodder Education, an Hachette UK Company, 338 Euston Road, London NW1 3BH

Contents

HODDER
EDUCATION
The Expert Choice

What does 'the expert choice' mean for you?

We work with more examiners and experts than any other publisher

● Because we work with more experts and examiners than any other publisher, the very latest curriculum requirements are built into this course and there is a perfect match between your course and the resources that you need to succeed. We make it easier for you to gain the skills and knowledge that you need for the best results.

● We have chosen the best team of experts – including the people that mark the exams – to give you the very best chance of success; look out for their advice throughout this book: this is content that you can trust.

More direct contact with teachers and students than any other publisher

● We talk with more than 100 000 students every year through our student conferences, run by Philip Allan Updates. We hear at first hand what you need to make a success of your A-level studies and build what we learn into every new course. Learn more about our conferences at **www.philipallan.co.uk**

● Our new materials are trialled in classrooms as we develop them, and the feedback built into every new book or resource that we publish. You can be part of that. If you have comments that you would like to make about this book, please email us at: **feedback@hodder.co.uk**

More collaboration with Subject Associations than any other publisher

● Subject Associations sit at the heart of education. We work closely with more Associations than any other publisher. This means that our resources support the most creative teaching and learning, using the skills of the best teachers in their field to create resources for you.

More opportunities for your teachers to stay ahead than with any other publisher

● Through our Philip Allan Updates Conferences, we offer teachers access to Continuing Professional Development. Our focused and practical conferences ensure that your teachers have access to the best presenters, teaching materials and training resources. Our presenters include experienced teachers, Chief and Principal Examiners, leading educationalists, authors and consultants. This course is built on all of this expertise.

How to get the most out of this book

Introduction

Welcome to OCR Design and Technology for A level. This book has been written to meet the specification requirements for OCR's GCE in Design and Technology: Product Design (specification numbers H053 AS level, and H453 A level).

Keep the book with you for easy reference as it will be an essential guide for the course. It contains clear guidance and support material for all of the units in the specification. It will be an extremely effective resource in helping you to prepare for assessment and will greatly increase your chances of being successful.

This book:
■ Has been designed to enable you to find information quickly and easily

■ Is focused on the OCR specification with further guidance, examples and extension resources supplied on the accompanying website

■ Has relevance and value to other GCE Design and Technology specifications

■ Has a page-by-page guide to clarify exactly what is expected of you in the coursework units.

A team of writers and contributors have brought their specialist knowledge and considerable experience in teaching Design and Technology to this book. The different approaches of the writers and the varied nature of the subject matter in each chapter have resulted in a range of styles throughout the book. It is hoped that this will make the book inspirational as well as informative.

The book gives an outline of the specification and explains how skills, knowledge and understanding will be developed and extended as you progress through the course.

Each of the four units is explained in detail. Your school or college will plan a programme of delivery and the units can be accessed at the appropriate time to support your learning. Further support, including assessment guidance sheets, Microsoft PowerPoint® presentations and exemplar material, can be found on the website.

Unit F521 Advanced Innovation Challenge

Chapter 1 includes details of the use of themes to instigate research and the timing of examined sessions. It explains the use of job bags to support design thinking and looks at the workbook section by section.

Unit F522 Product Study

Chapter 2 explains what you need to do, section by section, and includes details of exercises and introductory tasks to help to prepare you for your study. Specific reference is made to the assessment criteria requirements, such as moral implications including sustainability, user needs, manufacturer needs, testing and aesthetics. It suggests ways in which design activity can be recorded in real time.

Unit F523 Design, Make and Evaluate

Chapter 3 follows a similar format to Chapter 2. It explains the requirements of the unit section by section and includes exercises and introductory tasks to help prepare for this major coursework unit. Specific reference is made to the assessment criteria requirements including use of information, inspiration and influences to support design activity; preparing a marketing presentation; and completing the project with a review and reflection.

Unit F524 Product Design

Chapter 4 will help to prepare you for your written examination. It describes the style of the paper and gives examples of questions and mark schemes. Examiner tips are given to help you focus on the key requirements and employ specific examination techniques.

It is acknowledged that design and technology covers an extremely wide range of knowledge and understanding. The content section of the specification focuses on what is deemed to be reasonable for a student to be taught over a two-year course. Centres are able to introduce additional areas, develop specialist skills and extend the knowledge and understanding to deliver a course that meets the centre and its students' expectations.

Chapter 5 covers the core content of the specification, which all candidates follow. Chapters 6 to 13 cover the specific material content for the eight focus areas:

- Built environment and construction
- Engineering
- Food
- Graphic products
- Manufacturing
- Resistant materials
- Textiles
- Systems and control

Your centre will decide which focus areas that they will prepare you for.

You will need to have covered all of the content in at least one of the focus areas. You may wish to study the content for more than one focus area in order to give you more of a choice in the written examination.

Some focus areas, such as resistant materials, engineering and manufacturing, have some overlapping content. Where this occurs, follow the links and references between the chapters and sections.

Only the topics and information contained in the core and focus area content will be used to set questions in the June and January examination sessions.

Key Point

- Key points boxes list key aspects of a given topic.

Key Term

Key terms boxes clarify and define the technical terms used in a particular field of study.

Activities boxes suggest interesting and challenging tasks and activities, focused on the requirements of the specification, to support, enhance and extend learning opportunities.

Questions boxes provide practice questions, of the appropriate demand and rigour, to test key areas of the content of the specification. Specific guidance relating to question responses is given in Chapter 4.

EXAMINER'S TIPS

Examiner's tips boxes give specific tips to improve performance in each area of the specification. All tips are written by senior examiners and moderators of the OCR Design and Technology: Product Design course.

Website boxes are used where there is further information or resources on the website accompanying this book. The website is www.hodderplus.co.uk/ocrd&t.

Acknowledgements

Every effort has been made to trace and acknowledge ownership of copyright. The publishers will be happy to make arrangements with any copyright owners that it has not been possible to contact.

The authors would like to thank the following: Liz Barnett; Dave Barnwell; Terry Bream; Jason Davies; Doug Drane; Ken Edwards; Mike Facey; Doug Hales; John Houghton; Amanda Howard; Andrew Lees, Partner Edan Print Services; Jayne March; Johnsons Prams; Land Rover; Alan Padbury, Managing Director of Westdale Press; Practical Action; Meryl Simpson; Paul Tallett, Managing Director Kall Kwick; George Taylor; Traidcraft; Whitaker Engineering; Ashlawn School, Rugby; Bedford Modern School, Bedford; Churston Ferrers Grammar School, Brixham, Devon; City of London School for Girls, London; Colyton Grammar School, Colyton, Devon; East Barnet School, Barnet; John Leggott Sixth Form College, Scunthorpe; King Edward's School, Birmingham; King Henry VIII School, Coventry; Nottingham High School for Girls, Nottingham; Ounsdale School, Wolverhampton; Sherbourne School for Girls, Sherbourne; Silcoates School, Wakefield; St Mary's Catholic High School, Chesterfield.

The authors and publishers would like to thank the following for the use of photographs in this volume:

p.56 © AGphotographer- Fotolia.co; p.58 © KonstantinosKokkinis - Fotolia.com; p.59 (Fig 3.9) © Kwest - Fotolia.com; p.60 (Fig 3.12) © Sam Bailey/Hodder Education; p.65 (right) © PASQ - Fotolia.com; p.100 (Fig 5.3) Kim Steele/Workbook stock/Jupiter Images; p.100 (Fig 5.4) © Shaun Lowe/istockphoto.com; p.101 (Fig 5.5) TechSoft UK Ltd; p.103 (Fig 5.10) and p.319 (Fig 10.14) © Bulent ince/istockphoto.com; p.103 (Fig 5.11) Brand X Pictures/Photolibrary; p.117(Fig 5.19) © Beboy - Fotolia.com; p.117 (Fig 5.20) © Ali Taylor - Fotolia.com; p.118 © sallydexter - Fotolia.com; p.131 © CSeigneurgens - Fotolia.com; p.134 Paul Giamou/Workbook stock/Jupiter Images; p.135 (top) © Yuri Arcurs - Fotolia.com; p.140 (Fig 5.55) © TRADA; p.5.56 (Fig 5.56) © British Standards Institution; p.145 © Christian Delbert - Fotolia.com; p.179 (Fig 7.10) © Willian D Fergus McNeill/istockphoto.com; p.243 (Fig 8.1) © Crown Copyright material is reproduced with the permission of the controller of HMSO and Queen's Printer for Scotland; p.287 (Fig 9.29) © narvikk/istockphoto.com; p.287 (Fig 9.28) © Kall Kwick; p.292 (Fig 9.44) © narvikk/istockphoto.com; p.292 (Fig 9.45) © Olga Ermolaeva/istockphoto.com; p.294 (Fig 9.50) © Vladimir Caplinskij/istockphoto.com; p.295 (Fig 9.57) © Julie Deshaies/istockphoto.com; p.295 (Fig 9.58) © Johnny Scriv/istockphoto.com; p.296 (Fig 9.61) © Johnny Scriv/istockphoto.com; p.296 (Fig 9.66) © Johnny Scriv/istockphoto.com; p.297 (Fig 9.70) © Bart Sadowski/istockphoto.com; p.297 (Fig 9.71) © Alice Day/istockphoto.com; p.298 (Fig 9.72) © L. R. Kyllo/istockphoto.com; p.298 (Fig 9.73) © Asli Cetin/istockphoto.com; p.298 (Fig 9.75) © Edward Grajeda/istockphoto.com; p.299 (Fig 9.77) © Nicola Stratford/istockphoto.com; p.299 (Fig 9.78) © Valerie Loiseleux/istockphoto.com; p.299 (Fig 9.79) © Jamie Cross - Fotolia.com; p.299 (Fig 9.80) © Tom Mc Nemar/istockphoto.com; p.300 (Fig 9.81) © Joe Gough/istockphoto.com; p.300 (Fig 9.83) © Alina Pavlova/istockphoto.com; p.304 (Fig 9.106) © AVTG/istockphoto,com; p.304 (Fig 9.107) © fabphoto/istockphoto.com; p.305 (Fig 9.109) © dotshock - Fotolia.com: p.309 (Fig 9.121) © Nathan Winter/istockphoto.com; p.309 (Fig 9.122) © Tony Kwan/istockphoto.com; p.309 (Fig 9.132) © Don Bayley/istockphoto.com; p.311 (Fig 10.1) © Glen Bowden/istockphoto.com; p. 311 (Fig 10.2) © Eric Thompson/istockphoto.com; p.311 (Fig 10.3) © Kevin R. Morris/CORBIS; p.312 © kativ/istockphoto.com; p.313 (Fig 10.5) © caraman - Fotolia.com; p.313 (Fig 10.6) © Arno Massee/istockphoto.com; p.314 © PDL Design/istockphoto.com; p.315 (Fig 10.8)

© Kevin Jarrat - Fotolia.com; p.315 (Fig 10.9) © Thor Jorgen Udvang - Fotolia.com; p.318 (Fig 10.11) © Sheldon Kralstein/istockphoto.com; p.318 (Fig 10.12) © Milos Luzanin/istockphoto.com; p.319 (Fig 10.13) © Dainis Derics/istockphoto.com; p.319 (Fig 10.14) © Bulent Ince/istockphoto.com; p.356 (Fig 11.10) © Kennymac - Fotolia.com; p.357 (Fig 11.12) © Einar Bog - Fotolia.com; p.357 (Fig 11.13) © megasquib - Fotolia.com; p.358 (Fig 11.14) © pamtriv - Fotolia.com; p.359 (Fig 11.16) © Finishing magazine; p.359 (Fig 11.17) © Finishing magazine; p.359 (Fig 11.19) © Rohit Seth - Fotolia.com; p.404 © TIM MCCAIG/istockphoto.com; p.405 (Fig 13.2) © Jojo 100/istockphoto.com; p.405 (Fig 13.3) © Doug Cannell/istockphoto.com; p.405 (Fig 13.4) © Heidi Anglesey/istockphoto.com; p.407 (Fig 13,5) © B Sieckmann/istockphoto.com; p.408 (Fig 13.7) © Yasin GUNEYSU/istockphoto.com; p.408 (Fig 13.8) © Duncan Babbage/istockphoto.com; p.409 (Fig 13.11) © Udo Feinweber/istockphoto.com; p.411 (Fig 13.13) © P Wei/istockhoto.com; p.411 (Fig 13.5) © Terraxplorer/istockphoto.com; p.411 (Fig 13.16) © AVTG/istockphoto.com; p.413 (Fig 13.20) © Shaun Lowe/istockphoto.com; p.413 (Fig 13.21) © Ben Blankenburg/istockphoto.com; p.414 (Fig 13.22) © Jason Lugo/istockphoto.com; p.415 (Fig 13.24) © Jim DeLillo/istockphoto.com; p. 415 (Fig 13.25) © Marco Onofri/istockphoto.com; p.417 © Christopher Steer/istockphoto.com; p.420 (Fig 13.32) © Sergey Lemeshencko/istockphoto.com; p.421 (Fig 13.35) © Gabriela Abratowicz/istockphoto.com; p.426 © David H. Lewis/istockphoto.com; p427 (Fig 13.45) © Jacob Wackerhausen/istockphoto.com; p.427 (Fig 13.46) © ROBERTO CAUCINO/istockphoto.com; p.437 (Fig 13.77) © Tom Brown/istockphoto.com.

Advanced Innovation Challenge

Introduction to the Advanced Innovation Challenge

This paper is a seven-hour design challenge set by OCR. It is undertaken in two three-hour sessions over the period of a day and a one-hour session at a later date.

The Advanced Innovation Challenge provides opportunities for research but also allows you to demonstrate your creativity and innovation when reacting to a design situation (challenge).

All sessions are to be completed on dates set by OCR. The task assesses your ability to be innovative, demonstrate flair, work with materials and apply knowledge gained throughout the AS course.

Fig 1.1

Fig 1.2

A theme is released in the September prior to the examination. Each theme runs for a year, enabling you to research and gather resources to form a personal handling collection. You will work with materials in either a supervised examination room or a design workshop.

Fig 1.3

Fig 1.4

- Session 1 – with your handling collection as inspiration you will produce a specification and design brief using annotated sketches, notes and models to describe your ideas. Then you will choose one design to take forward into Session 2
- Session 2 – you will model your chosen design and further your ideas in a workbook, evaluating your final design and model against the original specification
- Session 3 – you will have the chance to reflect on the product designed in Sessions 1 and 2 in a written examination

There is a teacher script that guides the activities through Sessions 1 and 2.

The real benefit of this unit is that you are free to explore original ideas and demonstrate creativity and innovative skills more readily than in a project. Innovation is the key as the name suggests. You will also be working under pressure for a set time and this reflects what happens in the real world of design.

The workbook is linked to a teacher script and has sections that allow the challenge to be undertaken in a structured manner. This chapter

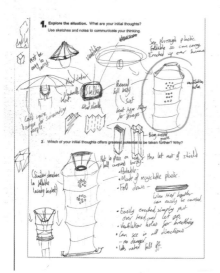

Fig 1.5, 1.6, 1.7 Example workbook pages

DESIGN AND TECHNOLOGY: UNIT F521 TEACHER GUIDANCE

SESSION 1	ORGANISATIONAL INSTRUCTIONS	MUST SAY	SAY IN YOUR OWN WORDS
THE FOLLOWING SECTION SHOULD BE DONE IN BOX 3, 4 AND 5 IN YOUR ANSWER BOOKLET			
3 mins	Box 3	Open out the last page to reveal boxes 3,4 & 5 Complete the following in box 3 A design brief.	Think before you fill out the boxes. Consider your options. Select a design brief that you will enjoy doing. One that will show your creative ability.
4 mins	Box 4	Examine the contents of your job bag remembering your design brief. Identify 'Key Points' which will help you write your specification.	Explain that Box 3, 4 and 5 will be visible throughout the challenge and they should refer to them from time to time.
12 mins	Box 5	Produce a detailed and justified specification.	

NB The Summary Sheet for F521 is for teacher/candidate use only.
The Unit is externally assessed by OCR

Session 1	Marks
1.1 Initial thoughts, design brief and specification	**9**
Clearly outlines initial thoughts in detail, responding with an open mind showing unexpected and/or challenging ways of thinking. Analyses problems in depth, responds in a way that allows scope for innovation. Identifies a user/market resulting in a clear design brief. Develops a detailed specification that identifies the key features of the product.	7-9
Outlines initial thoughts in some detail, some creative thinking. Analyses some aspects of the problem. Identifies a user/market, resulting in a design brief. Develops an adequate specification that gives some basic requirements of the product.	4-6
Outlines initial thoughts, these are predictable/non creative. Analyses the problem at a superficial level that lacks depth. Some consideration of a user/market resulting in a design brief. Produces a basic or superficial specification that is vague/generic.	0-3
1.2 Designing	**12**
Presents a wide range of innovative/creative initial ideas, using high quality annotated sketching showing full details of construction/materials. Presents a wide range of evidence to show the sources of inspiration and influences on the designing . Presents a detailed and objective evaluation of ideas against the design specification and justifies all decisions. Reflects on their chosen design and responds to feedback from others, making further improvements if necessary.	9-12
Presents a good range of innovative/creative ideas using reasonable quality annotated sketching showing some detail of construction/materials. Presents an adequate range of evidence to show the sources of inspiration and influences on the designing. Presents an adequate and objective evaluation of ideas against the design specification and justifies most decisions. Some reflection on their chosen design and response to feedback from others.	5-8
Presents only a limited range of innovative/creative ideas using annotated sketching at a limited level with little detail of construction/materials. Little or no reference made to the design specification. Presents a limited range of evidence to show the sources of inspiration and influences on the designing. Presents only a limited and mainly subjective evaluation of ideas with little or no justification of decisions. Limited reflection on their design and little/if any response to feedback from others.	0-4
1.3 Development and planning	**9**
Presents improvements, presents evidence of modelling, experiments, testing, making modifications their design to define and refine it, thorough consideration of materials, components or ingredients and methods of manufacture. Produce a detailed action plan for Making, to include a list of materials/ingredients/resources etc.	7-9
Presents improvements, presents some evidence of modelling, experiments, testing, making modifications their design, some consideration of materials, components or ingredients and methods of manufacture. Some consideration sustainability issues. Produce a reasonable action plan for making, to include a list of materials/ingredients/resources etc.	4-6
Presents limited improvements, and limited evidence of modelling, experiments, testing, and modifications their design, little if any consideration of materials, components or ingredients and methods of manufacture. Limited understanding of sustainability issues and how this affects their design. Produces a simplistic action plan for making, that shows limited awareness of materials/ingredients/resources etc.	0-3
Total	**30**

Fig 1.8, 1.9 Example teacher script page and mark scheme

will go through those sections and the mark scheme. It will also tell you what your teacher will say. The full mark scheme, teacher script and workbook can be found on OCR's website (www.ocr.org.uk).

Each centre can alter time allocations suggested in the script to suit their own preferences provided that the total time allowed is not exceeded for each session. It is worth practising example challenges.

You will model your most creative and exciting

idea using a range of easy to handle materials. Depending upon the activity, you can use paper, card, thin plastics, fabric, wire, foil, thin metal sheet, clay, polymorph, foam board, food ingredients, components and joining devices.

The tables overleaf are a summary of the activities in the workbook. They are taken from the specification. Your centre can adjust timings in the first two sessions as long as it's within the overall time allowed for each session. The last column shows alternative timings that could be used.

Fig 1.10, 1.11, 1.12

Table 1.1 Session 1

Reference to OCR workbook box number	Activity	Time allocation	Alternative timings
Member of staff gives introduction to the challenge		5	5 Context
The use of the job bag		5	5
1	Explore the situation	5	10
2	Initial thoughts	5	5
3 & 4	Your design brief – key points	10	10
5	Your design specification	10	12
6	Start designing product ideas	40	35
7	What do you think of your ideas so far?	10	10
8	What is your best idea?	5	5
Break (15 minutes)			
Member of staff gives introduction to presentations		2	2
9	Reflect and record Group presentation planning	10	7
Presentations	(minimum three minutes presentation, maximum five minutes)		10–15
10	Feedback Record suggestions made by others	10	8
11	Improvements and modifications	40	35
Introduction to modelling materials	Practice should mean you don't need this!		
12	Your model	10	9
13	Action plan for Session 2	10	8

Table 1.2 Session 2

Reference to OCR workbook box number	Activity	Time allocation	Alternative timings
14	Review	10	10
15	Modelling	40	40
Progress report 1 (photo)		10	5
16	Continue modelling	30	40
Progress report 2 (photo)		10	5

Table 1.2 continued

Reference to OCR workbook box number	Activity	Time allocation	Alternative timings
Break (15 minutes)			
17	Plan for last 60 minutes	10	5
Final modelling session		50	50
18	Evaluation	10	25
	Evaluation against specification	10	

Research skills

The Advanced Innovation Challenge requires you to research a set theme and then respond to a challenge under timed examination conditions.

A theme is released in the September prior to the examination. Each theme runs for a year enabling you to research and gather information and resources forming a personal handling collection. This is then taken into the sessions to provide inspiration when addressing the chosen challenge. An example of a theme might be:

SPORTING EVENTS

London often hosts large international sporting events. An organising committee of these events has stated that they wish to stage 'inspirational events that capture the imagination of young people around the world and leave a lasting legacy'.

The handling collection/resources will be used in Sessions 1 and 2 of the challenge – it is worth bearing in mind that a carefully selected and edited set of 'relevant materials will be easier to use during the challenge than masses of material selected without sufficient thought being given to its possible value in the challenge'.

Preparing for the challenge

It is worth producing a mind map with the rest of your group. Try to put yourself in the situation and think of possible products that would be needed and think of similar situations and products that are available. Mind maps are used to generate, visualise, organise and classify ideas.

Do not try and guess what the challenges could be! Keep an open mind and gather information and resources that will inspire and assist you when designing. Guessing will only narrow your research and make it hard for you to respond to the challenge on the day and be innovative.

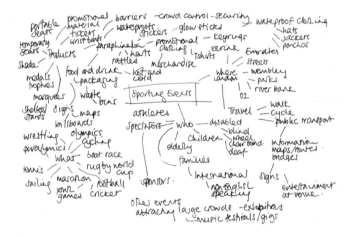

Fig 1.13 Mind map 1

Think about useful research and the techniques you could use to gather this. Research can be done at home, in school, or out and about. Think about things that may help you respond to the challenge.

Key Term

A **mind map** is a diagram used to represent words, ideas, tasks or other items linked to, and arranged around, a central key word or idea.

Primary and secondary information

It is useful to think in terms of primary and secondary information.

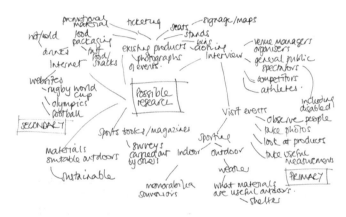

Fig 1.14 Mind map 2

Primary research

Primary research should record your observations of people in similar environments, using similar products, and opinions of the potential end users.

Think of places you could visit to gather information that may be useful. Think of materials and construction methods that may be suitable for use in the suggested environment, as well as safety and surface finishes. Collect samples that may be useful. Could you interview anyone to help you understand the environment and potential issues further? Can you visit specialist shops or venues?

Secondary research

Secondary research should include information from books and the internet. It may include

useful information such as anthropometric data, ergonomics, nutrition tables and so on, as well as photographs of products and information about materials.

Don't forget to think about social, environmental and moral issues too.

Any information from primary and secondary sources needs to be carefully selected and edited. You will need to think carefully about how to organise this research so that you can access it easily and quickly in the exam.

It may be that some information would be best in a mood or inspiration board. Pictures can be a good substitute for real products – group similar products together or things that have certain characteristics, for example materials, together. Lower cost or free products that are small may be worth taking into the exam in a job bag or ideas box.

You need to develop your own handling collection/resources and you will have to record how you used it later in the challenge. However, it is worth sharing ideas and experiences with your group and your teacher regularly.

SESSION 1

Section 1.1   Initial thoughts, design brief and specification

Responding to the challenge – initial thoughts

- Analyse the problem in depth. Respond in a way that allows scope for innovation
- Clearly outline initial thoughts in detail. Respond with an open mind showing unexpected and/or challenging ways of thinking

Teacher's instructions to candidates

Read through the challenge sheet. Select one situation from the list, which you will solve during this examination activity. Cross out the situations that you are not undertaking.

You should have your 'job bag' available for Sessions 1 and 2 of this examination. The first thing we would like you to do is to put some of your first thoughts down on paper.

Remember, we want you to be as creative as possible, so sketch and add notes of any ideas you have, even if they seem a bit risky or outrageous at this stage.

We really want you to feel able to 'let your mind go out to play'. In this box put down your initial thoughts. You have eight minutes, so work quickly. Try and remain focused.

Look at your initial thoughts. Think about the situation. Which of your initial thoughts has the greatest potential to be taken further? Why?

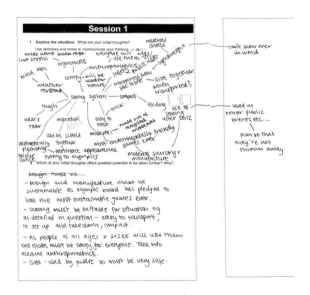

Fig 1.15

There are different ways of approaching this section as you can see from the examples. Mind mapping allows you to work quickly and get lots of ideas down; it will help you to think about information that could be useful to you and issues you will need to consider.

If you sketch, remember these are initial thoughts that you may have before you write your design brief.

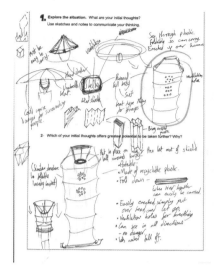

Fig 1.16

EXAMINER'S TIPS

You can show your thoughts in any format you want, for example a mind map or through sketches and notes. But don't be afraid to put things down, remember this is the innovation challenge.

Design brief

- Identify a user/market resulting in a clear design brief.

A design brief is a set of instructions for a designer and is written once the planning ideas and information have been consolidated. It should outline the extent of the product and confirm particular ideals and elements that are important. The design brief does not present the solutions for the product.

It should start with: 'I am going to design and model a…'; this needs to be followed by a general description of the type of products you feel will answer the challenge you have chosen.

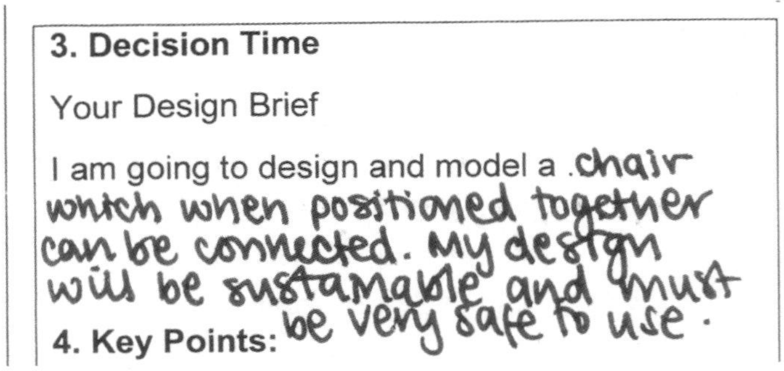

3. Decision Time

Your Design Brief

I am going to design and model a *chair which when positioned together can be connected. My design will be sustainable and must*

4. Key Points: *be very safe to use.*

Fig 1.17

The brief should be a general description that allows you flexibility regarding the type of product you intend to make. Do not be specific about materials but describe the materials or properties of product, for example strong, tough, flexible, natural, man-made, recycled, waterproof, healthy, or a similar general description. You could mention points such as safety, general size, what it will do (its functions), general properties of the materials needed, who it is for (for example children), basic cost of manufacture or a lower and upper cost limit, and other points you feel are important. This can be done in the **key points** section. The key points could be presented in the form of a mind map or as a bullet point list as in the example.

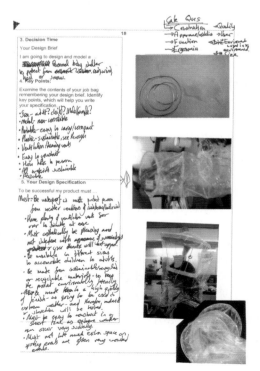

Fig 1.18

Teacher's instructions to candidates

Open out the last page to reveal boxes 3, 4 and 5. Complete a design brief in box 3.

Examine the contents of your job bag, remembering your design brief. Identify key points which will help you write your specification.

Fig 1.19

EXAMINER'S TIPS

- Think before you fill out the boxes
- Consider your options. Select a design brief that you will enjoy doing, one that will enable you to show your creative ability
- Boxes 3, 4 and 5 will be visible throughout the challenge: refer to them regularly when designing to ensure you are on track

Specification

- Develop a detailed specification that identifies the key features of the product.

After producing your design brief you are required to produce a design specification. This is a list of requirements that your design must meet. It may also contain a wish list of possible features. Below are some points you may find useful to include in your specification.

Function

What do you want the object to do? How well do you want it to do it? How often? How quickly?

Environment

Think about where your product will be used: what temperature range it will be exposed to, will it be used outdoors, might it suffer corrosion in contact with fluids, and so on? This will affect your choice of materials and finishes.

Environment (sustainability)

Consider your choice of material and manufacturing process, life-cycle analysis and ease of recycling. Could your product be reused? Can you reduce the number of components and different materials?

Quantity, construction, materials and finishes

The scale of production has an effect on the manufacturing processes and materials used. What quantity of your product is likely to be produced? Are certain properties of materials desirable? Are there reasons for not using a particular material or finish?

Product service life and maintenance

Are you expecting the product to perform its task 24/7 or once in a lifetime? Is regular maintenance needed?

Cost

Target costs can be checked against existing or similar products. Costs will influence your choice of materials and manufacturing processes.

Ergonomics

All products have a human interface. It is therefore important to take into account ergonomic considerations and anthropometric data, for example human sizes for ease of use, posture, reach, forces, colours, and so on.

Size and weight

There are likely to be size constraints for your product that must be considered. This could be due to where the product is going to be used. Weight is an issue for anything portable.

Aesthetics

We see it first, and try it later. Is it important that it stands out or is visually striking, or should it be discreet? Try to specify colour, shape, form and texture of finish.

User

The end user is important – what are their likes and dislikes, and specific requirements? In project work we usually carry out questionnaires and interview potential users.

Quality and standards

Consider the aspects of your product for which quality and reliability are important. Does the product need to meet British or European standards?

Safety

This links in with standards, but you should consider things like instructions and warning labels. You should consider what the implications could be if your product was misused.

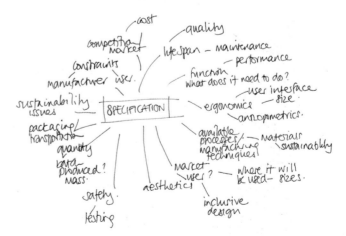

Fig 1.20 Mind map showing elements of a design specification. Models such as 'Pugh's Plates' can be used to structure your specification; the designer has to consider all aspects and try to continually balance all issues when designing

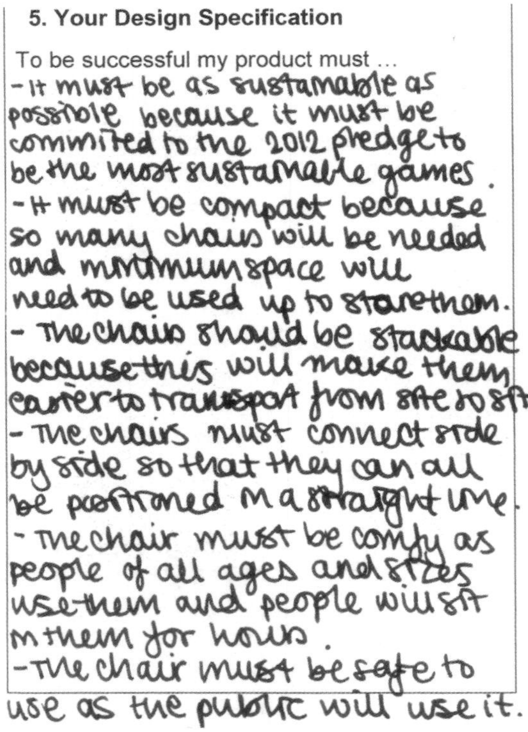

5. Your Design Specification

To be successful my product must …

- It must be as sustainable as possible because it must be committed to the 2012 pledge to be the most sustainable games.
- It must be compact because so many chairs will be needed and minimum space will need to be used up to store them.
- The chair should be stackable because this will make them easier to transport from site to site.
- The chairs must connect side by side so that they can all be positioned in a straight line.
- The chair must be comfy as people of all ages and sizes use them and people will sit in them for hours.
- The chair must be safe to use as the public will use it.

Fig 1.21 Example of a specification for a seat. Points are qualified (e.g. 'It must be compact because so many chairs will be needed and minimum space will be required to store them')

EXAMINER'S TIPS

- Avoid 'leading' the design and predicting the outcome even if the end product is clear in your mind: it will hamper your creativity in your initial ideas
- Use mnemonics to help you remember points, for example:
  - **C A F E Q U E S**: Construction, Aesthetics, Function, Ergonomics, Quality, User, Environment, Size
  - **Too Many Students Make Poor Errors And Critical Failures**: Target Market, Materials, Size, Manufacture, Performance, Ergonomics, Aesthetics, Cost, Function
- Avoid generic points that can apply to any product
- Qualify all points.

Key Term

Specification – A detailed, exact statement of particulars, especially a statement prescribing materials, dimensions, and quality of work for something to be built, installed, or manufactured.

Design ideas

- Present a wide range of innovative/creative initial ideas using high-quality annotated sketching, showing full details of construction/materials.

In this section you are required to produce a creative range of ideas. It is worth doing exercises to develop quick sketching techniques: it will help to give you confidence and to give you strategies for coming up with the big idea, and with being innovative and creative when generating ideas.

Teacher's instructions to candidates

In a moment you can start designing. You can sketch, make notes, make models and use photographs.

From now on you may use any of your resources at any time. Think about your job bag. Use this as a source of inspiration.

If you use 2D or 3D modelling, or trialling and testing to develop ideas, take photographs and stick them into Box 6. Ensure these photographs are fully annotated.

If there are materials you need that are not here, ask. Remember, from now on you are developing your own ideas.

Approaches/methods for designing

There is more than one way to do it! These approaches can help you think of alternative ideas that are different:

- User-centred design – this focuses on the needs of the end user of the product
- KISS principle – Keep It Simple, Stupid – which aims to keep the idea simple, avoiding unnecessary complications

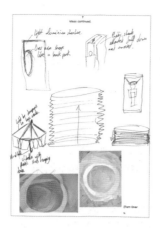

Fig 1.22

Fig 1.23

Fig 1.24

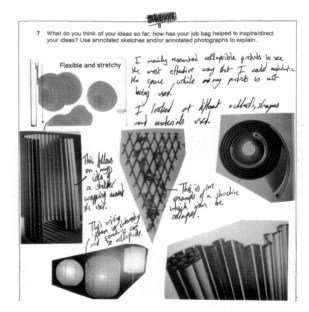

Fig 1.25

EXAMINER'S TIPS

- Focus on the word 'innovative'
- Practise quick sketching techniques and creativity exercises
- Use 3D and 2D sketching and quick modelling to communicate ideas – ensure sketches are detailed and annotated
- Show constructional detail where necessary

Inspiration and influences

- Present a wide range of evidence to show the sources of inspiration and influences on the designing

You also need to show evidence of how your resources/job bag has inspired you.

Teacher's instructions to candidates

Think carefully about your job bag and/or other resources. How have these resources helped with your designing? Use annotated sketches and/or annotated photos to explain how they have been used.

Evaluation

■ Present a detailed and objective evaluation of ideas against the design specification, and justify all decisions.

Your annotation of sketches, ideas and possible solutions should be evaluative as well as descriptive. How successful is the idea likely to be, and why? Explain the strong points of each design idea. Use the specification headings to guide your comments.

Reflection and response

■ Reflect on your chosen design and respond to feedback from others, making further improvements if necessary.

You are required to share ideas and respond to feedback from others. Here are some reasons why it's good to work with other people on a project:

■ To share ideas and have fun
■ When you need expertise or extra skills
■ You may have a better chance of success
■ Everyone sees things differently – you can bounce ideas off each other

Teacher's instructions to candidates

You will have the opportunity to tell others what you are designing (students only). You can present drawings and/or models if it helps to communicate your idea.

Students in your working group can ask you questions. Students in your working group can suggest improvements/developments to your idea.

You will need to plan what you are going to say.

Fig 1.26   Fig 1.27

Teacher's instructions to candidates

Record any suggestions made by students within your working group. How might you modify your idea in response to the feedback?

EXAMINER'S TIPS

● Be sensitive – you're talking about someone's work and they've spent time and effort on it.
● Be constructive – offer practical suggestions for improving it rather than just picking out the faults
● Be encouraging and positive
● Point out strengths as well as weaknesses. Try to begin and end on a positive note – this will make the person more open to the rest of the feedback
● Each person you work with needs to feel good about what they're doing: show you value their contributions

Creativity exercises

The following creativity exercises will help you to develop confidence in being innovative and creative when generating ideas.

4 × 4[i]

This idea originated from the Design And Technology Association (DATA) who, in a recent calendar, produced a range of pre-drawn ideas to use as a starting point. The first person draws a design and then passes it to a second person who uses the first sketch as inspiration and develops it further. This carries on until four boxes are complete.

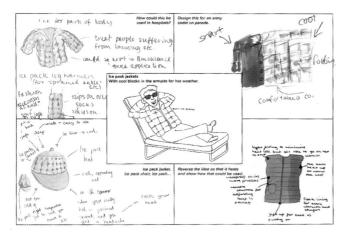

Fig 1.28 Example of 4 × 4[i]

Fig 1.29 Example of 4 × 4[i]

A time of four minutes is suggested for this activity. You may need more than four minutes to sketch when you start but decreasing time to ultimately produce a sketch in four minutes will develop your ability to sketch quickly. The final development can then be discussed. This activity can be done with either a chosen idea before design development or an existing product.

Dare to change – and find a hidden treasure of untapped ideas[ii]

What would happen if a car designer suddenly started producing mobile phones or mp3 players?

There is no doubt that car designers are innovative – they come up with many new ideas, such as side airbags – however, the longer you work in a certain field and the more expertise you have in your area, the harder it is to 'look with new eyes'. Quite simply, familiarity breeds contempt.

Use your own expertise to develop a product in a subject that you have no experience of. Best of all, take a couple of people from different subject areas and see what solution they come up with together. How would a resistant materials student look at a textiles or food problem? And, by the way, what would an mp3 player developed by a leading car designer be like?

> **Key Term**
>
> **Creativity** – 'The creative act is not an act of creation in the sense of the Old Testament, it does not create something out of nothing; it uncovers, selects, reshuffles, combines, synthesises already existing facts, ideas, faculties and skills …The more familiar the parts the more striking the new whole.' Arthur Koestler, *The Act of Creation*

Bring out more ideas

Many regard J. P. Guilford as the father of modern creativity. One of Guilford's first creativity tests was asking people to find as many uses for a brick as possible. Although simple, this is a good way of testing someone's creativity. Some just churn out an endless number of uses faster than you can write them down while others think for minutes before coming up with five uses. This is also a good way of kick-starting the creativity skills of a person or group.

So how many uses for a brick can you think of? Start by trying to come up with 50 different uses in ten minutes. Perhaps try this with a group. What about other everyday objects and their uses, for example a paperclip or a blanket?

Thinking hats[iii]

The six 'thinking hats' method developed by Edward De Bono is used by major companies such as IBM to improve the results of thinking and discussion.

Fig 1.30

The idea is to separate six different types of thinking to help us think more clearly. You should 'wear' each hat in turn, and analyse the idea or issue from six different aspects separately. In a group setting, the entire group is encouraged to use just one hat at a time. Each hat is marked with a different colour and represents the following type of thinking:

- The white hat represents facts and information, as well as identifying information that may be needed to solve the problem and thinking about how we can find it
- The red hat symbolises emotions and feelings. While wearing this hat, you are 'allowed' to express your gut feelings about the subject or idea. This hat does not require justification, as feelings are often subjective rather than rational

- The yellow hat is used to look at the positive aspects of a situation or idea and the benefit of it. Support for the idea should be justified and not simply stated without explanation
- The black hat is, in a way, the opposite of the yellow hat. It is used for discussing the negative aspects of the idea and criticism of arguments made in its support. As in the case of the positive hat, logical justification is expected
- The green hat stands for creativity and unconventional thinking. When wearing this hat, you are encouraged to think creatively. Creative thinking methods can used in order to search for unexpected developments of the idea or the discussion
- The blue hat is used for directing the discussion, for switching hats when necessary, for summarising the major points of the discussion, as well as for making decisions.

Modelling ideas through talking and discussing

This can be used at any stage of designing and making to develop and move ideas or products forward. Get into groups of three or four with your design and/or practical work. You have three minutes to describe the current situation of your work. The rest of the group then have three to five minutes to comment, suggest, evaluate and constructively criticise. Now write down at least three action points to describe what you need to do immediately to develop your work. This technique can be useful in the reflection session of the challenge.

EXAMINER'S TIPS

Try using 'thinking hats' for discussion topics when revising issues for Session 3 of the challenge. They could also be used for A2 'discuss'-type questions or when deciding which idea to develop in a project.

Development

Present your improvements with evidence of:

- Modelling
- Experiments
- Testing
- Making modifications to your design to define and refine it
- Thorough consideration of materials, components or ingredients
- Thorough consideration of methods of manufacture

Teacher's instructions to candidates

You have 40 minutes to develop your idea. You may use any of the resources at any time. Use annotated sketches and/or annotated photographs to communicate your thoughts. Make clear reference to your specification. If you use 2D or 3D modelling, or trialling and testing, to develop ideas, take photographs and stick them into Box 11. Ensure that these photographs are fully annotated.

You now have to develop your idea, refining and defining it to move towards a final solution. Some ways of approaching this could be:

- Use the SCAMPER technique to help generate twists to an idea:
  - **Substitute** – materials, components, and so on
  - **Combine** – mix or combine parts of other ideas
  - **Adapt** – alter the design, change its function or the way it functions, use part of another idea/element
  - **Modify** – reduce or increase the size of the whole design or part or it, or change the shape
  - **Put to another use** – is there an alternative application for your design? Could it have more than one function?
  - **Eliminate** – remove and reduce parts, simplify to basics
  - **Reverse** – turn inside out or upside down
- Use other people's suggestions to adapt your idea
- The mind map below may also help you develop ideas in the challenge or during project work; it can be worked through in any order:

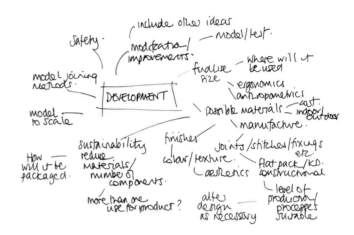

Fig 1.31

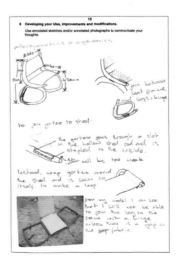
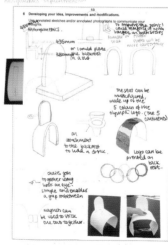

Fig 1.32, 1.33 Examples of student work

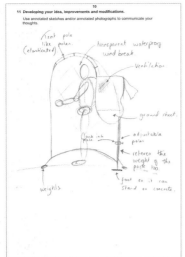

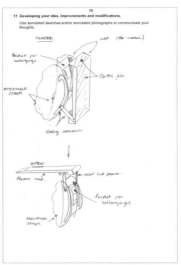

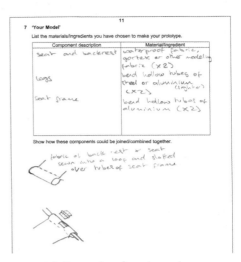

Fig 1.34, 1.35 Examples of student work. Note the different approaches to breaking the product down into parts, modelling and looking at the design as a whole.

Fig 1.36 Example of student plan

Planning

> ■ Produce a detailed action plan for making, including a list of materials, ingredients, resources and so on

Teacher's instructions to candidates

Think about your design proposal. What are the main components? What modelling materials/ingredients are required to model your prototype product? What tools/equipment will be required? How will these components be joined/combined together?

Teacher's instructions to candidates

In Session 2 of the innovation challenge you will be modelling your design proposal. Plan what you need to do in the second session. Do staff need to get anything for you? Do you need to find out anything that will help you next time?

This is not a teaching opportunity. Candidates may find things out for themselves – they must not be taught at this stage

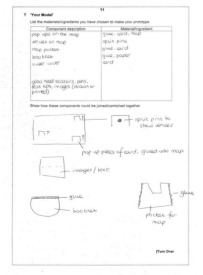

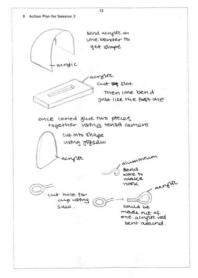

Fig 1.37 Example of student plan

Fig 1.38 Example of student plan

Fig 1.39 Example of student plan

Modelling techniques

During the challenge you will need to model your most creative and exciting idea using a range of easy-to-handle materials. Depending upon the activity you could choose from paper, card, thin plastics, fabric, wire, foil, thin metal sheet, clay, polymorph, play dough, Plasticine, Mod-Roc, foam board, corruflute, food ingredients, components and joining devices, and so on.

During project work you will also need to model ideas in 2D and 3D. The examples here and on the website show some modelling techniques and may help you to think of alternative materials you could use. Scale models can be used to communicate the aesthetics and function of a product.

EXAMINER'S TIPS

- Remember you are modelling – techniques need to be quick and easy. Don't attempt to model using mild steel bar
- In food don't worry if you don't have the exact ingredients, or in textiles the right fabric. Remember it's a model, and the purpose is to communicate your ideas.

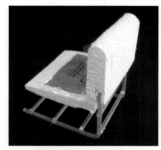

Cardboard and paper can be used to communicate ideas quickly. Why not try to re-use old cardboard boxes and the like?

Modelling can also be used to test principles full size, as in the case of the wind-powered light and spaghetti packaging on the previous pages. Use it to make sure your idea will work before beginning manufacture in a project.

Scale modelling can also be useful, as in the case of the 'folding chair in a carry case' made from corruflute and laser-cut plywood on page 18, and the final scale model of a coffee table from laser-cut plywood and aluminium on page 17.

EXAMINER'S TIPS

Although CAM can achieve a high-quality finish, it may be best to avoid it during the challenge, when working under pressure. It is advisable not to put yourself in a situation where you have to wait to use a machine or piece of equipment.

Textiles modelling can be done in card, newspaper, tissue paper and plastic – both full size and to scale. Styrofoam, wire, card etc. can all add structure.

Don't just think of Styrofoam and card for modelling in resistant materials or graphics: it was used to plan the cake design on page 18.

EXAMINER'S TIPS

Don't worry if your model is crude. If it communicates your idea then it's successful!

SESSION 2

Section 2.1   Making

Making: progress reports

- Record and reflect on progress in detail at various stages

Teacher's instructions to candidates

In Session 2 of the Advanced Innovation Challenge you will be modelling your design proposal. You will have longer periods of uninterrupted time so that you can model your idea. You will be modelling your design so that you and the examiner can see your ideas.

Although you are not working in an examination hall you must still treat this activity as an examination. You should not talk unless you are told to do so by a member of staff.

We will continue to explain everything as we move through each stage of this challenge. You will be using tools and equipment, so your normal workshop risk assessments must apply. Photographs will be taken at the end of each modelling session. Additional photos can be taken at any time and included in your workbook. Additional photographs should be annotated.

Fig 1.40 Photograph showing progress – modelling in Styrofoam and acrylic rod

You have two 35-minute sessions to start modelling your design proposal. A photograph will be taken by a member of staff at the end of each 35 minutes to record your progress. These will be stuck into Boxes 1 and 2 and are not annotated.

Additional annotated photographs may be stuck in to your progress reports and annotated. Be honest about the problems you have experienced. What decisions have you made? What were the solutions to your problems? What has been successful?

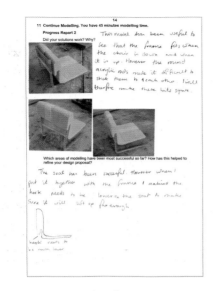

Fig 1.41 Example of Progress report 2

Fig 1.42

Making: the final modelling session

- Select and use materials innovatively and creatively and further develop ideas to define and refine them
- Complete a product/model to a high standard
- Demonstrate a range of making skills and/or complexity

With limited time you will need to plan carefully what you intend to do, what materials you will need and how you will model. You could do this in a flow chart or block diagram. Working to a plan will mean that, if when you need to use a drilling machine someone else is doing so, you can get on with the next part of your model and use the equipment once it is free.

Teacher's instructions to candidates

Plan what you need to do in the last 50-minute time allocation. You have 50 minutes to finish modelling your product. A photograph will be taken by a member of staff at the end of the 50 minutes to record your final product.

Fig 1.43

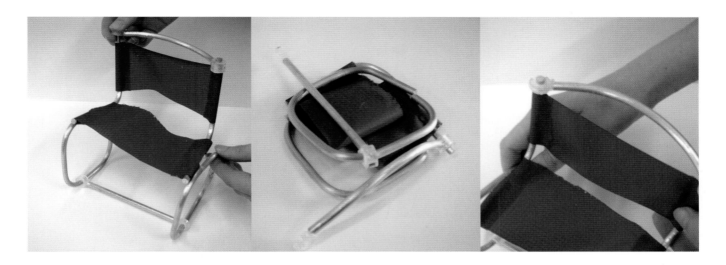

Fig 1.44–46 Planning and final modelling pictures; final scale model made from aluminium rod, acrylic rod and rip-stop nylon

EXAMINER'S TIPS

Although only one photograph is required, extra pictures showing close-up features or details of mechanisms and so on can be used in your evaluation.

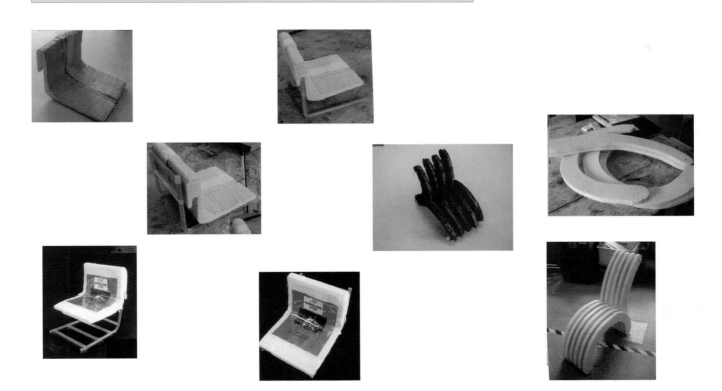

Fig 1.47–48 Examples of seating models: scale models in Styrofoam, acrylic rod, acetate sheet and foam board; full-size model in Styrofoam and dowel rod

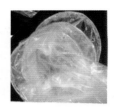

Fig 1.49–50 Examples of personal shelter models

Section 2.2 Evaluation

Modifications and evaluation of the final product

- Present realistic and detailed modifications to your idea, using annotated sketches. Improvements should be creative
- Produce a detailed evaluation of your product, identifying strengths and weaknesses, and show good consideration of the users/market
- Evaluate your design thoroughly against your product specification

Teacher's instructions to candidates

Describe the effectiveness of your developed design proposal and how it meets the needs of the original design situation. You may use annotated sketches and/or annotated photographs.

Look at your original specification in Box 5. Evaluate your final proposal against your original specification.

At the end of Session 2 you are required to evaluate your work. When evaluating any piece of work it is worth working through a checklist. The mark scheme breaks the evaluation down into three areas.

Key Term

Evaluate – e·val·u·ate (i-văl'yū-āt') – To ascertain or fix the value or worth of. To examine and judge carefully; appraise.

The evaluation should not simply be a written document. It should consist of notes, sketches and, where appropriate, photographs to show evidence.

Identify what you think are the good aspects of your design – where it is successful. What do you think are the weaker aspects of your design and how could they be improved?

Evaluate the end product in terms of the user. How easy is it to use? Is it safe to handle? Could ergonomics be improved? Is it aesthetically pleasing? Think of all the ways that it will be used or affect the user.

Evaluate the way that it was modelled. Explain why you used these materials.

Explain what method of manufacture would be used if it was to be produced in larger numbers. Describe in detail any changes that you would need to make to your design to make it better suited to larger scale production

How easy is your product to recycle? Could it be more sustainable/environmentally friendly?

Look at the design specification produced earlier in your booklet. Go through it point by point and identify how well each need and requirement has been met. It is not necessary to re-write each point, but it must be clear that you have considered each one.

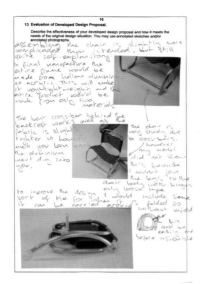

Fig 1.51 Example of evaluation against specification

Fig 1.52 Example of evaluation showing improvements and describing how the product would be made in larger numbers

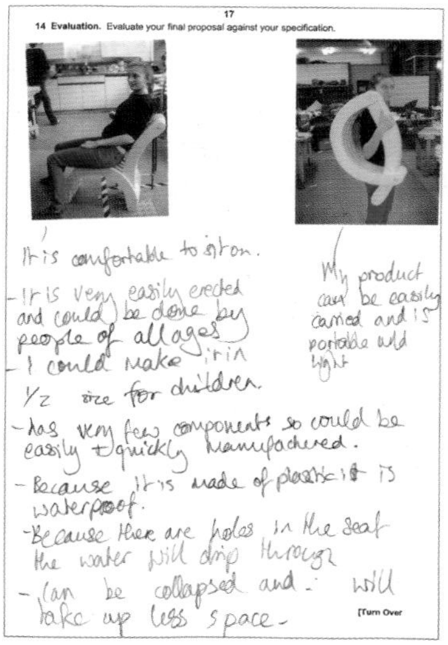

Fig 1.53 Example of evaluation against specification

EXAMINER'S TIPS

Don't be afraid to point out weaknesses in your design and how they can be improved. Use notes, sketches and extra photographs if they help to clarify points.

SESSION 3

Reflection

- Produce two outline presentation plans aimed at an expert panel
- Answers will be in the form of written material supported by annotated sketches and will be assessed on:
  - Relevant points/issues raised (**P**)
  - Quality of explanation of these issues (**Q**)
  - Supporting examples and sketches (**S**)
  - Quality of written communication (**QWC**)

Session 3 is a one-hour examination paper. You will have the opportunity to reflect on the challenge by answering questions that require you to consider your product. Questions will be derived from a design, manufacturing or marketing perspective including the following:

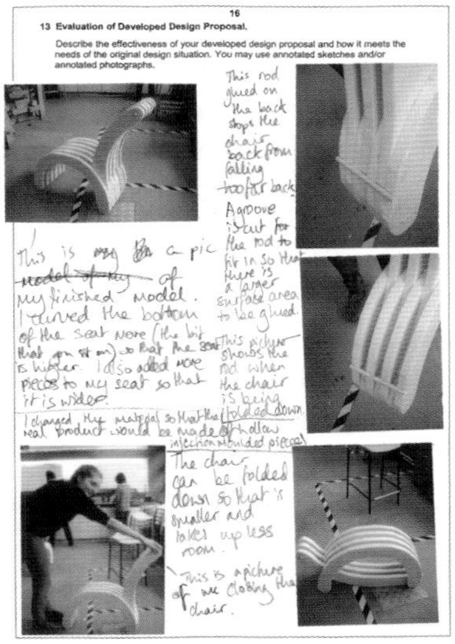

Fig 1.54 Example of evaluation showing improvements that could be made

- Sustainability and the environment
- Product life
- Social, moral and cultural issues
- Environmental issues
- Inclusive design
- The human interface
- Aesthetics
- Scale of production
- Production technologies
- Fashion
- Marketing
- Commercial issues

Fig 2.1 Analysis of pasta salad

You will find the first assessment criterion below. It starts with the statement: 'Select a specific single product, which has a focus within your area of expertise.' This sounds a bit complicated but all it means is that you need to select a product from an area that is familiar to you. If you are studying food, choose a single food product for study; if you are a construction student choose a product used in the construction industry for study.

You should choose one product rather than a range of products. Try to select a product that is not too complicated and which you believe may have some scope for possible improvement.

A suitable product for a student studying resistant materials would be a single coat hanger rather than 'coat hangers' in general. A textiles student could pick one specific coat rather than 'coats'. You don't have to redesign the whole product: the coat's hood or the arm of a chair is OK.

Select a product from one the following focus areas:

- Built environment and construction
- Engineering
- Food

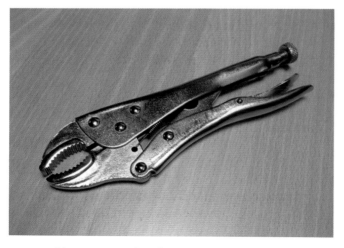

Fig 2.2 Choose a product from an area you have studied and think about how it could be improved

- Graphics products
- Manufacturing
- Resistant materials
- Systems and control
- Textiles

Section 1 — Product focus and analysis

What you need to do

This section is given a total of eight marks and should be presented on two sheets of A3 or two Microsoft PowerPoint® pages. You need to:

- Select a specific single product, which has a focus within your area of expertise suited to development within a prescribed time scale.
- Examine and give the intended purpose of the selected single product, including the needs of both the manufacturer and consumer. Identify the original key criteria against which the selected product was developed.

- Examine it, use it and, if possible, take it apart – detail its function. Record what you do in real time.
- Identify the needs of the manufacturer that relate to this specific product – list them.
- Identify the needs of the consumer that relate to this specific product – list them.
- Identify the original key criteria against which the selected product was developed. Many candidates do this by annotating a picture of their selected product with key points.

To access all of the marks you must cover all of the above areas of study.

This piece of coursework encourages the use of digital technology to record aspects of the product study in real time. Use annotated photographs, videos and sound bites to record things as they actually happen.

Some candidates may choose to complete all of these areas on one sheet of A3 or one Microsoft PowerPoint® page. This is acceptable provided that all of the areas are covered. Another approach is to cover the key criteria on the second page. There is an example of this in Figure 2.7.

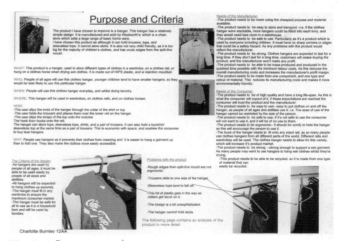

Fig 2.4 Purpose and criteria

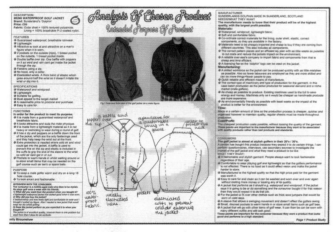

Fig 2.3 Analysis of chosen product

There are five distinct areas associated with this assessment outcome:

- Select a single product which has a focus within your area of expertise: if you are a textiles student, choose something made from textiles! State what it is (the exact type/model/manufacturer) and show it clearly.

Key Points

- Choose one product – name it
- Only one product!
- Use it – analyse it, state its purpose
- Show it in use – use digital technologies such as photographs and video
- List key design criteria
- Detail the needs of the consumer and the manufacturer
- Do all of this to access all marks!

Fig 2.5 Chicken noodle packet

Fig 2.7 Criteria for designing the original product

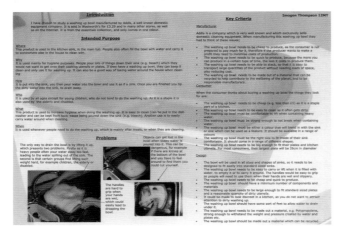

Fig 2.6 Intro and key criteria

The advantage of taking two pages to respond to these assessment criteria is that you will have more space to:

- Show the product in use (a video clip with commentary could replace one of the photographs shown in Figure 2.6)
- Clearly identify the intended purpose
- Identify the needs of the user
- Identify the needs of the manufacturer

Use video clips with sound bites and annotated digital photographs where possible to record things in real time – as they actually happen!

A whole page dedicated to the key criteria used for designing the original product will enable you consider this in detail. The example in Figure 2.7 identifies 20 key criteria some with up to five sub sections. This is a very good response following on from the previous page (Figure 2.3).

Section 2 — Comparison of strengths and weaknesses

What you need to do

This section is given a total of 12 marks and should be presented on two sheets of A3 or two Microsoft PowerPoint® pages. You should:

- Analyse the strengths and weaknesses of the single selected product
- Compare the product to other similar products in terms of:
  - Function
  - Suitability of materials, components or ingredients
  - Manufacturing processes used
  - Ergonomic suitability
  - Aesthetics
  - Cost

The most important thing to realise here is that you are required to do two quite separate things, which are then cross-referenced. You should clearly identify your selected product; if possible show it in use and analyse its strengths and weaknesses.

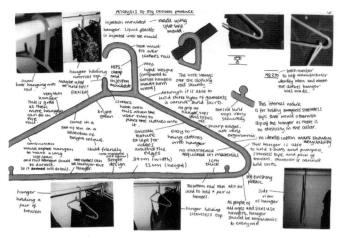

Fig 2.8 Coat hanger

You will then need to compare it to other similar products in terms of function; suitability of materials, components or ingredients; manufacturing processes used; ergonomic suitability; aesthetics and cost. You can only do this if you specifically refer to these points.

The example in Figure 2.9 has clearly identified these points as headings and used them in the analysis of the selected product, and to compare it to similar products (Figure 2.10). The example in Figure 2.8 uses a more creative approach and then summarises the results in a formal table.

Whichever style you choose it is important to end with a conclusion that cross-references your product to similar products.

Fig 2.10 Strengths and weaknesses comparison

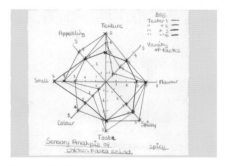

Fig 2.11 Spider diagram

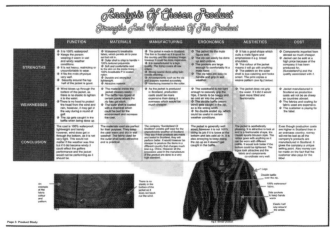

Fig 2.9 Strengths and weaknesses of jackets

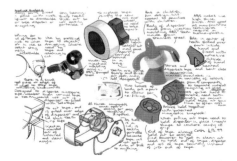

Fig 2.12 Product analysis – Sellotape dispenser

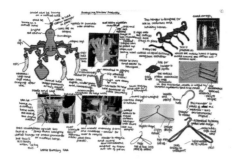

Fig 2.13 Product analysis coat hanger

Good responses to this section can include creative analysis by annotating images, photographs and diagrams of similar products. Some candidates choose to do this before they record their findings in a table.

Better responses to this section show actual images of similar products actually being used – video clips would be good for this.

The candidates whose work is shown in Figure 2.12 and 2.13 chose to add an additional page here – they could afford to do this as they considered section 1 as a single page.

In this section you can also analyse similar types of product outside the immediate group of products you are considering; for example if you have selected a particular type of folding ironing board for study it is quite acceptable, and good practice, to consider some other folding devices such as a deck chair or a folding clothes rack.

Use digital technology to record strengths and weaknesses in real time as you analyse products.

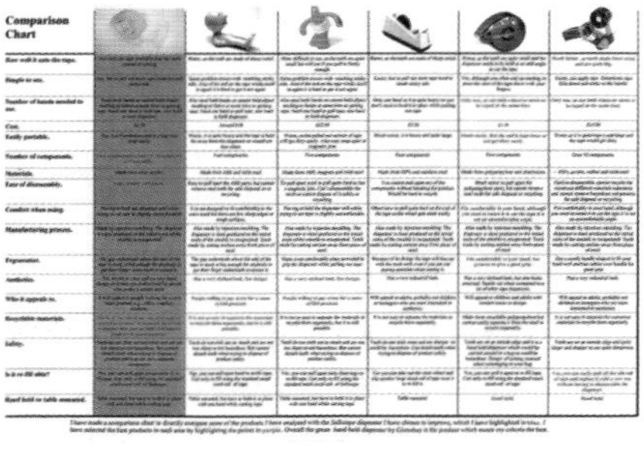

Fig 2.14 Product comparison

Key Points

- Make it clear which is your selected product. In Figure 2.14 for example, the selected product is highlighted in blue
- Make sure you write a conclusion or summary that cross-references your selected product to the similar products

Section 3 Moral implications

What you need to do

This section is given a total of eight marks and should be presented on one sheet of A3 or one Microsoft PowerPoint® page. You should:

- Identify and analyse the relevant moral implications associated with the:
  - environmental
  - social and
  - economic

  issues in the design and use of a product.

The Sustainability Handbook
for D&T Teachers

The Sustainability Handbook has been produced by Practical Action in collaboration with the Centre for Alternative Technology and Loughborough University. Funded by the European Community and the Countryside Council for Wales.

This publication has been produced with the assistance of the European Union. The contents of this publication are the sole responsibility of Practical Action and can in no way be taken to reflect the views of the European Union.

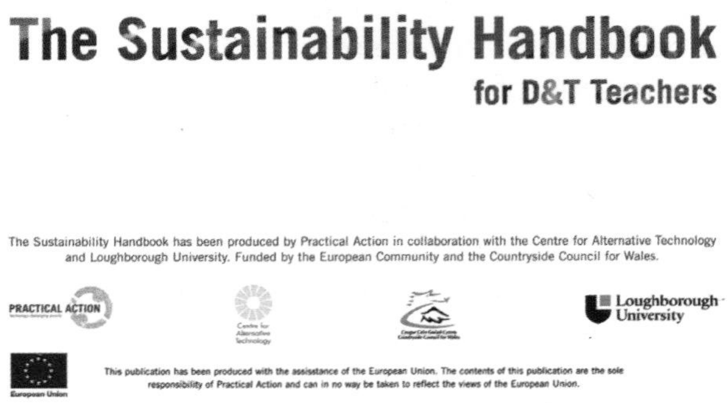

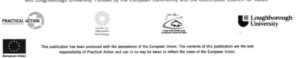

Fig 2.15 Sustainability handbook

A recent news item estimated that up to 80 per cent of presents that are given to children at Christmas are made in China, while a high street, brand-name store recently destroyed a whole warehouse full of clothes after a news programme revealed they had been made using child labour.

- Do you know what the average wage in China, India or Sri Lanka is?
- Did you know that some hand-loom weavers in India have been put out of business by Chinese competitors?
- Did you know that polluting chemicals could constitute over 25 per cent of a cotton product?

The answers to some of these questions and many other ethical considerations can be found in the excellent resource produced by Practical Action shown in Figure 2.15. This book might already be in your school or college. OCR and Practical Action have worked together for many years and some sections of the new specification were written to mirror exactly their excellent resources. Further information on sustainability can be found in Chapter 5 of this book.

This section can not be completed without some serious ethical debates – you need a good resource!

OK, so what have gorillas got to do with your mobile phone?

- Demand for new mobile phones
- Increased demand for columbite-tantalite (CT, a material used in the production of high-tech capacitors for mobile phones)
- Groups including militia from neighbouring countries fight to control CT supplies in the Democratic Republic of Congo
- Land cleared to make mining easier, leading to destruction of natural habitat
- Local population forced off the land to allow mining to take place
- Farmers no longer able to grow food so look for alternative food, supplies and work
- Gorillas either killed or leave because their habitat has disappeared

Ian Capewell, Practical Action

And another thing: if you are persuaded to buy a bar of chocolate, check it is fair trade! Environmental, social and economic responsibility – make sure that you understand what this means.

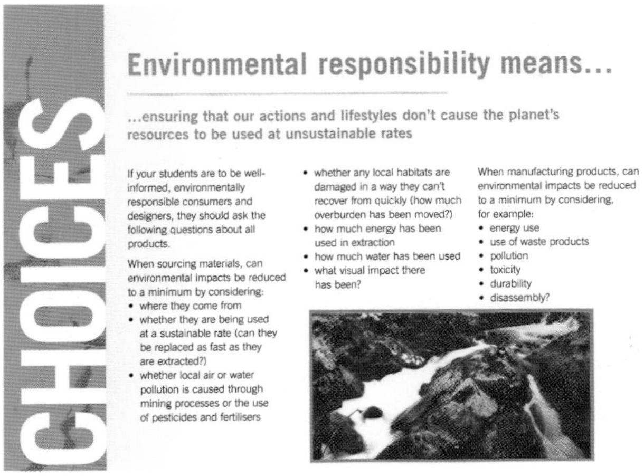

Social responsibility means...

...ensuring that our own and other people's quality of life and human rights are not compromised to fulfil our expectations and demands

If your students are to be well-informed, socially responsible consumers and designers, they should ask the following questions about all products:

- Does the product improve the quality of life for its users?
- Is this product appropriate for the society and culture in which it will be used?
- Does the product encourage the maintenance of traditional knowledge and skills, or could traditional knowledge and skills be lost over time as a result (e.g. home cooking)?

- Does the making of the product (e.g. material or energy used) have a positive or negative impact on the quality of life for some people, including those living elsewhere in the world, sometimes in poverty?
- Does the product help to maintain valuable social or cultural traditions, e.g. the food we eat, clothes we wear, our music, leisure activities?
- Does the product encourage us to be sociable, to enjoy the company of others, when we want to?

- Does the product meet the needs of people today without limiting the ability of future generations to meet their needs satisfactorily?
- Does the making of the product infringe any basic human rights, e.g. fair pay, decent working conditions?

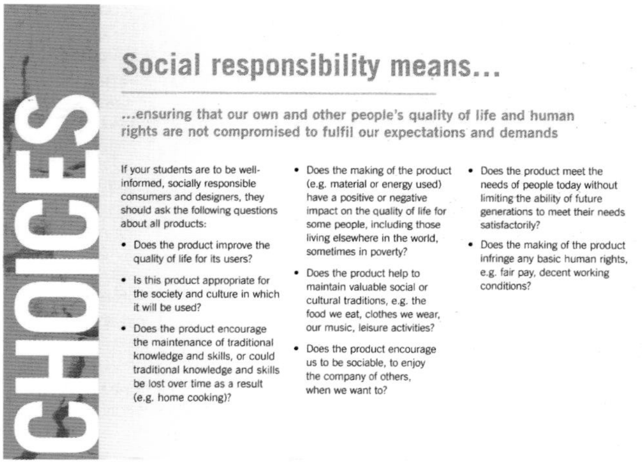

Fig 2.17 Extract from the Sustainability Handbook

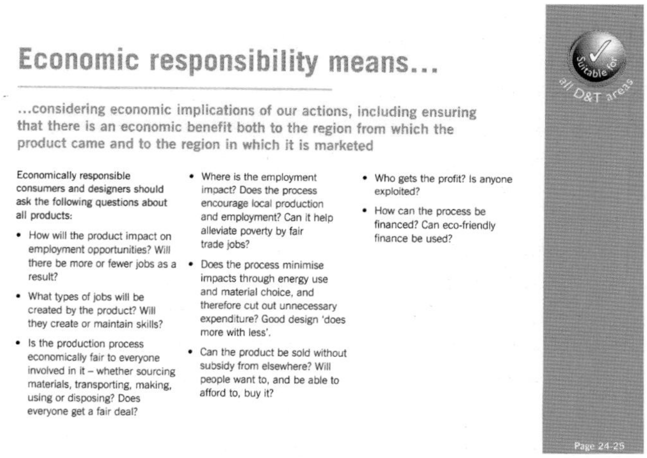

Economic responsibility means...

...considering economic implications of our actions, including ensuring that there is an economic benefit both to the region from which the product came and to the region in which it is marketed

Economically responsible consumers and designers should ask the following questions about all products:

- How will the product impact on employment opportunities? Will there be more or fewer jobs as a result?
- What types of jobs will be created by the product? Will they create or maintain skills?
- Is the production process economically fair to everyone involved in it – whether sourcing materials, transporting, making, using or disposing? Does everyone get a fair deal?

- Where is the employment impact? Does the process encourage local production and employment? Can it help alleviate poverty by fair trade jobs?
- Does the process minimise impacts through energy use and material choice, and therefore cut out unnecessary expenditure? Good design 'does more with less'.
- Can the product be sold without subsidy from elsewhere? Will people want to, and be able to afford to, buy it?

- Who gets the profit? Is anyone exploited?
- How can the process be financed? Can eco-friendly finance be used?

Page 24-25

Fig 2.18 Extract from the Sustainability Handbook

Environmental responsibility means...

...ensuring that our actions and lifestyles don't cause the planet's resources to be used at unsustainable rates

If your students are to be well-informed, environmentally responsible consumers and designers, they should ask the following questions about all products:

When sourcing materials, can environmental impacts be reduced to a minimum by considering:
- where they come from
- whether they are being used at a sustainable rate (can they be replaced as fast as they are extracted?)
- whether local air or water pollution is caused through mining processes or the use of pesticides and fertilisers

- whether any local habitats are damaged in a way they can't recover from quickly (how much overburden has been moved?)
- how much energy has been used in extraction
- how much water has been used
- what visual impact there has been?

When manufacturing products, can environmental impacts be reduced to a minimum by considering, for example:
- energy use
- use of waste products
- pollution
- toxicity
- durability
- disassembly?

Fig 2.16 Extract from the Sustainability Handbook

Economic implications

There are worldwide implications for the marketing of most products. When you complete this section it is essential that you make the content particularly relevant to your selected product.

If you are studying a food product go to great lengths to source ethical implications relating specifically to your area. The Stop the Traffik website – www.stopthetraffik.org – has an excellent and very thought-provoking story about child exploitation and trafficking.

If you are studying a construction product or a resistant material product, make sure you use appropriate, specific resources. The Practical Action book has excellent sections on most material areas – it even has a section on 'ethical electronics'. There are also many excellent sections relating to sustainable textiles and graphical products. Engineering and manufacturing students will also find many relevant areas. The most useful aspect of this recommended resource is the number of websites it lists on nearly every page (there are 12 references to cotton alone).

Ethical considerations are complex; there are tensions. Buying fair-trade goods from abroad could mean a large increase in air miles. There is a scheme in Middlesbrough to grow tomatoes all year round from waste energy from an ammonia plant. There is an increasing lobby to support these developments. Most fair-trade goods mentioned in the Values section of this book are items we can't produce in quantity.

Fig 2.20 Moral implications

Fig 2.21

Monda African Art is owned by Carol Monda and her Ghanaian husband Martin. They design and make intricate pieces of jewellery and other Kenyan and Ghanaian crafts. Increasingly they have found it more and more difficult to compete against Asian businesses.

'We cannot succeed unless buyers are prepared to invest in Africa and recognise that things are more expensive here. We will never be able to compete on price with countries like China and India.

'We were displaying some products at a trade fair. One of our pieces was a beautiful bag with quite an intricate design, hand-sewn beads, etc, and we had priced it at $12. But a buyer picked it up and told us he could buy a similar bag in China for $5. How can we compete against that?'

Carol Monda, April 2007
Traidcraft Exchange Development Review 2007

Fig 2.22 Monda African Art

Key Points

- Visit the Practical Action website (www.sda-uk.org). There are also many website links and other useful references in the Sustainability Handbook
- See also The Eco-Design Handbook by Alastair Fuad-Luke (Thames & Hudson)

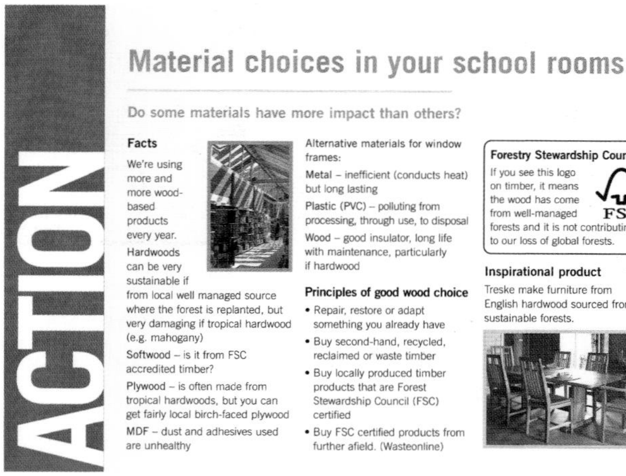

Material choices in your school rooms

Do some materials have more impact than others?

Facts

We're using more and more wood-based products every year. Hardwoods can be very sustainable if from local well managed source where the forest is replanted, but very damaging if tropical hardwood (e.g. mahogany).

Softwood – is it from FSC accredited timber?

Plywood – is often made from tropical hardwoods, but you can get fairly local birch-faced plywood

MDF – dust and adhesives used are unhealthy

Alternative materials for window frames:

Metal – inefficient (conducts heat) but long lasting

Plastic (PVC) – polluting from processing, through use, to disposal

Wood – good insulator, long life with maintenance, particularly if hardwood

Principles of good wood choice
- Repair, restore or adapt something you already have
- Buy second-hand, recycled, reclaimed or waste timber
- Buy locally produced timber products that are Forest Stewardship Council (FSC) certified
- Buy FSC certified products from further afield. (Wasteonline)

Forestry Stewardship Council

If you see this logo on timber, it means the wood has come from well-managed forests and it is not contributing to our loss of global forests.

Inspirational product

Treske make furniture from English hardwood sourced from sustainable forests.

Fig 2.23

Fig 2.24

(www.sda-uk.org)

equo e solidale
fair trade

Section 4 — Brief and specification for improving the product

What you need to do

This section is given a total of eight marks and should be presented on one sheet of A3 or one Microsoft PowerPoint® page. You should:
- Write a detailed design brief for improving the selected product in some way
- Develop and justify an objective design specification

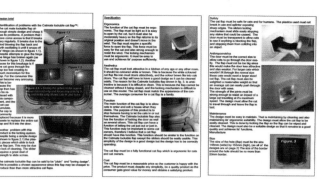

Design Brief and specification

Fig 2.25 Catflap design brief and spec

Design brief

It is a good idea to stop here and remind yourself that this is a product study and not a product design exercise where you develop a complete product from an open context. You have studied a selected product and have analysed in detail its strengths and weaknesses. From the study of other similar products you should now have a clear idea of how your selected product could be improved in some way. Now is your chance to state quite simply and clearly what you are going to improve.

- Use statements like: 'I am going to improve …… by ……'
- Avoid statements like: 'I am going to redesign a torch,' or: 'I have been asked to redesign a mobile phone.' Briefs written in these terms or that talk about the specification requirements will be awarded zero marks

Design specification

Better specifications identify design criteria as points and then fully justify them.

- A different font style, colour or size helps to differentiate the justification and makes it clear
- Specifications without clear justification will always be marked in the lower bands
- Better specifications are written under 'classification headings', such as 'sustainability'

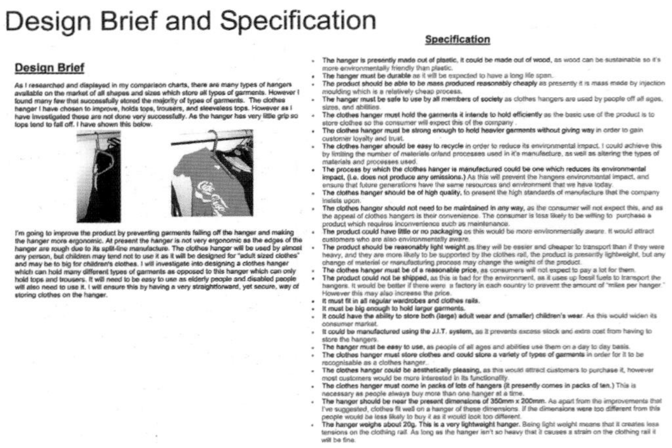

Design Brief and Specification

Design Brief

As I researched and displayed in my comparison charts, there are many types of hangers available on the market of all shapes and sizes which store all types of garments. However I found many few that successfully stored the majority of types of garments. The clothes hanger I have chosen to improve, holds tops, trousers, and sleeveless tops. However as I have investigated these are not done very successfully. As the hanger has very little grip so tops tend to fall off. I have shown this below.

I'm going to improve the product by preventing garments falling off the hanger and making the hanger more ergonomic. At present the hanger is not very ergonomic as the edges of the hanger are rough due to its split-line manufacture. The clothes hanger will be used by almost any person, but children may tend not to use it as it will be designed for "adult sized clothes" and may be to big for children's clothes. I will investigate into designing a clothes hanger which can hold many different types of garments as opposed to this hanger which can only hold tops and trousers. It will need to be easy to use as elderly people and disabled people will also need to use it. I will ensure this by having a very straightforward, yet secure, way of storing clothes on the hanger.

Specification

- **The hanger is presently made out of plastic, it could be made out of wood,** as wood can be sustainable so it's more environmentally friendly than plastic.
- **The hanger must be durable** as it will be expected to have a long life span.
- **The product should be able to be mass produced reasonably cheaply** as presently it is mass made by injection moulding which is a relatively cheap process.
- **The hanger must be safe to use by all members of society** as clothes hangers are used by people off all ages, sizes, and abilities.
- **The clothes hanger must hold the garments it intends to hold efficiently** as the basic use of the product is to store clothes so the consumer will expect this of the company .
- **The clothes hanger must be strong enough to hold heavier garments without giving way** in order to gain customer loyalty and trust.
- **The clothes hanger should be easy to recycle** in order to reduce its environmental impact. I could achieve this by limiting the number of materials or/and processes used in it's manufacture, as well as altering the types of materials and processes used.
- **The process by which the clothes hanger is manufactured could be one which reduces its environmental impact, (i.e. does not produce any emissions.)** As this will prevent the hangers environmental impact, and ensure that future generations have the same resources and environment that we have today.
- **The clothes hanger should be of high quality,** to present the high standards of manufacture that the company insists upon.
- **The clothes hanger should not need to be maintained in any way,** as the consumer will not expect this, and as the appeal of clothes hangers is their convenience. The consumer is less likely to be willing to purchase a product which requires inconvenience such as maintenance.
- **The product could have little or no packaging** as this would be more environmentally aware. It would attract customers who are also environmentally aware.
- **The product should be reasonably light weight** as they will be easier and cheaper to transport than if they were heavy, and they are more likely to be supported by the clothes rail, the product is presently lightweight, but any change of material or manufacturing process may change the weight of the product.
- **The clothes hanger must be of a reasonable price,** as consumers will not expect to pay a lot for them.
- **The product could not be shipped,** as this is bad for the environment, as it uses up fossil fuels to transport the hangers. It would be better if there were a factory in each country to prevent the amount of "miles per hanger." However this may also increase the price.
- **It must fit in all regular wardrobes and clothes rails.**
- **It must be big enough to hold larger garments.**
- **It could have the ability to store both (large) adult wear and (smaller) children's wear.** As this would widen its consumer market.
- **It could be manufactured using the J.I.T. system,** as it prevents excess stock and extra cost from having to store the hangers.
- **The hanger must be easy to use,** as people of all ages and abilities use them on a day to day basis.
- **The clothes hanger must store clothes and could store a variety of types of garments** in order for it to be recognisable as a clothes hanger..
- **The clothes hanger could be aesthetically pleasing,** as this would attract customers to purchase it, however most customers would be more interested in its functionality.
- **The clothes hanger must come in packs of lots of hangers (it presently comes in packs of ten.)** This is necessary as people always buy more than one hanger at a time.
- **The hanger should be near the present dimensions of 350mm x 200mm.** As apart from the improvements that I've suggested, clothes fit well on a hanger of these dimensions. If the dimensions were too different from this people would be less likely to buy it as it would look too different.
- **The hanger weighs about 20g. This is a very lightweight hanger.** Being light weight means that it creates less tensions on the clothing rail. As long as the hanger isn't so heavy that it causes a strain on the clothing rail it will be fine.

Fig 2.26 Design brief and specification

Key Points

- Make sure the brief clearly identifies an improvement to the selected product
- Specification points should be fully justified
- Digital images make things more dynamic – it's great to see the cat in Figure 2.25; a video would be even better!

Section 5 Development of improvement

What you need to do

This section is given a total of 56 marks and should be presented on 10 sheets of A3 or 10 Microsoft PowerPoint® pages. You should:

- Use annotated sketches, real-time digital images and interactive dialogue to generate and record a wide range of **initial ideas** that explore possible improvements
- **Photograph, record and comment** as improvement actually takes place
- Make sufficient appropriate **prototype models** to establish the validity of the proposed idea in terms of:
  - Physical requirements, for example construction, movement, stability, composition and strength
  - Aesthetic qualities and/or taste as appropriate
  - Suitable manufacturing processes
  - Suitability of materials, components or ingredients
- **Test and evaluate** developed ideas against the specification in real time and justify the choice of one idea worthy of being taken forward.

Key Points

- Integrate all aspects of the development together in 10 pages
- Record things in real time as they actually happen
- Work in an appropriate way for your focus area
- This is the most important section!

Most marks are awarded here; this one section alone is worth nearly half of the marks for the whole study. It is very important that you understand what is required.

Different candidates working in different focus areas will approach this section in their own way. The most important thing is to ensure that ideas, appropriate prototype modelling and ongoing evaluation are presented in an integrated way.

The key word here is *appropriate* – what is appropriate to a graphic products focus may not be suitable for a food focus. One focus area may require more 2D modelling, another area more 3D prototypes; ongoing menu development and sensory testing may be the main feature of another candidates work. The common element is that ideas, appropriate prototype modelling and evaluation must be developed in an integrated form and recorded in real time. This means that any ideas, models and ongoing testing should be recorded as they actually take place.

Clearly justify one idea worthy of being taken forward.

Use video clips, sound recording and annotations to show results and to record the views of others.

Candidate examples

The next few pages show the work of five different candidates. The work was taken from the previous OCR specification where there was no requirement to integrate ideas, modelling and evaluation, so not all candidates have done this.

Ideas and modelling – the coat hanger

Initial Ideas

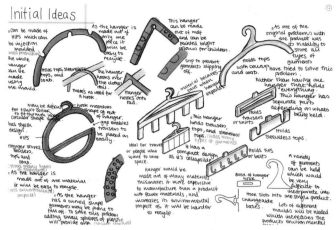

Fig 2.27

Initial Ideas

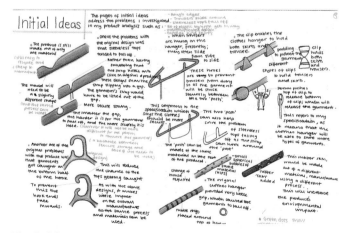

Fig 2.28

Modelling

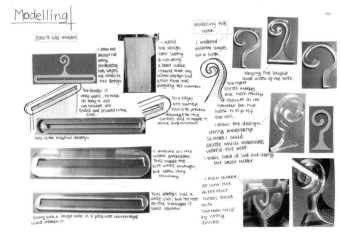

Fig 2.29

Initial Ideas

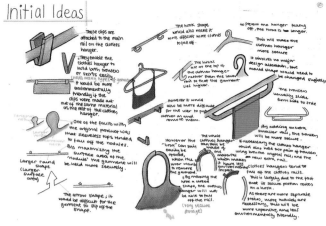

Fig 2.30

Modelling

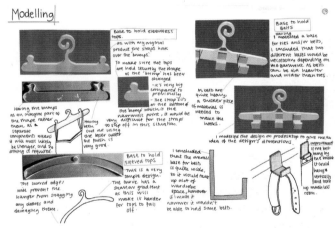

Fig 2.31

Initial Ideas

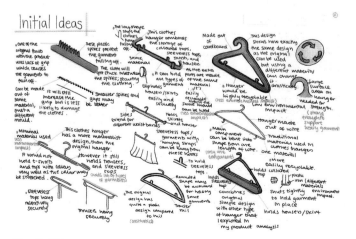

Fig 2.32

Modelling

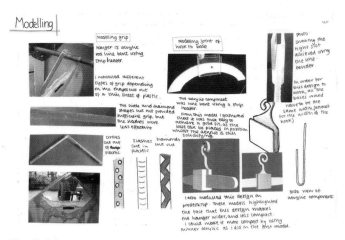

Fig 2.33

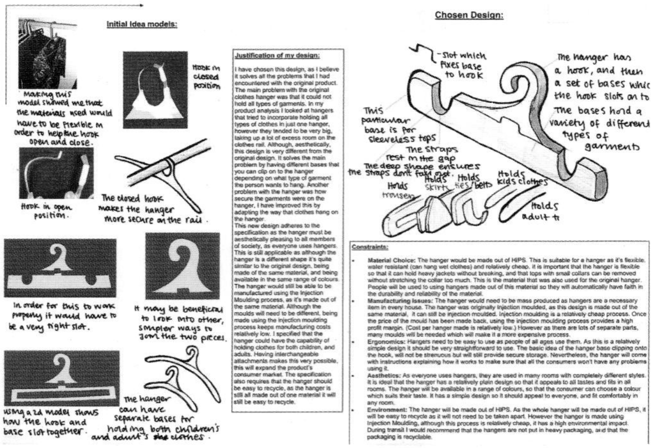

Fig 2.36

Final Model

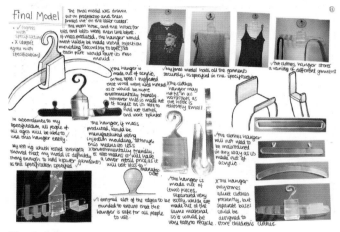

Fig 2.34

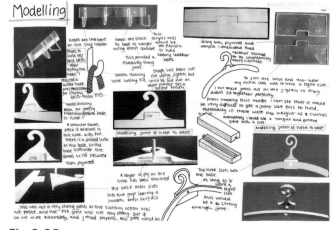

Fig 2.35

Fig 2.37 Product suitable for a built environment and construction focus

Textiles – waterproof jacket

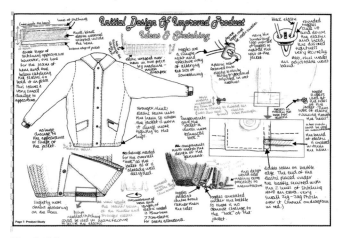

Fig 2.38

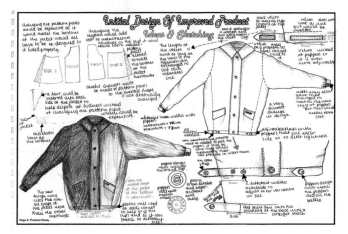

Fig 2.39

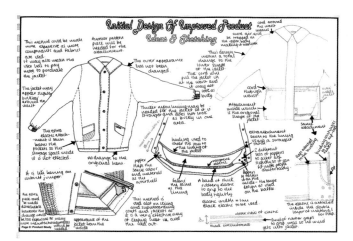

Fig 2.40

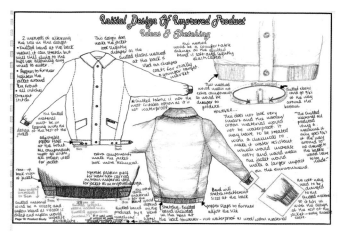

Fig 2.41

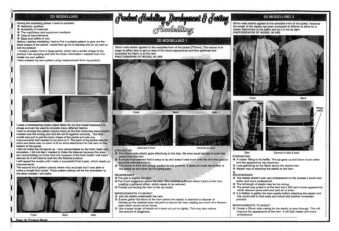

Fig 2.42

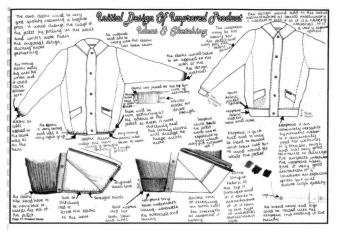

Fig 2.43

Fig 2.44 Products suitable for an engineering focus

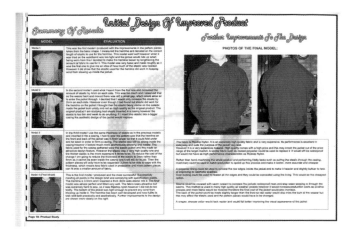

Fig 2.45

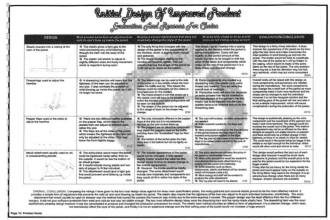

Fig 2.47

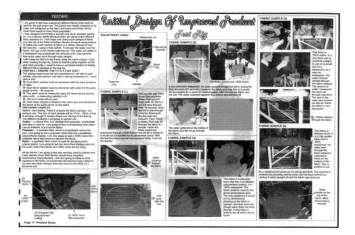

Fig 2.46

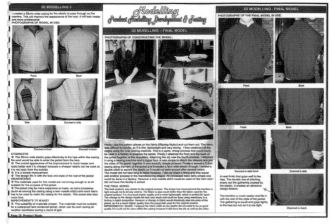

Fig 2.48

Fig 2.49 Products suitable for a food focus

Ideas and modelling – sticky tape dispenser

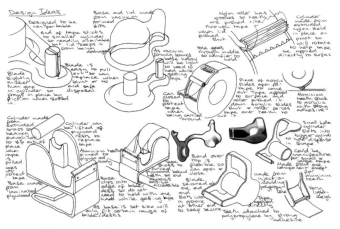

Fig 2.50

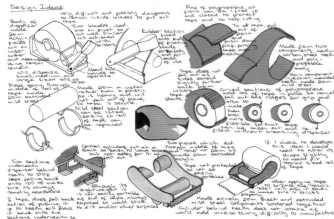

Fig 2.52

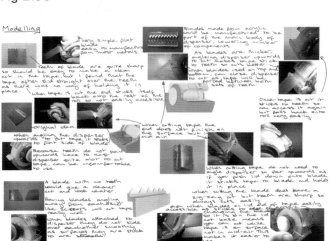

Fig 2.51

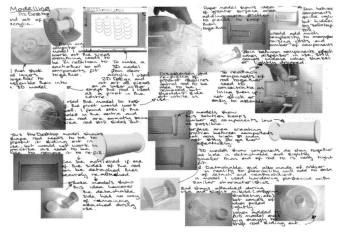

Fig 2.53

Design Ideas

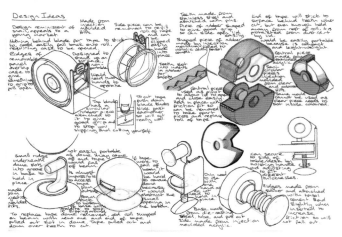

Fig 2.54

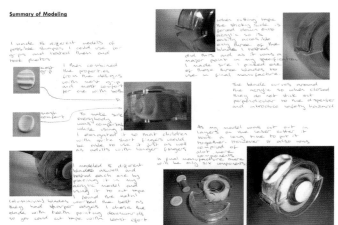

Summary of Modeling

Fig 2.56

Fig 2.55 Product suitable for a graphics products focus

Chosen Idea

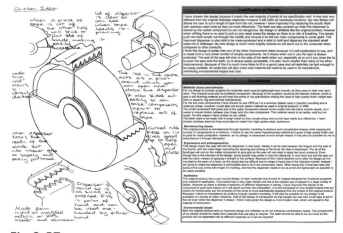

Fig 2.57

Modelling

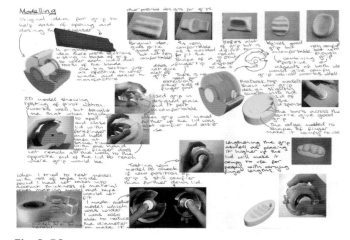

Fig 2.58

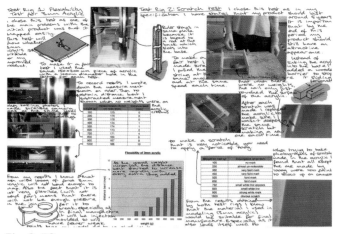

Fig 2.59

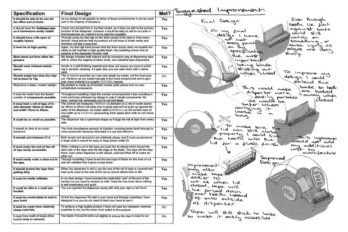

Fig 2.60

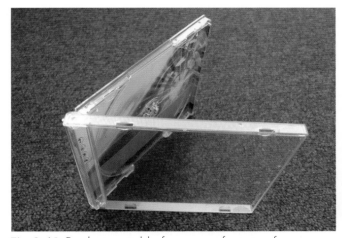

Fig 2.61 Products suitable for a manufacturing focus

Systems and control – laptop cooling device

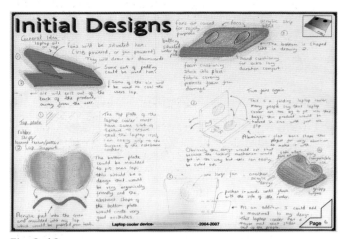

Fig 2.62

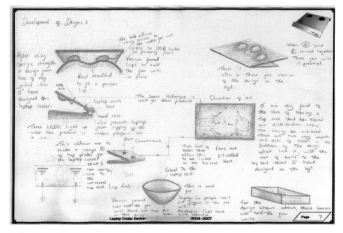

Fig 2.63

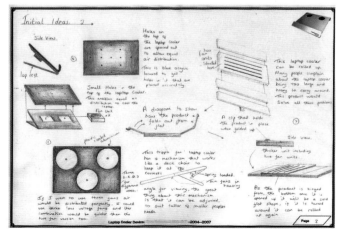

Fig 2.64

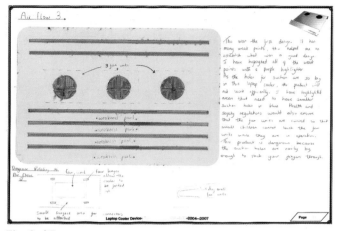

Fig 2.65

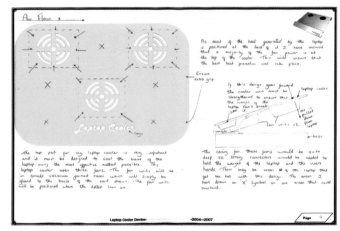

Fig 2.66

Fig 2.67 Product suitable for a resistant materials focus

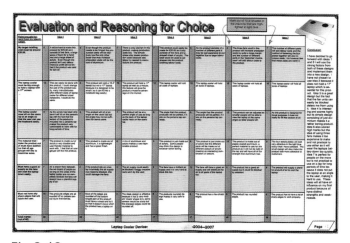

Fig 2.68

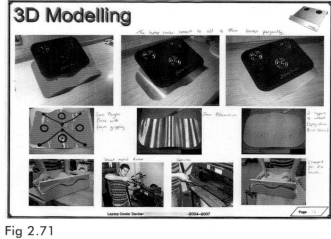

Fig 2.71

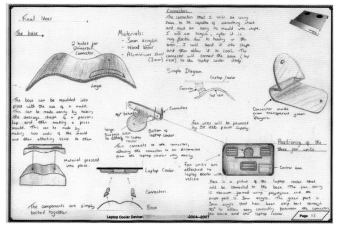

Fig 2.69

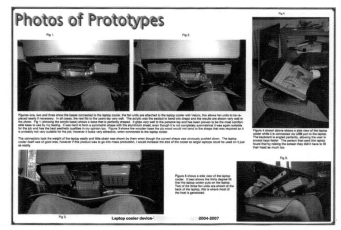

Fig 2.72

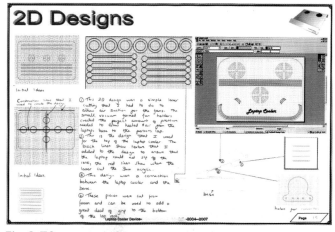

Fig 2.70

Key Points

- The final prototype is shown in use
- A video of this would be good
- Sound bites of users' comments could be useful

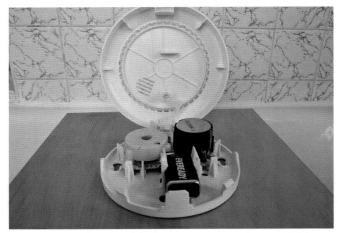

Fig 2.73 Product suitable for a systems and control focus

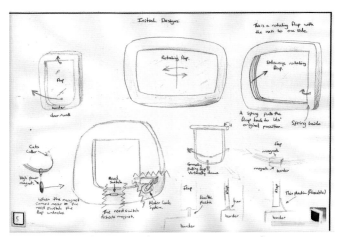

Fig 2.75 Development of improvement to catflap

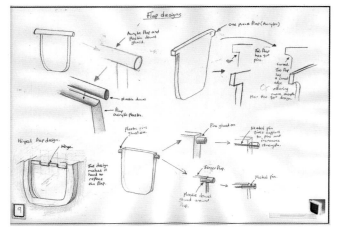

Fig 2.74 Development of improvement to catflap

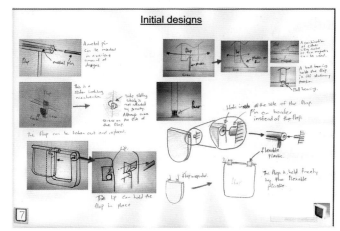

Fig 2.76 Development of improvement to catflap

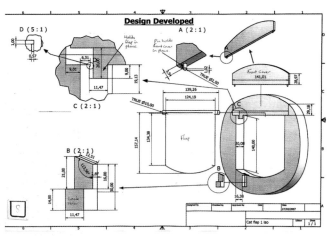

Fig 2.77 Development of improvement to catflap

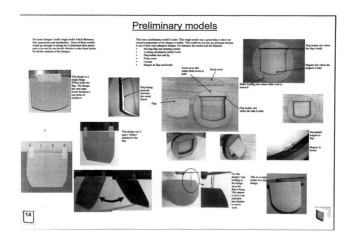

Fig 2.78

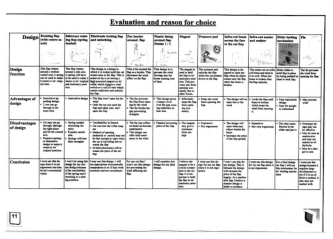

Fig 2.80

Key Points

■ This sophisticated presentation starts off with very basic ideas drawings. Don't throw these away at the expense of better presentation: your marks will go up, not down, if you show early development work.

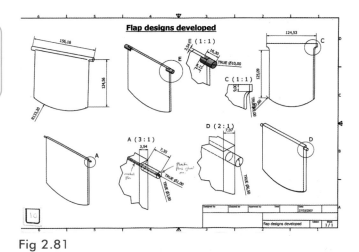

Fig 2.81

Fig 2.79 Products suitable for a textiles focus

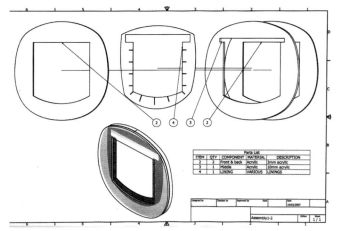

Fig 2.82 Development and modelling of improvement to catflap

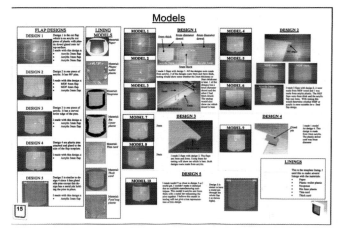

Fig 2.83 Development and modelling of improvement to catflap

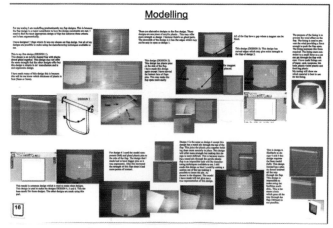

Fig 2.84 Development and modelling of improvement to catflap

If you want marks at the top of the top band you need to produce work of this quality, which shows clear evidence of mixing ideas/modelling and ongoing evaluation so that your ideas develop naturally.

Most work shown has excellent quality of presentation with good use of photographs. You will need to add video clips and sound bites where appropriate. A good time to do this would be during ongoing evaluation. You could show models actually in use and record responses to their use by others.

Work is not included from all focus areas, but whichever focus area you choose you will need to produce work of the same quality for marks in the top band.

As work becomes available in other focus areas it will be added to the website.

Whatever product you choose for analysis and development, try to make it exiting, both in content and presentation. If you are developing something for cats, let's see one! Show clothes on hangers and videos of problems. Show food products with real people sampling them with recordings made as things actually happen in real time.

The best projects are produced by candidates who have obviously enjoyed themselves in the process!

Look on the website for examples from your focus area (www.hodderplus.co.uk/ocrd&t).

Section 6 — Testing of final developed idea

What you need to do

This section is given a total of 12 marks and should be presented on two sheets of A3 or two Microsoft PowerPoint® pages. You should:

- Use an appropriate method or system to formally test and evaluate your final developed idea, or the suitability of the proposed materials, components or ingredients
- Present results in real time, clearly and concisely

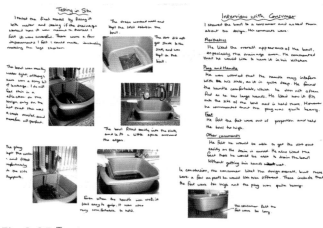

Fig 2.85 Testing

happened. Use digital technology, such as photographs and videos with sound bites.

In the previous section you tested and evaluated your developed ideas and justified the choice of one selected idea to be taken forward. This justified and selected idea is the one you now need to formally test.

You can do this by:

- Actually testing the final outcome in an appropriate way for your focus area, or
- Testing the suitability of the proposed materials, components or ingredients

There is a wide scope of opportunities. You choose the appropriate method of testing – you can wear it or get others to wear it, use it or get others to use it, or eat it and get others to eat it. You can carry out formal systems tests on components or sensory tests on food products.

Whichever method you choose, you must present the results in real time – as they actually

Carry out testing in whatever way is appropriate to your focus area. You choose!

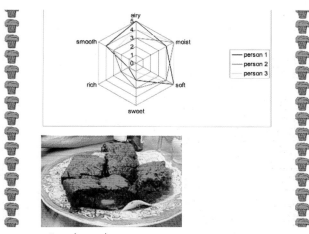

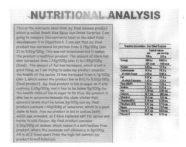

Fig 2.87 Food product testing

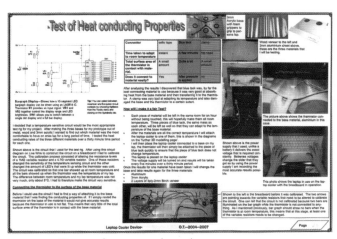

Fig 2.86 Testing laptop coding device improvements

Fig 2.88 Nutritional analysis

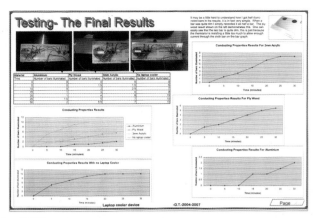

Fig 2.89

Whatever method you choose to test your final outcome you must:

- Present your results in real time (show photographs or video clips of the test actually being carried out)
- Obtain sufficient good-quality data to enable you to make sound conclusions
- Record sound bites of user/testers comments as they actually say them
- Present results clearly and concisely
- Refer to the design specification

Some formal presentation of results will be needed to tie together your real-time recording.

Fig 2.90

Key Points

- Show results in real time
- Record things as they actually happen
- Use digital technology: photographs, videos, sound
- Present results formally in a clear and concise manner

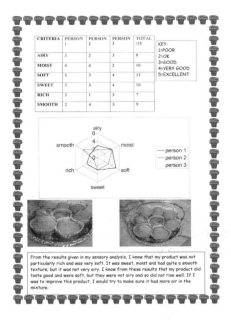

Fig 2.91

Section 7 — Summary of results

What you need to do

This section is given a total of eight marks and should be presented on two sheets of A3 or two Microsoft PowerPoint® pages. You should:

- Produce a summary of the results of the development and prototype modelling which includes:
  - Analysis of information gained from the prototypes
  - Details and analysis of the results gained from the testing
- Provide suggestions for further improvements to the proposed idea

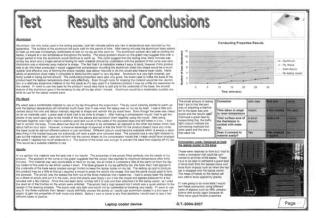

Fig 2.92

Summarise the results of your prototype modelling. You need to provide evidence that you have actually analysed and gained information from your prototypes.

Leading on from your formal presentation of results you need to summarise your findings and analyse them in depth.

A conclusion would be a good outcome. The example in Figure 2.92 shows this.

The purpose of this section is to tie things together in order to see whether the final selected idea meets the intended requirements. How realistic are your proposed improvements?

Fig 2.93

Key Points

- What have you learnt?
- Analyse test results – what do they tell you?
- Write a conclusion to lead you on to the next section
- Does your final analysis highlight any faults that need to be addressed?

Specification	Final Design	Met?
It should be able to be use din the office and at home.	As my design is not specific to either of these environments it can be used well in the majority of situations.	Yes
It should have the Sellotape logo on it somewhere easily visible.	I have not included this in my final model, as it does not add to the primary function of the dispenser, however it would be easy to add to any side in final production as I intend it to be injection moulded.	Yes
It should have a life span of roughly 5years.	Through using my test rigs on the 3mm acrylic to be used in final manufacture I have proven that my product will not snap or break easily and therefore will last a long time.	Yes
It must be of high quality.	Again, my test rigs have proven that the 3mm acrylic does not scratch too easily to will maintain a high quality finish. My modeling shows that all aspects of the dispenser function well.	Yes
Must stand out from other dispensers.	My design includes bold colours and an innovative way of dispensing tape with is unlike the majority of other small, non-industrial tape dispensers.	Yes
Should need minimal maintenance.	Acrylic is a self-finishing material and does not require any kind of polishing or periodic cleaning. If it gets dirty you can wipe clean with a damp cloth.	Yes
Should weigh less than the original product at 16g.	This is hard to ascertain as I can only weigh my model, not the final product. However as my model includes a few more components and in general, more material it is roughly 45973... heavier.	No
Must have a sleek, modern design.	My product is made up of smooth curves, bold colours and no over-complicated components.	Yes
It must be made from the fewest number of components possible.	Throughout modelling I kept the number of components I was including in mind and have achieved my design in only 5 simple components: lid, base, central rod, removable end of rod and blade.	Yes
It must hold a roll of tape of inside diameter 26mm to 30mm and width 15mm to 20mm.	The central rod measures 18... in diameter so a roll of inside diameter 26mm to 30mm will slide over it easily and will not push up against the sides of the dispenser. Its inside width is 85... so will contain tape of any width up to 42432mm (accounting 2mm space each side so can move easily).	Yes
It could be as small as possible.	The dispenser has a cylindrical shape so it hugs the roll of tape from every side.	Yes
It should be able to be mass-produced	The final manufacture process of injection moulding lends itself strongly to mass-production because otherwise it is not cost-effective.	Yes
It should cost between £1-5.	Both acrylic and aluminium are relatively cheap, and if mass-produced on a large scale it should be easy to keep prices under £5	Yes
It must make the end of the roll of tape easily accessible.	When making a cut in the tape you push the lid closed which forced the stick side of the tape onto the flat edge of the blade. The tape will the stay there, even when dispenser is left closed, until next lifted off to make another cut.	Yes
It must easily make a clean cut in the tape.	Through modelling I have found the best type of blade for this kind of cut and am satisfied that it gives a neat finish.	Yes
It should protect the tape from getting dirty.	When the dispenser is not in use the end of the roll of tape is covered and kept quite close to the rest of the roll so cannot attract dust or dirt.	Yes
It must be easily refillable.	In my final design I have included the instruction "pull" of the end of the central rod you need to remove to refill. Once this has been done refilling is self explanatory and quick.	Yes
It could be able to e used one handed.	You can operate the dispenser easily with only your right or left hand.	Yes
It must be comfortable to hold in your hand.	At first the dispenser fits well in your hand and through modelling I have designed it so you do not need to twist your hand to use it.	Yes
It must be made form relatively cheap materials.	To achieve a high quality product I have not used the cheapest materials available to me, but the ones most suited to the purpose	No
It must have teeth of blade either tucked away or covered.	The blade it forced to stick out slightly to ensure the tape is easy to cut	No

Fig 2.94

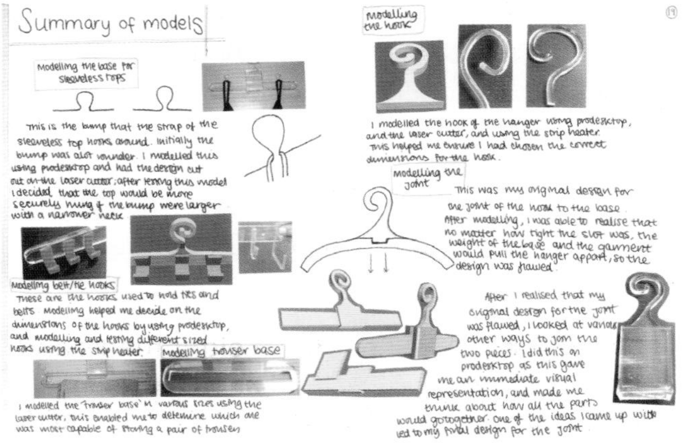

When you have looked at and analysed all of your prototypes, and analysed the results of your final testing, this will inevitably lead to the identification of final points that need to be addressed.

This section gives you the final opportunity to consider what refinements you could make to your selected final idea to make it even better for its selected purpose.

The excellent example in Figure 2.94 considers whether the original specification has been fully met. The example in Figure 2.95 leads on from the analysis of models to identify a further improvement.

Fig 2.95

Key Points

- From the analysis of models and consideration of the test results, check the specification to identify whether further possible improvements are necessary.

Section 8 Communication

What you need to do

This section is given a total of eight marks; marks are awarded for the communication skills you demonstrate throughout your product study. You should:

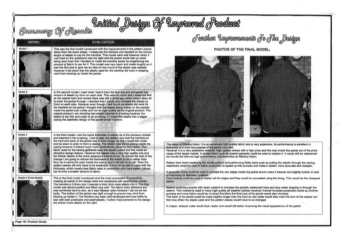

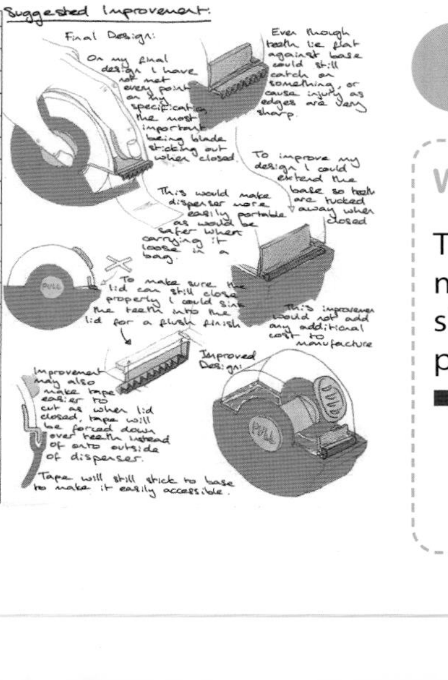

- Use a combination of text, graphical techniques, digital technology, real-time digital images and interactive dialogue as appropriate to present information

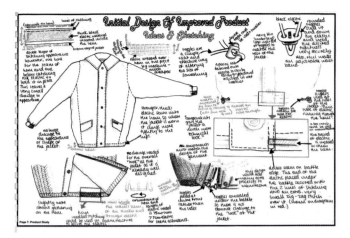

Fig 2.96

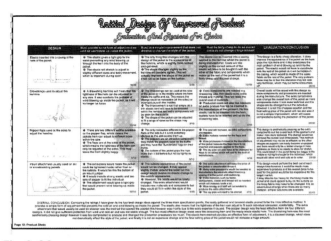

Fig 2.98

For marks in the top band you must check whether your work has all of the required features:

- Text
- Graphical techniques
- Digital technology
- Real-time digital images
- Interactive dialogue

You need a combination of all of these to access marks in the top band.

Better work will have a balance that shows good aspects of all of the above. Do not rely too heavily on the use of scanned images – there is still room for creative, first-hand presentation in the digital age!

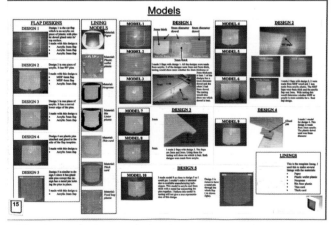

Fig 2.97

Key Points

- Do not over-enhance your background – the one in Figure 2.98 is OK but some projects are difficult to read
- If yours is an A3 paper folder:
  - Bind your project securely
  - Do not use plastic sleeves or folders

Checklist

- Have you selected only one product, not a group of products for analysis?
- Have you identified the needs of the manufacturer and the needs of the consumer?
- Are the key design criteria clear?
- Have you shown the product in use?
- Have you shown similar products in use?
- Have you identified the strengths and weaknesses of your selected product?
- Have you identified strengths and weaknesses of similar products?
- Have you written a summary or conclusion that cross-references similar products to your product?
- Have you researched moral/ethical implications relevant to your product?
- Is there a reference to ethical implications of economic issues? Not just how much it costs!
- Does your brief identify a clear improvement?
- Have you justified your specification points?
- Have you integrated your design ideas, 2D modelling and 3D modelling?
- Have you given evidence of ongoing evaluation on your development of ideas sheets?

- Have you justified one idea as being suitable for taking forward?
- Have you used an appropriate method or system to test and evaluate your final developed idea?
- Are your results presented in real time (that is, as they actually happened)?
- Have you produced a summary that includes analysis of information gained from the models and the testing?
- Have you provided suggestions for further improvements to the proposed idea?
- Have you met the requirements for a combination of communication techniques?
- Have you used digital technology?

And finally:

- If you are submitting your product study as an A3 paper version, have you bound your project securely and is your candidate name and number, and your centre name and number, on your front cover?
- If you are submitting your product study as Microsoft PowerPoint® file on a CD, is your candidate name and number, and your centre name and number, on your CD case cover *and* on your CD?

Achieving your potential

Remember that to have a chance of passing this unit you must check that you have addressed all of the assessment criteria.

The work shown in this chapter is of an exemplary standard at the top end of the A grade. Although the majority will not reach this standard, the important thing is to do everything to the best of your ability and try for the highest assessment band possible in each section. If you do better in some sections it might help to balance out some of the sections that you find more difficult.

An example of three assessment bands is shown below in Table 2.1; this is for the development of improvement' section, worth 56 marks in total.

To achieve an A grade you must meet the minimum competence at the bottom of the top band. If you are determined to achieve a grade A, make sure you read every word and check you have met all requirements.

If you are an average student you will probably be working at a standard around the middle band. Be careful because you will be unlikely to pass if you miss one section out!

To pass you must achieve a standard at the top end of the lower band. If you find AS-level work challenging, you could still pass if you try as hard

Table 2.1 Assessment bands for the development of improvement section

Development of improvement	Marks awarded
Presents a wide range of innovative/creative initial ideas, which demonstrate a high level of development using high-quality annotated sketching, real-time digital images and interactive dialogue. Makes a wide range of appropriate prototype models. Presents a detailed and objective evaluation of ideas against the design specifications in real time and justifies all decisions.	44–56
Presents a good range of innovative/creative ideas with varying levels of development using reasonable-quality sketching, digital images and interactive dialogue. Makes a good range of appropriate prototype models. Presents an adequate and objective evaluation of ideas against the design specification in real time and justifies most decisions.	25–43
Presents only a limited range of innovative/creative ideas, which are developed only to a simplistic level or not at all using annotated sketching at a limited level. Makes a more limited range of moderate prototype models. Presents only a limited and mainly subjective evaluation of ideas with little or no justification of decisions. Little or no reference made to the design specification.	0–24

as you can in each section and make absolutely sure you have made a response to every assessment criteria. If you know you are working at this level, you will not pass if you miss anything out!

See the website for the full assessment criteria for this coursework unit.

Timings

This unit represents 75 hours' work, 45 hours of which should be allocated to the learning of skills. This leaves 30 hours in which to complete the study, which is worth 120 marks.

In general terms it should take about one hour to earn four marks. (eight marks = two hours, 12 marks = three hours, and so on). This should mean you complete the brief and specification section in two hours, but you should spend 14 hours developing your improvement. Testing of the final developed idea should take about three hours.

Remember this is actual production time – there is a considerable amount of time (45 hours) to prepare and practice any skills you may need.

Choose a project that interests you – time will fly!

CHAPTER 3

Design, make and evaluate

Introduction to the A2 coursework project

This unit of the A level course will take up a major slice of your design technology time in your A2 year. It requires a total of 75 hours, divided into:

- 35 hours developing a range of designing, making and evaluating skills
- 40 hours producing the final project for assessment

It represents 30 per cent of the total A level GCE (60 per cent of the A2 course).

The coursework will use and develop skills you have acquired during your AS coursework and consists of designing, making and evaluating a product to meet a need. A marketing presentation and a review and reflection are also included.

You will use and develop skills learnt in the two AS units to produce a coursework portfolio and product that will fully demonstrate your skills. Creativity, flair and innovation are important elements.

For the Product Design A2 written examination you will select a focus area from the list below. If you choose a product that majors in the same focus area, you are likely to achieve the best coursework outcome through working in an area of study you are familiar with:

- Built environment and construction
- Engineering
- Food
- Graphic products
- Manufacturing
- Resistant materials
- Systems and control
- Textiles

The completed project, presented in either an A3 portfolio or in Microsoft PowerPoint® format, will be marked by your teacher and moderated by OCR using the assessment criteria for this unit.

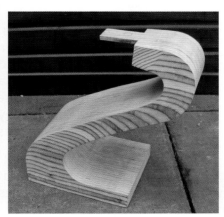
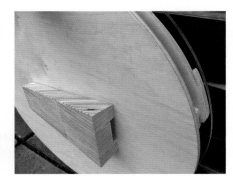

Fig 3.1 Feature table with interchangeable top and CD storage

Important guidelines and procedures

You should make your own judgements and decisions, and direct your own project work, but your teacher will advise, support and assist you by suggesting approaches, alternatives and possibilities, and by directing you to appropriate resources.

Throughout the project you must remember that you will be expected to sign a declaration saying that the work is your own original work. Where some of the coursework is carried out outside the centre, it is important that your teacher is able to confirm the work you have carried out. Sufficient work must be carried out under the direct supervision of your teacher for the whole of your work to be verified.

The nature of some coursework projects at Advanced level may involve significant input from industry or involvement by companies and individuals. These external links and support are encouraged. You must, however, ensure that you yourself have carried out sufficient in-depth and challenging designing, making and evaluating to satisfy the assessment criteria. You can obviously only be awarded marks for your own work, and for the way in which you manage and integrate the contributions of others.

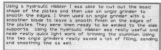

Fig 3.2 Contact with industry

If your centre sets a theme for all projects in your teaching group, you need to identify a design need or opportunity and present a design brief that is individual to you. If you are working together with other students on a group project, you must identify and take responsibility for a uniquely definable aspect of the overall product. Ultimately, your work must constitute a complete project in its own right and provide unique evidence for assessment against each of the assessment criteria.

QCA (Qualifications and Curriculum Authority) have published guidelines for the completion of coursework, indicating the advice and help that may be given. Visit the QCA website (www.qca.org.uk)[1] for clarification.

It is important that you acknowledge the sources of all information and assistance clearly, either at the appropriate point in your record of designing and making, or in a separate list or bibliography. This includes extracts from newspapers, magazines, catalogues, websites, CD-ROMs, photocopied materials and practical assistance with making tasks. By doing this you emulate the approach of professionals who use the knowledge and expertise of others to ensure the highest quality outcomes.

Assessment and submission of coursework

The project is not just about assessment and helping you to attain a qualification; it is as much about your learning and development as an individual, and you will undoubtedly enjoy 'learning by doing' through this coursework.

You should structure your work to follow the assessment criteria, and present your work in section number order. A summary or contents page with a numbering system should be included to aid organisation.

The intention is that the assessment framework should not restrict, interrupt or influence the natural flow and progression of your designing and making of a product to meet a need. The assessment criteria are a means of assessing your approach to key elements in that process: the appropriateness, depth and quality of your work.

Quality of work is more important than quantity. You should aim to present focused, appropriate, high-quality design and supporting written work. Ensure that you include sufficient evidence of your work to enable accurate assessments to be made. Include clear, well-framed photographs or digital images, together with close-up details where appropriate.

If submitted in paper form, a maximum of 25 A3-size pages presented in landscape format is recommended. Sheets should be hole-punched (preferably using a four-hole punch along the left-hand edge) and secured together with treasury tags. Sheets larger than A3 may be included if appropriate, but should be folded down to A3 size and punched to secure them with the A3 sheets. Individual plastic sleeves must not be used for paper sheets.

If submitted in Microsoft PowerPoint® form, it is expected that a maximum of 50 slides will be sufficient, containing text and graphics, scanned-in/digital images and audio/video clips.

The importance of the right project choice

The purpose of the project is for you to show your abilities in designing, making and evaluating. If you are to do this, the project you choose will need to:

- Match your ability and potential – to enable you to achieve the grade you are capable of
- Cover an adequate range and depth of designing, making and evaluating activities, in order for you to satisfy the assessment criteria at Advanced level
- Meet a real and specific need or problem – a client-based project extending beyond your personal needs is most likely to provide you with the best opportunities to display your capabilities throughout the project
- Display creativity, flair and innovation
- Mirror and use industrial processes
- Result in a complete, high quality, marketable product that can be fully tested by the intended user
- Be realistic and manageable within the time and resources available, the facilities in your centre, staff expertise and your access to specialist assistance

Table 3.1 Assessment headings, marks and recommended number of A3 pages

Section		Marks	Pages
1.	Design brief	3	½
2	Information, inspiration and influences	9	2½
3	Design specification	3	1
4a	Design, design development and making	57	13
4b	Innovation	15	
5	Testing and independent evaluation of the final product	9	3
6	Marketing presentation	15	3
7	Review and reflection	9	2
	Total	120	25

Fig 3.3 Snow bike: a marketable product

- ■ Have I thoroughly checked that I have covered all the aspects that I need to?
- ■ Should I ask someone else to look at this?
- ■ Train yourself not to do the minimum or to just meet the minimum standard to satisfy your teacher. Be self critical and always ask: 'How can I improve this further?' Enjoy exceeding other people's (and your own) expectations and achieving the 'wow' factor

Clear and concise communication and presentation

Communication is central to your project. At various stages of the project you will be communicating with:

- ■ Your client and potential users of your product
- ■ Your teacher
- ■ Other students
- ■ The general public
- ■ Those from whom you require advice or information
- ■ Those who will be giving opinions on your designs
- ■ Those giving expert or specialist help
- ■ Those testing and evaluating your final product
- ■ Potential manufacturers or purchasers of your product

You will need to decide at the start of your project whether you will be presenting your work for assessment in Microsoft PowerPoint® or in paper format. You must choose one or the other – a mix is not acceptable. Your choice may be determined by the facilities and expertise available to you at your centre.

You will have various forms of presentation and communication available to you, which may include some or all of the following:

- ■ Text, drawings, diagrams and images on paper
- ■ CAD on paper or in animation/video format
- ■ 2D and 3D models, mock-ups and representations
- ■ Digital images
- ■ Microsoft PowerPoint® with video and audio
- ■ Verbal – face to face or by telephone
- ■ Email, text message, letter, questionnaire…

It is important that you choose the most appropriate means so that you present and communicate what you need to clearly, concisely and fluently. Use bullet points, tables, charts and images. (How many words is a picture worth?) Summarise and keep your work succinct and to the point.

Think about the most appropriate means of presenting an early interview with your client. A video or audio recording would seem to be ideal, with a list of bullet points. This provides a list of key details for the designing together with evidence of what you have done for assessment purposes. If you are presenting your portfolio in paper format then a number of 'still' images from a video, or photos, along with a list of key points, should be sufficient evidence.

It is crucial that you maintain a clear focus on the specific product you are designing, making and evaluating. This means not getting side-tracked into things that have no direct relevance to what you are designing, making and evaluating. You must avoid any generic work that does not directly relate to the specific project.

Fig 3.7 A design team

Collaboration and consultation

The designing and making of products often requires specialist knowledge and skills. In industry and business, design teams include specialists in the various materials and technologies involved, and each brings their own area of expertise to the project.

At this level you are expected to make use of a wide range of people, companies and organisations to support your project work. There will be people within and beyond your centre who will be able to give you specialist help and assistance.

'Bounce' your thoughts and design ideas around with friends and other DT students. Obtain honest opinions. Their suggestions may just give

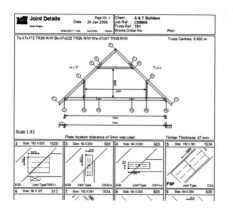
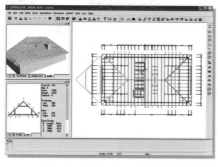

Fig 3.8 Mitek Industrial CAD modelling and drawings of a roof structure and the roof trusses and joint details

you the breakthrough that you need. Don't forget to record and acknowledge their input!

Use of industrial and commercial practices

The intention of the coursework is that, as far as possible, you perform all tasks involved in the project in a way that reflects how they would be carried out in the commercial world. Some of the specific areas to consider are:

- Project management and planning
- Modelling and prototyping
- IT systems
- Production in large quantities
- Manufacturing aspects
- Economic issues
- Style and fashion trends
- Product marketing and retailing
- Quality control
- Health and safety regulations
- Standards, for example testing methods

Understanding society and the environment

Consider the wider effects of design and technology activities. What impact will your product or the way it is produced have on society and the environment?

Use of digital technology

This is now a big part of everyone's lives and should be an integral part of all work in design

Fig 3.9 Digital technology

technology, in particular the use of CAD/CAM and other digital applications related to the specific materials and type of product.

Any time you spend developing your IT skills is time well spent. The use of digital technology is here to stay, and its use in your DT coursework will considerably enhance your capability and attainment.

Examples of projects for each focus area

Built environment and construction

A permanent installation of a ramp and handrails giving wheelchair access to a building. It has appropriately constructed foundations and below-ground preparations, anchorages for handrails, and high-quality finishes to all surfaces.

Fig 3.10 Built environment and construction

Engineering

These athletes' starting blocks required extensive modelling and use of jigs during designing and manufacturing. A range of machining processes including milling, turning and threading were used, with CAD and CAM being used extensively throughout. A range of standard parts were used effectively.

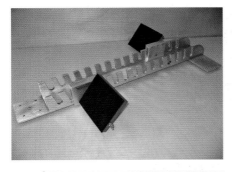

Fig 3.11 Engineering

Food

This project focused on range of products suitable for a special occasion or festival meal

that can be combined to cater for a large group of people, including assorted canapés on crackers and toasts. Skills included savoury biscuit making as well as presentational skills. A gelatinised and non-gelatinised mousse was prepared for the toppings and a number of alternatives trialled.

Fig 3.12 Food

Graphic products

A point-of-sale display for a range of new cosmetics products. It includes development of graphic design of the net through to camera-ready artwork for the card packaging, as well as a laser-cut, folded and assembled acrylic table stand with logo and brand details. Promotional leaflets can be stored underneath the display. A range of graphic design skills are evident.

Fig 3.13 Graphic products

Fig 3.14 Manufacturing

Manufacturing

A rifle light for night shooting: a multi-material, multi-process product. The prototype was produced in glass-reinforced moulded plastic lamp casing, with a cast and machined aluminium bracket, fabricated polystyrene battery holder and carrying case in textiles. The project included consideration of the manufacturing context in terms of the materials and standard/bought-in components, production planning, manufacturing processes, quality control techniques, assembly, and handling and storage requirements.

Resistant materials

A 'back to driving' physiotherapy aid/leg exerciser for use in hospital (see Fig 3.16 on next page). The flat-pack, heavy-duty hospital version was constructed in MDF and mild steel tubing with fittings; the 'take-home' economy version was constructed in laser-cut recycled card and clip-together plastic fittings. It uses 'theraband' strapping available in different forms.

Systems and control

A 'park-safe' ultrasonic parking sensor: the circuit was developed from first principles using an astable oscillator at ultrasound frequency with a band-pass filtered receiver and phase-lock loop. It has a wailing siren output and is self-calibrating; it has a vacuum-formed casing with laser-cut apertures and engraved brand name.

Fig 3.15 Systems and control

Textiles

This student designed an interactive headboard for two- to five-year old children to ease the transition from a cot to a bed (see Fig 3.17 on p63). It included a wide range of fabrics, hand

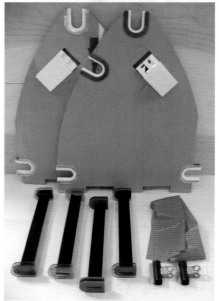

Fig 3.16 Resistant materials

and machine processes and techniques, and included the use of CAD/CAM. It incorporates an amplifier and speaker for the MP3 music player.

Choosing your project

Possible starting points to help you find a suitable project

- Your sports and hobbies
- Your workplace, work experience, business or industry visit
- Your home and neighbourhood
- Your leisure time
- Your subjects and activities at school
- Your interests and career aspirations
- Your family, relations, neighbours, friends, their places of work or study
- Things you find difficult or frustrating – or see others struggling with
- People with special needs, the young, the elderly, the disabled

Fig 3.17 Textiles

- The community
- Local businesses
- Public/emergency services
- Environmental issues
- Design for the developing world

See www.hodderplus.co.uk/ocrd&t
List of possible contexts

Start by listing things you are interested in, things you enjoy or are keen to be involved in, places you like to be, your possible career.

Your commitment over the months of the project is more likely to be maintained if you have 'ownership' of your project in this way. Your project should be your personal choice, but agreed in conjunction with your teacher.

Think of the many activities and tasks that are involved with the interests you have listed and investigate possible design projects arising from them.

You also need to think of the project as your opportunity to 'make a difference' in the world. You will be spending time using your increasingly wide range of skills to look into real needs and problems and come up with meaningful proposals to solve them, which will potentially change and transform

a situation. At this very moment, wherever you are, stop and look at the products that you are surrounded by. Consider where the ideas might have started, the stages of design that may have been involved, and the impact the final products have on the way people live and work.

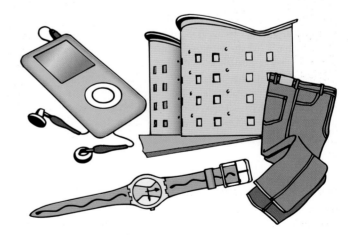

Fig 3.18 Everyday designed products

Talking to others

Talking to others about possible needs and problems is invaluable. This can be with other students, family, friends and neighbours. Talk to them about their situations and possible project choices that might arise.

In turn, talk through your typical weekday and a typical weekend day with another student. As you do, make a list of things you find difficult or frustrating, or things you remember others struggling with. Are there possible design projects here?

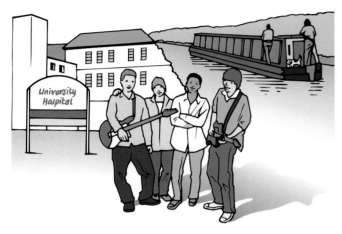

Fig 3.19 Identifying a need

Make a list of people you know or have contact with – people from different work situations, people from different backgrounds and with different interests, who you can talk to about possible design projects.

Maybe someone who works in a shop, keeps lots of pets, is a keen musician, works in a hospital, or someone who enjoys boating holidays? All of these people may have quite a different outlook on life, and may well help you identify a genuine need for your project.

Find a suitable time and go and chat to some of those people! Try and do this as early as you can with the aim of finalising your project choice and making a prompt start to your coursework project.

Making your decision

Following your exploration of possibilities, propose three final possible project choices.

The 'Coursework Project Proposal Form' is useful to give the details ready for you to discuss with your teacher, including:

- Proposed title/product
- Opportunities for innovation/original design/new thinking

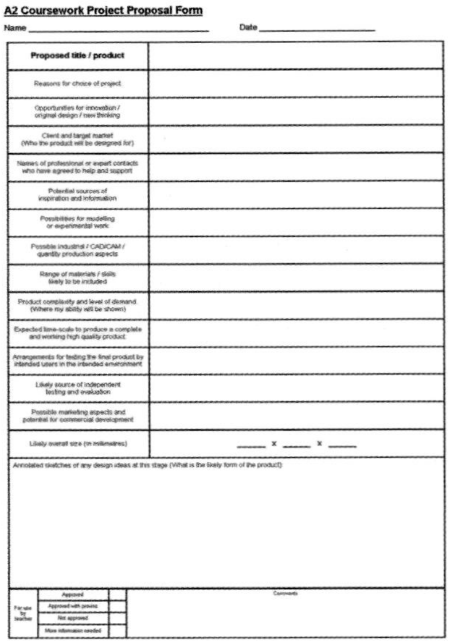

A2 Coursework Project Proposal Form

Name ______________________ Date ______________

Proposed title / product	
Reasons for choice of project	
Opportunities for innovation / original design / new thinking	
Client and target market (Who the product will be designed for)	
Names of professional or expert contacts who have agreed to help and support	
Potential sources of inspiration and information	
Possibilities for modelling or experimental work	
Possible industrial / CAD/CAM / quantity production aspects	
Range of materials / skills likely to be included	
Product complexity and level of demand (Where my ability will be shown)	
Expected time-scale to produce a complete and working high quality product	
Arrangements for testing the final product by intended users in the intended environment	
Likely source of independent testing and evaluation	
Possible marketing aspects and potential for commercial development	
Likely overall size (in millimetres)	____ x ____ x ____

Annotated sketches of any design ideas at this stage (What is the likely form of the product)

For use by teacher			Comments
	Approved		
	Approved with provisos		
	Not approved		
	More information needed		

Fig 3.20 The 'Coursework Project Proposal Form'

- Client and target market (who the product will be designed for)
- Names of professional or expert contacts who will help and support
- Possible industrial/CAD/CAM/quantity production aspects
- Range of materials/skills likely to be included
- What form the product is likely to take
- Possible marketable features to be considered

Be prepared to present your proposals to your teaching group (with your teacher of course!) for discussion and comment. The opinions of others will be extremely valuable and will help you identify potential problems and key focus areas for your designing. Make sure you record these comments as they will be useful in the formation of your design brief.

At this stage you need to do two things before continuing and making a start on the project:

- Carry out an internet/catalogue search to ascertain the availability of products that already meet your identified need or problem. It could be that you are proposing to design a very common item (for example a desk tidy) but with a new arrangement or new feature of some sort (such as revolving with holders for an MP3 player and a mobile phone). Your search should concentrate on the new elements
- Make contact with individuals or organisations

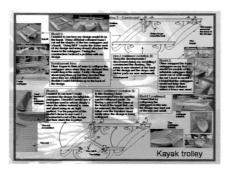

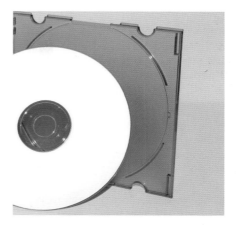

Fig 3.21 Choose your means of presentation

who may be involved in the project, to confirm their support

The results of your search should help you to clarify your proposal. They may prompt some revision to your proposed project, which should be carried out in conjunction with your teacher, but it is essential that you are fully aware of these issues before you begin in earnest.

Getting started

Presentation

Choose your means of presentation for the portfolio – Microsoft PowerPoint® or A3 paper format. Either way, you will need to decide on your slide or page layout.

Choosing Microsoft PowerPoint® will give several additional benefits, including:

- The ability to integrate real-time video or audio recordings of:
  - The problem or need in context
  - Group discussions and presentations
  - Talks with clients and user groups
  - Interviews
  - Surveys
  - Product disassembly
  - Materials, components or ingredients
  - Making processes
  - Development and modelling work
  - User trials
  - Product testing and evaluation
  - Marketing presentations
  - Project diaries

(Many of these would otherwise require lengthy writing up as text and diagrams in the portfolio, thus there is a significant time saving.)

- The ability to include animated sequences of CAD assemblies and CAM graphics
- The facility for indexing and using hyperlinks within the program to enable easy navigation
- The ability to include hyperlinks to websites to show sources of information and inspiration
- Reduced use of paper (and trees) and associated storage and handling
- All work is recorded electronically/digitally, giving the facility to backup all work regularly and easily

If you choose to present your work for assessment in Microsoft PowerPoint®, ensure you have the necessary skills, or that there are people who can support you when using the software. You also need to follow the examination board's requirements, for example:

- The work must be one single coherent piece – portfolios comprising of multiple documents are not acceptable
- 'Pack and Go' or 'Package for CD' must be used when copying to CD/DVD to save all related files (for example video clips)

Remember to check that your Microsoft PowerPoint® will run exactly as required from the CD/DVD on another computer before you submit it for assessment.

3D items such as materials and samples should not be included in paper format portfolios – digital images or photographs of such items are sufficient evidence for assessment.

Look at examples of previous students' A2 coursework portfolios. Some copies may be available in your centre. You will find there are invaluable in terms of both presentation and content.

Examine a range of professionally produced reports, portfolios, articles, brochures, catalogues, advertisements, websites, and other presentations and documents (printed and digital). Look closely at the layout and use of text, colour, diagrams, symbols and images. Assess the impact and effectiveness of the different methods and media used.

What can you learn from the design and detail of these items about how to present your own work?

Having a written plan for the project, in which you record successes and failures along the way, will enable you to 'review and reflect' on the effectiveness of your designing and making processes, which will be in Section 7 at the end of the project.

Produce a spreadsheet or time line to indicate key stages and target dates for the coursework project. Leave spaces for your notes and comments.

Fig 3.22 Project planning

Planning

A time plan is not a requirement in this specification, but you do need to think about the various stages of the coursework and what will be involved. The success of any project of this nature is dependent on careful planning ahead and being prepared.

Sections 1, 2, and 3 total 15 marks, 12.5 per cent of the marks for this unit. It is therefore important that you do not spend too long on these first three sections. The largest amount of marks is for Section 4 – the design, design development and making, and innovation – which has a total of 72 marks, or 60 per cent of the total marks for this unit. This is the main section, and where your energies should be focused.

Remember that there are a number of tasks to be completed following the completion of the making of your product, and 27.5 per cent of the marks are available for these in Sections, 5, 6 and 7. Ensure you leave sufficient time to do justice to these sections. Arrangements for the testing of your product may need to be made in advance.

Section by section

We will now look in detail at each section:

- What you need to do – the requirement for the section
- The assessment criteria – with the marks and suggested number of pages
- Key points – how to maximise your marks
- Examples of students' coursework showing the type and quality of work you should aim to produce – look at the aspects of these extracts that make them successful
- Activities to help you gain the skills and knowledge you need

Remember that you must keep all coursework tightly focused in the time available. It is crucial that you keep on track. You will not be expected to incorporate every guideline and suggestion given in each section, but to carefully consider and follow those that are most appropriate to your personal approach to the project and those most relevant to the product you are designing.

Section 1   Design Brief

What you need to do

Present a design brief for a marketable product

Table 3.2 Assessment criteria, marks and pages for Section 1

	Marks	Suggested Pages
Design brief	3	½
Presents a clear and precise design brief for a marketable product.	3	
Presents a reasonable design brief for a marketable product.	2	
Presents a superficial design brief for a marketable product.	0–1	

By the time you get to this stage you will have already done a great deal of preparatory work, thinking and exploration. This is vital if your design brief is to be clear and precise.

You need to state what you are going to design and to give a clear direction for the project.

Your design brief should include many of the following:

- The need or problem – the situation as it is now
- Images and details of the context/situation/need/problem
- An outline description of the product (for example outdoor public seating, savoury snacks, packaging and display, sports equipment container)
- Why existing products are not suitable or why your design concept may be superior to existing products
- Your client or user group – who the product is for, what specific groups of people will use it (for example families on holiday, vegetarians, young people on the sports field…), how will it help, the difference it will make
- The key stakeholders (those with a direct interest, involvement or investment in the project)
- Why the project is worth doing. Why it is needed now. The desired end result of the project
- The product's key functions and features
- A breakdown of the known key aspects of the product at this stage; for example in a window-opening device: the electronics, the mechanism, the casing, the controls. This will help structure the designing. You could do this in a 'thought shower' format
- The key focus areas in the design that should ensure a marketable product – the aspects that will need particular attention if it is to be an attractive and desirable product
- The extent of the project – things that will be included and things that are outside the scope of the project
- The aspects of the project that you anticipate needing specialist help with

Key Points

- Use your completed 'Coursework Project Proposal Form', approved by your teacher, to guide you
- Remember that the brief is not a solution or even a list of requirements, but a starting point for the design process
- Be specific to your own product and avoid statements that could apply to any project

Examples

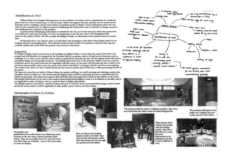

Fig 3.23 Physiotherapy aid – details of location, context and current situation; specific exercise requirements and existing (makeshift) solutions (see also Figure 3.16)

Fig 3.25 Leaflet stand for museum – contact with client English Heritage from the outset; details of context and specific design issues to be tackled

Examine design briefs from previous coursework projects that are available at your centre. Consider and comment on the importance of those briefs in providing a clear direction for the project.

Look closely at any existing product. Make a list of the main points that you think would have been included in the design brief.

Fig 3.24 High-fibre meal for teenagers – context for healthy eating; key issues to be addressed; clear brief

Section 2   Information, inspiration and influences

Table 3.3 Assessment criteria, marks and pages for Section 2

	Marks	Suggested pages
Information, inspiration and influences	9	2½
Obtains all significant information relevant to the design of the product. Presents a wide range of evidence to show the sources of inspiration and influences on the designing.	7–9	
Obtains some information relevant to the design of the product. Presents an adequate range of evidence to show the sources of inspiration and influences on the designing.	5–6	
Obtains limited information relevant to the design of the product. Presents a limited range of evidence to show the sources of inspiration and influences on the designing.	0–4	

When embarking on a project, professional designers use a variety of methods to collect the information they need and to gain inspiration for their designs.

When embarking on a project, professional designers use a variety of methods to collect the information they need and to gain inspiration for their designs.

You will follow some of these methods and present a collection of evidence to show the information you gathered and your sources of inspiration. You need to show the relevance and influence of the items you include.

Your collection of evidence may include a selection of the following:

- A notebook – notes on meetings with your client, user group survey results
- A diary – a progress record showing your planning for key aspects
- Photographs of industrial visits, existing and similar products
- A scrapbook
- A paper version – magazine and newspaper cuttings, photographs
- A digital version – images, screenshots, audio clips, video clips
- A sketchbook – drawings and diagrams of existing products, innovative features, design ideas or possibilities

- A 'mood board' or collage of images relating to your product, showing colours, styles and other inspirational material
- A 'job bag' – designers use these to contain everything relating to a project – cuttings, sketches, CDs, samples, components, products
- An 'inspiration' or 'ideas' box – colour samples and swatches, materials, existing products

You are likely to be involved in some of the following:

- Consultations with your client or user group regarding their requirements
- Establishing the needs and requirements of the key 'stakeholders' in this product, such as the consumer, manufacturer, distributor, retailer, society
- Contact with specialists who may be involved in the project
- Personal examination or disassembly of existing and similar products
- Investigating the work of other designers
- Looking at historical and traditional influences
- Obtaining details of constraints such as sizes relating to the product and its location and use
- Reference to documentation published by regulatory bodies and authorities relating to your product and the materials, components or ingredients you may use
- Analysis of results from surveys, questionnaires and interviews
- Use of internet forums to gain opinion from a broad range of potential users
- Exploration of current market trends and the influences of fashion and lifestyle
- Consideration of relevant social, moral, environmental, cultural and sustainability issues

Fig 3.26 An inspiration box

Fig 3.27 Horse box trailer storage – use of a website forum to obtain information and ideas from the target market

Key Points

- Focus on the key information needed to support your designing
- Remember that the purpose of this section is to refine your brief and enable you to compile a detailed list of requirements and features for your product in the next section
- Obtain information from primary and secondary sources on a 'need to know' basis
- Record in real time, not retrospectively
- State the value and relevance of the information and of the other items you have obtained
- Acknowledge all sources

Examples

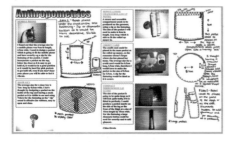

Fig 3.28 Physiotherapy aid – interview with client to obtain the various exercises required

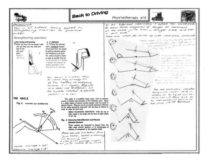

Fig 3.29 Leaflet stand for museum – relevant standards, ergonomics, anthropometric data and sizes

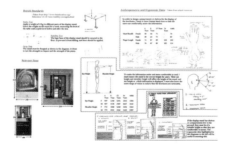

Fig 3.30 Outfit for clubbing, incorporating storage for handbag items – shapes and sizes of items to be stored; design possibilities

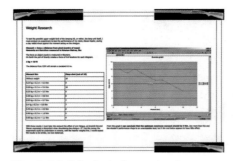

Fig 3.31 Rifle light for night shooting– tests to determine the effect of additional weight on shooting performance; maximum weight for lamp established (see also Figure 3.14)

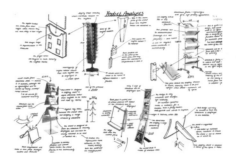

Fig 3.32 Leaflet stand for museum – information and inspiration from existing products: aesthetics, materials, construction, sizes

Fig 3.33 Food products suitable for cystic fibrosis sufferers – disassembly of sausage products

Fig 3.34 Children's clothing – possible directions for the project; adapting garments

Fig 3.35 Fashion clothing – inspiration board with notes

Fig 3.37 Children's clothing – historical influences

Fig 3.36 Fashion clothing – current trends and predictions

- Examine a successful product and list the reasons for its success. List the key information that would have been obtained by the designer in order to design that product.
- Study the design of a product closely. List the possible sources of inspiration for its design.
- Examine the products of a well-known designer, from any design field, and discuss the likely influences on their work.

Section 3 Design specification

What you need to do

Produce a design specification for the product.

Table 3.4 Assessment criteria, marks and pages for Section 3

	Marks	Suggested Pages
Design specification	3	1
Produces a detailed design specification.	3	
Produces an adequate design specification.	2	
Produces a superficial design specification.	0–1	

The design specification is a list of design requirements and should follow on directly from Section 2. It needs to be a list of detailed must haves, must bes, and must dos; a conclusion to all the work on the project so far – giving a reference point for the designing of your product and a framework for its testing and evaluation.

Look at everything you have gathered and the information you have found out. What are the key issues and influences identified that will guide your designing?

It is helpful to have a checklist or a series of headings in this section. These must then be applied to your own product. Chapter 1 makes some suggestions that you may have used during the advanced innovation challenge.

You may find it valuable to follow the practice of designers who often set priorities for the design of the product by stating the relative importance of the specification points. This is done by applying a numerical ranking, by categorising essential and desirable criteria, or by stating primary and secondary requirements.

It is important that the design specification covers all key areas relating to your product. It should include all of the following that relate to your product:

- Measurable targets – optimum, maximum or minimum sizes, weights, capacities; quantities and costs; nutritional details
- Performance criteria – specific details of the performance required; properties of material and finish; aspects of quality, standards and reliability; expected life span
- Technical detail
- Aesthetics, texture and taste aspects
- Marketing and commercial aspects – key factors affecting the distribution and presentation of the product to potential purchasers, clients or user groups. These will include reference to brand image, packaging, documentation, aesthetics, fashion and current trends
- Reference to user factors – ergonomics, safety, legislation, handling, storage and maintenance
- Aspects relating to manufacturing – use of standard or bought-in components and ingredients, scale of production factors and economics
- Consideration of environmental aspects – disposal, recycling and reuse

Key Points

- Make sure your requirements are realistic
- Your specification should be agreed by your client or user group
- A clear specification will maximise the likelihood of your product meeting the original need

Examples

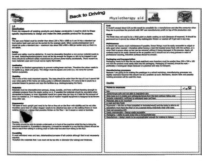

Fig 3.38 Physiotherapy aid – clear and specific requirements including quantities and technical detail; 'points to overcome' checklist

Fig 3.39 Savoury party snacks – summary of information and influences; design specification in the form of essential and desirable criteria

Fig 3.40 Folding electric guitar – detailed specification including tolerances and requirements for both electrical and mechanical components (see also Figure 3.5)

Examine closely and analyse any existing product to determine the likely design specification. List the key design requirements under a series of suitable headings. Justify each of the requirements.

What you need to do

Demonstrate competence in the design, design development and making of the product, to include the following package of evidence:

- The generation and exploration of design possibilities
- The use of digital technologies
- Experimenting and modelling
- The refining and defining of a final design through ongoing evaluation
- The planning and making of the product

Table 3.5 Assessment criteria, marks and pages for Section 4a

	Marks	Suggested pages
Design, design development and making	57	13
Demonstrates a high level of competence in the design, design development and making of the product.	45–57	
Demonstrates a sound level of competence in the design, design development and making of the product.	23–44	
Demonstrates a low level of competence in the design, design development and making of the product.	0–22	

Section 4 (4a and 4b together) accounts for 60 per cent of the marks for this unit, so most of your time and energy should be focused on this section.

You will use your creative skills to generate and explore a range of design possibilities and to develop and refine a final design solution. You will then make the product using materials, components, ingredients and appropriate technologies.

Remember that a relatively simple product involving little real difficulty and challenge must include a considerable level of detail and depth if the coursework is to satisfy the examination requirements at this level.

In a similar way to Section 2, you need to present a collection of evidence for assessment.

Your collection of evidence is likely to include many of the following:

The generation and exploration of design possibilities

- Sketches, drawings, and diagrams
- Annotation and notes, descriptive and evaluative
- Ranges of possibilities and ideas

The use of digital technologies

- CAD/CAM
- Video, audio, photography
- Image manipulation software
- Simulation and analysis software

Experimenting and modelling

- 2D and 3D models – to scale and/or actual size
- Mock-ups, prototypes, trials and tests, sensory analysis, use of charts and diagrams to aid analysis and comparisons
- Models focusing on specific details of the solution and/or models of the final proposal
- Experimentation with techniques to improve and refine possible processes
- Simulations and computer modelling
- All with a purpose and value, recorded in real time, and including analysis and evaluation

Refining and defining a final design through ongoing evaluation

- Progression, evolution, refinement, increasing depth and detail (a definition of development)
- Recorded changes and modifications to the design, referring to the design specification
- Final CAD working drawings showing clear details of the whole product and the parts
- For food items, the manufacturing specification should include:
  - Detailed recipes with specific quantities, function of ingredients used and nutritional breakdown
  - Sizes of components, quantities and depths of toppings, sauces and fillings
  - A HACCP (Hazard Analysis and Critical Control Points) chart to identify food safety issues in production and storage
  - Allergy concerns and special claims (for example organic, vegetarian)
  - Packaging and labelling details, including heating instructions where appropriate
- For items to be printed, camera-ready artwork or print masters using suitable software; a 3D outcome is required for part of the making for graphics products
- For textiles items, a detailed lay plan using suitable software. This should include overall sizes of the pieces, and the layout and positioning for the cutting of the fabric from the roll. Other important elements to include are: grain direction arrows, marks to show where parts match, seam allowances and indication of scale
- Exploded or cross-sectional views, details of patterns, jigs or templates.

- A clearly defined final design which, if it was sent to a distant manufacturer, would enable that manufacturer to produce the item exactly as intended

The planning and making of the product

- Key stages and a clear, logical sequence for the making
- CAD-animated assembly sequences
- The sourcing and selection of bought-in/pre-manufactured components as appropriate
- Use of hand and machine processes, use of CAM
- Use of industrial processes where possible
- Video clips of key processes
- Quality control checks – ensuring the product outcome is the best it can be
- A real-time record of progress – a diary or log, a video diary

Evidence of the following will maximise your marks:

- An appropriate range and depth of skills
- Initiative, innovation and enterprise
- An integrated approach, combining many of the designing elements listed above
- Clear and fluent progression toward the final proposal
- Consideration of the economic use of materials, components and ingredients
- A planned and structured approach to problem solving (rather than random trial and error)
- Regular consultations with client/user groups at key stages of design and development, and consumer testing and evaluation, with feedback reflected upon and used to influence product development. This mirrors professional practice where designers evaluate their design ideas and proposed solutions through contact with a representation of their target market. For example, at 'car clinics', ideas and models for new car designs are presented to specific sectors of the car-purchasing public for their opinions on a range of design aspects
- Collaboration with others where required,

team working, managing and integrating the input of others

- The incorporation of peer group evaluation
- Care, precision and attention to detail
- Technical and scientific detail including sizes, quantities, capacities, nutritional information
- A high-quality, complete outcome, suitable for its intended market, and safe and easy to use and maintain
- A detailed knowledge of the working properties and functions of materials and components/ingredients
- Correct use of tools and equipment
- Regular reference to, and evaluation against, the list of design requirements (namely your design specification)
- Additional information gathered on an 'as required' basis during development, for example ergonomics, safety, fittings, ingredients, components, materials
- Consideration of marketing aspects such as packaging (see Section 6)
- Consideration of commercial and manufacturing issues

Key Points

- Record your designing and making in real time; maintain a 'live' record
- Organise and manage your time and resources effectively
- Forward thinking and planning is crucial
- Benefit from the opinions and expertise of others
- Respond positively to challenges and changes
- Ensure clear evidence of all stages is presented, including close-up photographs or video, and 'print screens' to show detail
- Acknowledge all sources of assistance received (teacher, specialist, technician, another student…)

Examples

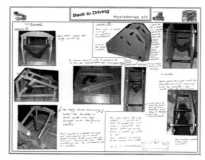

Fig 3.42 Physiotherapy aid – extensive full-scale modelling and refining of ideas

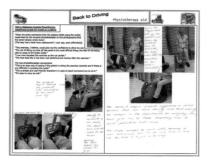

Fig 3.43 Physiotherapy aid – testing of the model during design development; feedback from client and patient to inform further development

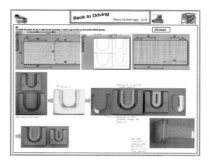

Fig 3.44 Physiotherapy aid – modelling using CADCAM/laser cutter, and trials for 'clip joint'; Use of simple MDF jig to assemble components accurately

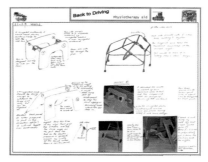

Fig 3.41 Physiotherapy aid – annotated ideas and early modelling

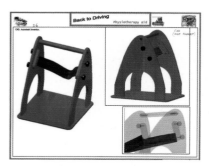

Fig 3.45 Physiotherapy aid – CAD images of final solution

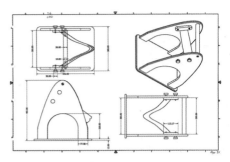

Fig 3.46 Physiotherapy aid – CAD assembly drawing of final solution

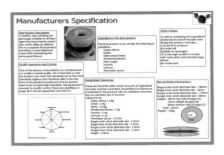

Fig 3.50 Healthy meal for teenagers – final solution including full details of ingredients and manufacture with details of QA and QC (quality) and tolerances

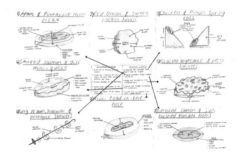

Fig 3.47 Savoury party snack – ideas and possibilities with reference to essential and desirable criteria

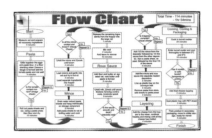

Fig 3.51 Food product flow chart – planning for making

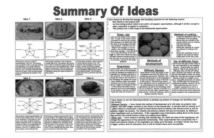

Fig 3.48 Sausage-based casserole – ideas developed and tested with ongoing evaluation

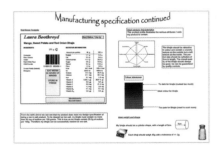

Fig 3.52 Low-salt food product – nutritional and sensory analysis; manufacturing details and tolerances, label for food packaging

Fig 3.49 Food product HACCP analysis

Fig 3.53 Children's outfit – ideas and possibilities explored for several items

Fig 3.54 Textiles product – exploration of colours, fabrics, styles, buttons/ties

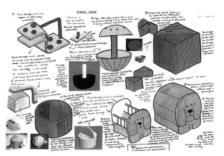

Fig 3.58 Child's storage item – integrated sketching, card modelling and CAD modelling

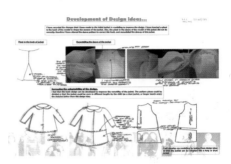

Fig 3.55 Jacket – ongoing development and refinement

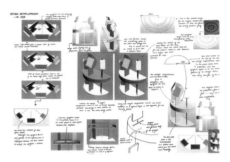

Fig 3.59 Leaflet stand for museum – use of CAD in design development to show alternatives

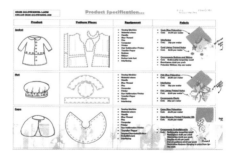

Fig 3.56 Jacket – final product details: pattern pieces, equipment and costs

Fig 3.60 Recycling bin – ideas inspired by the Bauhaus and car styling

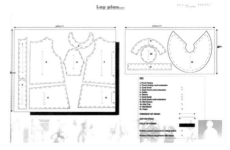

Fig 3.57 Jacket – lay plan showing grain direction and overall sizes

Fig 3.61 Stirring device for use in a microwave – ideas, CAD modelling and testing of 3D model inside microwave

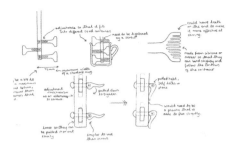

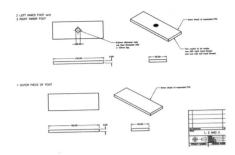

Fig 3.62 Stirring device for use in a microwave – possible 'paddle' shapes and clamping mechanism to allow for different sized dishes

Fig 3.66 Stirring device for use in a microwave – one of a series of parts drawings

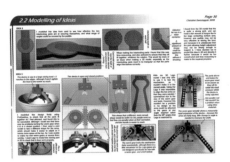

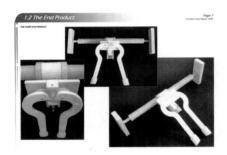

Fig 3.63 Stirring device for use in a microwave – development of the paddle and mechanism design using modelling in card, CAD and LEGO®

Fig 3.67 Stirring device for use in a microwave – images of final outcome

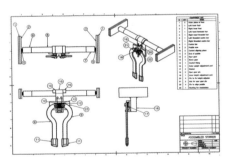

Fig 3.64 Stirring device for use in a microwave – CAD assembly drawing with components list

Fig 3.68 Stirring device for use in a microwave – record of the making processes

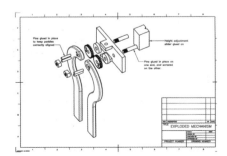

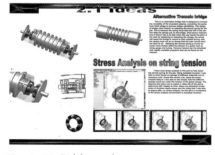

Fig 3.65 Stirring device for use in a microwave – exploded drawing showing assembly details

Fig 3.69 Folding electric guitar – stress analysis feature within CAD software used to calculate spring tension required

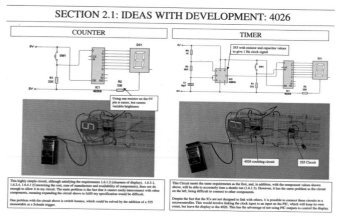

Fig 3.70 Fitness timing device – design development, CAD, breadboard modelling

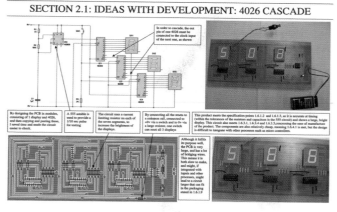

Fig 3.71 Fitness timing device – design development, CAD, PCB trials and modelling

See hodderplus.co.uk/ocrd&t

Further images of these projects and further examples of students' work in all focus areas/materials

Generate a range of alternative solutions to one of the following problems:

■ Fastening the lining of an item of clothing to enable quick and easy removal

■ A means of fixing the lid of a container without the use of a hinge or other additional fitting

■ Alternative combinations of spices for inclusion in an Indian food product

Look closely at a small component or a part of a product (for example a small shelf support bracket or a nylon pulley wheel). Think creatively and quickly to generate/explore as many alternative methods of manufacture for that part as possible. This is best carried out in discussion with another student, with the use of a whiteboard or sketch pad.

Examine a product involving a wide range of components or parts. List some of the standard components used. Explain why the designer has chosen to make use of these 'bought-in' or 'ready-made' components.

Some of the activities suggested in Section 6 will be helpful for Section 4.

Section 4b Innovation

What you need to do

Show innovation.

Table 3.6 Assessment criteria and marks for Section 4b

	Marks
Innovation	15
The designing/making shows clear evidence of innovation.	12–15
The designing/making shows some evidence of innovation.	7–11
The designing/making shows little or no evidence of innovation.	0–6

Note that separate work is not required in this section. Section 4b assesses the level of innovation you have demonstrated in Section 4a.

You should aim to show some of these characteristics of innovation:

- Exploration beyond the obvious and immediate
- Risk-taking, willingness to 'have a go'
- Thinking 'outside the box'
- Pushing back the boundaries
- Taking ideas and possibilities 'further'
- An unconventional or unusual approach to problem solving
- The ability to disseminate a design task and identify aspects where a creative approach is required
- A passion for getting things right
- Originality

- Creativity
- Flair
- Responding positively to challenges
- Perseverance
- An entrepreneurial approach

Key Points

- Your ability to show innovation may be evident in the design of the final product, for example the use of materials and technology, aesthetics and styling
- Innovation may be evident in the process of designing
- Innovation may be evident in aspects of manufacture – the methods used in the making

- **Study the work of an inventor or creative practitioner.**
- **Compare a range of similar products (for example mobile phones, point of sale displays, pairs of gloves). Identify and describe the innovative features.**

Section 5   Testing and independent evaluation of the final product

What you need to do

Show evidence of the testing of the product against the specification:
- Identify and state strengths and weaknesses in the product
- Respond to independent evaluation

Table 3.7 Assessment criteria, marks and pages for Section 5

	Marks	Suggested pages
Testing and independent evaluation of the final product	9	3
Shows evidence of thorough testing of the final product against the specification. Identifies and clearly states the strengths and weaknesses in the product. Responds positively to in-depth independent evaluation of the product.	7–9	
Shows reasonable evidence of testing of the final product against the specification. Identifies some strengths and weaknesses in the product. Shows a reasonable response to some independent evaluation of the product.	5–6	
Shows limited or no evidence of testing of the final product against the specification. Identifies few strengths and weaknesses in the product. Shows a superficial response to limited independent evaluation of the product.	0–4	

Now that you have completed the making of your product, you need to test it to see how well it fulfils all the requirements and performance criteria you identified in your design specification.

Remember that in this section you are testing the product. You are not commenting on the project as a whole or the process of designing and making.

The product needs to be tested by the intended user or consumer, in its intended location. It should ideally be subjected to all the expected conditions that it would face during the phases of its life – its suitability in all situations and conditions in which it may be placed, used, consumed, stored, packaged, or transported.

You should devise a series of tests to determine in a methodical and systematic way just how well it performs. Refer to standard tests used commercially including those that are specified by regulatory bodies and authorities such as British Standards and the Food Standards Agency.

Some things may be difficult to test (for example how long a product might last, how well it might keep when stored) so the opinions of experts and those with experience in the appropriate field will be important.

Arrange evaluation of your product by those qualified to give opinion and comment because of their knowledge, qualifications and expertise in the specific design field. Your response to their comments should include modifications needed to your product to satisfy any concerns or deficiencies.

Direct contact with the product is required for testing and evaluation to take place. The evaluation of a product by any other means (for example a digital image sent by email) is likely to be very limited.

In a similar way to earlier sections, you will need to present a collection of evidence for assessment.

Your collection of evidence is likely to include many of the following:

- Video, audio and photography recording the testing and evaluation being carried out
- Charts or tables showing clearly the results of testing and the strengths and weaknesses of the product
- Surveys or questionnaires conducted with your target market
- Original feedback direct from your client or user group and independent evaluators
- Sketches, drawings and diagrams showing possible improvements that are a result of the testing carried out

Key Points

- Evaluate the product, not the project
- Carry out testing of the product in a 'real life' situation
- Record the testing of the product in real time, not retrospectively
- State clearly the strengths and weaknesses identified
- Obtain objective and impartial comments from others
- Be positive and openly consider the suggestions made by specialists
- Propose modifications to the product that address the issues raised in the testing and evaluation

Examples

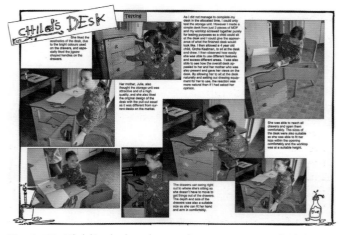

Fig 3.72 Child's desk – thorough user testing including comments from parent/client

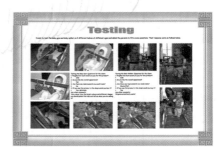

Fig 3.73 Baby gym/walker – testing with two babies of different ages and questions answered by parents

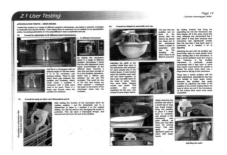

Fig 3.74 Stirring device for use in a microwave – product testing in a range of microwaves; strengths and weaknesses identified with reference to specification points

Fig 3.75 Leaflet stand for museum – full testing of the product in use at the museum; strengths and weaknesses identified; letter with comments from English Heritage and response by student

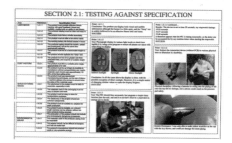

Fig 3.76 Fitness timing device – testing to specification points

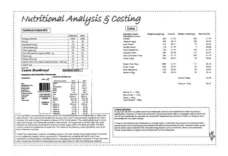

Fig 3.77 Raspberry and chocolate cheesecake – evaluation of final product including nutritional analysis and costing

Fig 3.78 Raspberry and chocolate cheesecake – expert opinion, peer group presentation and consumer target group testing/questionnaire

- **Explain the difference between subjective and objective evaluation**
- **Choose a product in your chosen material focus. Identify the opportunities and possibilities for user testing/sampling of these products**
- **Choose a product in your chosen material focus. Identify the opportunities and possibilities for obtaining independent evaluation**
- **Carry out an investigation into the legislation, regulations and national/international standards that apply to a product in your chosen material focus. Devise a series of tests to measure the performance and suitability of aspects of that product based on your findings**
- **Investigate the work of consumer organisations that report on the quality and suitability of ranges of products and give ratings and recommendations to prospective purchasers, for example Which? (www.which.co.uk). How do they go about testing and evaluating products? Who carries out the tests? What criteria do they use? How are their results reported?**

Marketing presentation

What you need to do

Using appropriate techniques create a marketing presentation suitable for the final product.

Table 3.8 Assessment criteria, marks and pages for Section 6

	Marks	Suggested pages
Marketing presentation	15	3
Creates a competent marketing presentation suitable for the final product.	12–15	
Creates an adequate marketing presentation suitable for the final product.	7–11	
Creates a weak marketing presentation for the final product.	0–6	

In this section you will show how your product will be marketed – how your product will be presented to potential customers or clients. A good starting point is to identify the key features and characteristics of your product that will appeal to the customer. Marketing professionals call this the 'Unique Selling Proposition' or 'USP':

- What is your product's USP?
- What is distinctive or unique about your product?
- What will make people buy your product rather than another?

The answers might include some of these aspects of your product:

- Its versatility
- Its reliability
- Its performance
- Its ease of use
- Its efficiency
- Its relevance to lifestyle and fashion
- Its superiority to other products

Your presentation should include proposals from the following:

How your product will be promoted and advertised

- How will you inform potential customers about your product?
- How they can acquire it? Consider the ways that your target market is likely to purchase such items
- Which media and advertising methods will be the most appropriate? Consider the lifestyle of your target market. Here are some of the possibilities:
  - Video/film/audio – television, cinema, radio…
  - The internet – websites, banners, pop ups…
  - Posters – roadside, public transport…
  - Printed leaflets and flyers
  - Newspapers
  - Magazines – general interest or those designed for specialist topics and careers, for example sport, housekeeping
  - Sponsorship – celebrities, sporting events
- What will the advert or promotion include?

Fig 3.79 Innovative advertising

How your product will be presented to potential buyers, and how it will be distributed and sold

- The 'product identity'. What is the name of your product?
- The 'logo' or trademark, font, style or images to be used in the product's promotion and packing
- Consider the 'brand image' that you want to create; this should be relevant to the target market
- In what form will it be sold (ready to use, flat pack, multi-pack, retail or bulk packs; in different sizes, weights or capacities; to be installed by professionals…)?
- Important features of the packaging – suggested design and layout, possible variations for different market sectors
- Selling may be direct to customers, through a catalogue, over the internet, or in a shop and store on the high street and elsewhere. Which would be the most appropriate?
- Pricing – likely selling price. Positioning of the product in the market place: are free trials or samples appropriate?

Some of the following may be appropriate as part of your presentation:

- Photographs or video clips of the product in use or being tested
- 2D or 3D modelling of marketing aspects such as packaging and point of sale displays using CAD software, suitable image-editing software, or drawings and sketches

- An audio clip explaining how the product works and its benefits
- Designs for posters or for other promotional aspects
- Testimonials/reviews from those who have used the product

Key Points

- Ensure you have an understanding of the key aspects and principles of marketing before you start
- Your presentation must show clearly your marketing plan and strategy
- Refer to the supply chain and the needs of the manufacturer, distributor, wholesaler, retailer and consumer in the marketing plan

Examples

Fig 3.80 Seating and tables for bar area – images of furniture in situ suitable for marketing and publicity

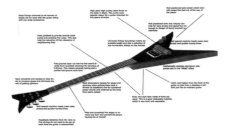

Fig 3.81 Folding electric guitar – design features of the product shown in publicity

Fig 3.82 Electric guitar – promotional aspects: company logo/brand name, poster for bus shelter, advertisement for *Guitar* magazine

Fig 3.86 Children's clothing – different colours/combinations to be marketed

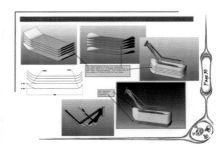

Fig 3.83 Snow bike (see also Figures 3.6 to 3.8) – technical information for buyers

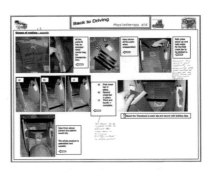

Fig 3.84 Physiotherapy aid – instructions for assembly and use (item supplied flat pack)

Fig 3.87 Fashion clothing – photo shoot for use in marketing

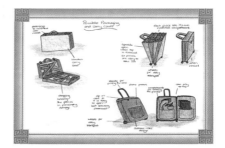

Fig 3.85 Baby gym/walker – packaging possibilities explored (item supplied flat pack)

- Examine in detail a product in your chosen material focus area. Identify key features that contribute to its success in the market place.
- Choose an everyday product without a known 'identity'. Devise possible names for the product that might increase awareness of the product and increase sales. Illustrate, using drawings or digital images, design changes to the product to reflect the new 'identity'. Design changes should specifically relate to colour and style.
- Find products made by the same manufacturer and describe how similar styling has been applied to each. Identify other common features and aspects that contribute to the brand identity.
- Study closely a product with distinctive styling (for example a car, a pair of trainers, a mobile phone). Identify the styling features (visual features such as colours, rounded or square edges, surfaces and lines, and so on). Make notes and sketches of these.
- Collect a range of printed advertisements and/or other adverts (such as radio adverts) and explain how they work. Refer to the information given, the product personality/identity, the target market, the use of the brand image, and the influence of lifestyle/fashion.
- As above, but do this for a range of product packaging.
- Explore and make a list of innovative forms of advertising, for example the advertising around a bus for an MP3 player.
- Make a study of the different ways that products in your chosen material focus are displayed for sale or displayed for presentation to clients. Identify features of the displays that are likely to attract customers and increase sales.
- Make a study of the ways in which product designers respond to the different requirements of manufacturers, wholesalers, retailers and consumers.

Section 7   Review and reflection

What you need to do
- Review and reflect on the effectiveness of the designing and making process that led to the final product
- Consider the possible wider implications and impact of the product, including possible future developments

Table 3.9 Assessment criteria, marks and pages for Section 7

	Marks	Suggested pages
Review and reflection	9	2
Presents a thorough and detailed review and reflection of the effectiveness of the designing and making process that led to the final product. Considers in detail the possible wider implications and impact of the product, including clear details of possible future developments.	7–9	
Presents an adequate review and reflection of the effectiveness of the designing and making process that led to the final product. Considers in reasonable detail the possible wider implications and impact of the product, with some details of possible future developments.	5–6	
Presents a superficial review and reflection of the effectiveness of the designing and making process that led to the final product. Considers in limited detail the possible wider implications and impact of the product, with few details of possible future developments.	0–4	

Your coursework project has occupied a significant amount of time, over several months. This section involves three specific tasks in which you:

- Look back and think about all that has been involved – the review and reflection
- Think about the implications of what has been done – the wider implications
- Consider further developments – the future

Choose the most appropriate methods of presentation – text, graphic, audio, photo or video.

Review and reflection

Review the project as a whole and in particular the process of designing and making. This is your 'self-assessment' of the stages of designing and making that culminated in your final product. What worked well? What are the aspects that caused difficulties? What might be done differently next time? Your review should refer to aspects such as:

- How decisions were made – was it the right approach?
- Specific setbacks and how they were overcome
- The value of inspiration gained from other sources
- The usefulness of information gathered
- The importance of the various stages of designing and making (for example specification, modelling, use of templates, testing)
- The working relationship between you and your client or user group
- The value of input from others in developing and refining your design
- The effectiveness of your project management and time planning
- Quality control in your making processes
- How opportunities for innovation and creativity were realised
- The significance of the use of ICT and digital technologies in the process

Wider implications of your product

You need to consider the effects of your product beyond its immediate function and use. Look at all aspects of the product. Answer such questions as:

- What are the moral and ethical implications?
- What are the sustainability issues – use of natural resources?
- What are the economic implications of decisions about choice of materials?
- What are the implications of choice of process for consumption of raw materials and energy?
- What are the maintenance and product life issues?
- What are the possibilities for recycling, reducing, reuse or repair?
- Life-cycle analysis can be used to evaluate the complete life of a product and its wider impact, from its conception through manufacture, distribution and use, to its recycling or disposal. Assess factors such as energy consumption, use of raw materials and ingredients, water consumption, and creation of waste.

The future of your product

View your final product as a prototype with design still ongoing. What would the next stages be?

- Identify possible further developments
- Identify key quality control issues
- Are there possible spin-off products?
- Comment on the likely success of your product in the marketplace
- Compare your product with products already available
- Does your product have a place alongside existing products?
- What aspects make it suitable or unsuitable?
- Make an honest appraisal of your product's commercial potential
- What modifications might be needed to enable quantity manufacture?
- What would be an appropriate scale of production?

- Describe different methods of finding possible solutions to a design problem
- Examine a familiar product and assess the impact of that product in its widest sense, for example socially, morally, environmentally and economically. Include a life-cycle assessment
- Summarise the role and responsibilities of a designer
- Make a study of products that have been redesigned in order to reduce their environmental impact. What innovations and features have been incorporated?

Examples

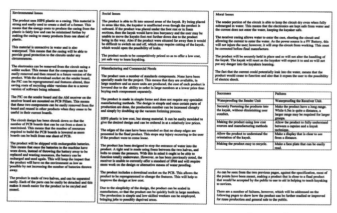

Fig 3.88 Fitness timing device – assessing the wider implications of the product

REFERENCES

[1]http://www.qca.org.uk/libraryAssets/media/qca-06-3403-csewk-parents.pdf

By the end of this section you should have a clear understanding of the nature of the written paper and style of questions. You should also have a better understanding of the standards required to be successful and be able to prepare for your examination with confidence.

F524: Product design

This is a written paper that consists of two components. You will be able to select questions across the focus material areas. Components 1 and 2 should both be available for the full two-and-a-half hour session.

Product design: component 1

Component 1 consists of eight questions, and each question follows a common format. You are required to answer one question.

The content of this unit focuses on products and applications and their analysis in respect of: materials, components and their uses, manufacturing processes, and industrial and commercial practices.

It is vitally important that materials and components are studied from the perspective of analysing modern consumer products that are designed to meet identified consumer needs, their design and manufacture, and product development.

You should be familiar with a range of materials, components or ingredients as used in the manufacture of commonly available products, and be able to make critical comparisons between them.

This unit is designed to provide a framework for analysing existing products that will enable you to make considered selections of appropriate materials and manufacturing processes when designing for making.

It brings together the knowledge, understanding and skills acquired in the study of F521 and F522 and should also directly contribute to work undertaken in F523.

The question paper consists of eight questions. Each question follows a common format: 24 of the 36 marks are drawn from the core content and relate to the material focus; 12 marks are allocated to the specific material content from each of the eight focus areas.

Product design: component 2

Component 2 assesses your ability to make immediate design thinking responses to a given situation. The question paper consists of eight questions. You will answer one question only on pre-printed OCR A3 sheets.

There are a total of 54 marks for the question. Each question will follow a common format and a generic mark scheme will be used for all questions.

You are required to respond to a given design situation. You will start by producing a specification and then produce a range of developed ideas. Ideas should be innovative and you should consider construction techniques, materials, components or ingredients and include appropriate measurements.

Your ideas should be evaluated with reference to your specification and volume production, and you should use sketches and appropriate annotation to show a final developed outcome. Specific features should be identified along with justification for your choices.

The mark breakdown for component 2 is:

- Three-point specification – 6 marks
- Range of innovative ideas with development – 33 marks
- Final developed outcome – 9 marks
- Efficient communication – 6 marks

Component 1

Every question starts the same from part (a) to part (d), for example:

6 Resistant materials

Figure 4.1 shows a push-along toy.

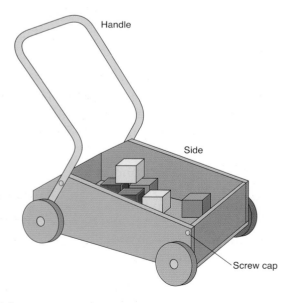

Fig 4.1

(a) Give **four** justified design requirements for the push-along toy. [4 × 1 mark]

An example of an acceptable answer, achieving one mark for each point, would be:

- The wheels must rotate freely to enable a child to manoeuvre the trolley easily
- The trolley must be robust to withstand the wear and tear of possible misuse by children
- All parts must be securely fixed to avoid the chance of small parts coming loose, causing a choke hazard
- The finish on the sides of the trolley must be resistant to damage and marking when blocks are thrown in

Marks will not be awarded to candidates who give generic or unjustified requirements such as:

- Ergonomically correct
- Value for money
- Aesthetically pleasing
- Safe to use

(b) Describe **two** examples of how modelling can be used in the design of the push-along toy. [2 × 2 marks]

Examples of acceptable answers would be:

- Ideas for the trolley can be modelled using CAD to explore a range of shapes for the handle
- 3D scale models can be used to check the overall proportions and test stability of the trolley

Marks are not awarded for repetition such as;

- 3D models can be made in polystyrene to check balance
- 3D models can be made in MDF to check balance

(c) Describe **two** quality control checks that may be carried out **during** the **manufacture** of the push-along toy shown in Figure 9. [2 × 2 marks]

Examples of acceptable answers would be:

- Visual check on the quality of finish applied/colour consistency
- Dimensional check on parts, for example alignment of holes drilled to connect handle to sides

Marks would not be awarded for statements relating to the quality of material or final product testing checks. The question states 'during manufacture'.

(d) Explain **two** benefits to the manufacturer of using standardised components in the production of the push-along toy. [2 × 2 marks]

Acceptable answers achieving two marks would be:

- Suppliers will provide quality-assured components, no need to carry out your own checks
- The manufacturer has a choice of suppliers and can make selections on cost, quality and reliability of service

Stating 'quality-assured components' would gain one mark; the justification 'no need to carry out your own checks' is also awarded one mark.

The next 12 marks are allocated for specific material content.

For resistant materials:

(e) Figure 4.2 shows parts of the push-along toy.

Choose **one** of the parts shown in Figure 10.

(i) State a suitable material for the part you have chosen. [1 mark]

(ii) Give **two** properties or characteristics that make the material suitable for this use. [2 × 1 mark]

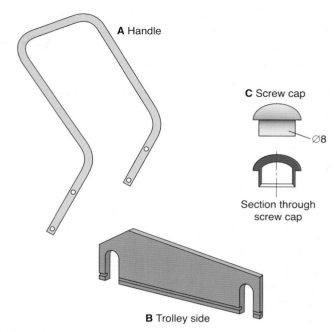

A Handle

C Screw cap

Ø8

Section through screw cap

B Trolley side

Fig 4.2

Acceptable answers achieving one mark to part (e) (i) would be:

- For part A, mild steel tube, aluminium tube or other appropriate
- For part B, plywood, MDF, beech or other appropriate
- For part C, HDPE, polypropylene or other appropriate

Plastic, wood or metal would not receive credit, and neither will any inappropriate materials.

Acceptable answers achieving one mark to part (e) (ii) would be:

- For part A, will resist force of child's pushing without deforming, can be easily formed in required shape, accepts high-quality finish
- For part B, machines easily giving a good finish, dimensionally stable, does not split/splinter
- For part C, flows easily when molten so suitable for injection moulding, flexes slightly to enable push fit, can be coloured as required

Unjustified statements such as 'resists force' would not achieve a mark.

(iii) Describe, in detail, how the part you have chosen would be manufactured. Use annotated sketches to support your answer. [9 marks]

Stages for A	Stages for B	Stages for C
Mark out and cut to length; Bend top curves with pipe bender; Bend other curves with jig; Crimp ends/braze? drill; Shape; Finish.	Mark out using template; Cut groove, router, plough plane; Cut slots, forstner, router; Profile edges, plane spindle moulder; Drill holes; Finish.	Prepare mould; Heat mould; Heat plastic; Inject; Cool; Eject; Remove sprue.

Marks are awarded for stages identified (a maximum of six marks) and quality of description and communication (a maximum of three marks).

(f) Discuss the issues that determine the commercial success of a product. [8 marks]

Clear instructions on how to answer 'discuss' questions are given in the instructions on the front page.

- Identify three relevant issues/points raised by the question [P]
- Explain why you consider three of these issues/points to be relevant [Q]
- Use two specific examples/evidence to support your answer [S]

Acceptable discussion issues would be:

- Effective marketing
- Value for money
- Advertising – celebrity usage/TV coverage
- Quality of design and manufacture

Each relevant points relating to the question is awarded one mark (up to a maximum of three marks); points/issues must be explained for one mark per point (up to a maximum of three marks); the suitable use of examples/supporting evidence is awarded one mark per example (up to a maximum of two marks).

Key Points

Discussion questions cannot be answered by a list of bullet points. Plan your response to this question with a scatterchart. Marks are lost when candidates focus on only one issue and fail to include appropriate supporting examples or evidence.

Component 2

Carefully select your choice of question. Examples from resistant materials and textiles are shown below.

Resistant materials

A chain of nurseries has decided to expand the age range that they accept by opening an 'early years' section for children from six months to two years. A product that will help the children to crawl or walk is required.

Data:

- The product must be able to be folded away or easily disassembled for storage when not in use
- Market research has indicated that there is a potential demand for an initial minimum of 5000 products

Textiles

A local nursery allows their 'first steps' children to go outside to play in good weather. A product, other than a baseball cap, that will protect the children from the sun when outside is required.

Data:

- The product must be able to fit children from the ages of one to three years old
- Market research has indicated that there is a potential demand for an initial minimum of 5000 products

Specification points

You are expected to produce three specification points. Specifications must be clear, specific to the product, and justified. Acceptable specification points achieving two marks would be:

■ The handle must have a good grip texture to ensure that the children do not slip off and hurt themselves
■ The structure must be rigid and stable so that it does not overbalance and tip over in use
■ The product must not move too quickly when the child is pushing it when learning to crawl or walk as they may fall over

A clear relevant statement would achieve one mark; a clear relevant justification would achieve one mark.

A relevant point repeating information already given would achieve a maximum of one mark. A generic non-specific statement would not achieve a mark.

After writing your three specification points, read them and check that you can tell what the product is.

You would then use annotated sketches to generate a range of initial ideas. It is essential that you work quickly, and ensure that your communication is clear and easy to understand.

Do not waste time producing high-quality rendered designs; you will only have approximately 90 minutes to spend on component 2.

Food students could use sketches where appropriate, but they would focus more on different combinations of ingredients and processing methods.

The examples shown in Figures 4.3 to 4.5 show a selection of design ideas for a device to transport garden waste. They demonstrate good quality design skills and would be marked in the top band of the criteria outlined below.

Examiners will mark ideas using a grid on page three. The letters refer to the criteria for assessing design thinking.

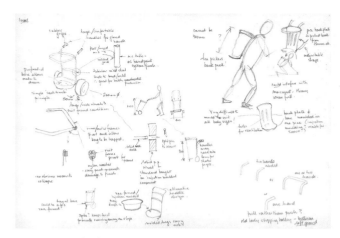

Fig 4.3 Initial ideas

R – range of ideas

There are **15 marks** for the range of ideas. You will need to generate a wide range of significantly different innovative ideas that are each developed as far as possible to achieve top band marks.

C – construction techniques

You will need to show evidence of appropriate manufacturing/construction techniques. Up to **six marks** are awarded for evidence of appropriate, detailed construction methods. Reference to volume production is required, both in the design features proposed and in construction methods. Simply stating injection moulding would not received credit: showing how the design features would be suitable for the process or giving details of the mould would be needed.

> **EXAMINER'S TIPS**
>
> Use a simple colour code for evaluative comments to ensure they are explicit and can be easily identified by an examiner.

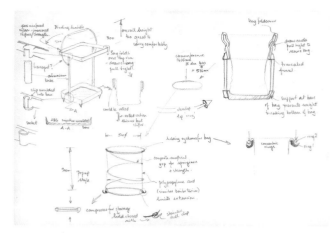

Fig 4.4 Developing initial ideas

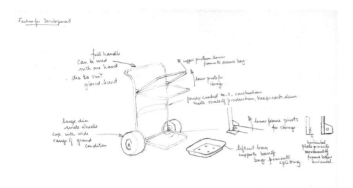

Fig 4.5 Final developed outcome

M – materials and components

Three marks are awarded for the justification of appropriate materials and components. Simply stating materials will not achieve full credit. Generic terms such as wood, metal, card, and so on, would not gain marks. Be specific.

E – evaluation

You must show clear evaluation of ideas with reference to your specification and to volume production. The scale of production will be given in the question. **Six marks** are awarded for detailed evidence of objective evaluation.

F – chosen features for final developed outcome

You are to identify the main features from your design ideas and propose a final developed outcome on the last sheet. Up to **six marks** are awarded for a detailed description of the identified features.

J – justification of choices made with reference to specification

Up to **three marks** are awarded for a detailed justification of your choices. The example in Figure 4.5 identifies the key features of a device to transport garden waste and justifies some of the choices.

Core knowledge

Learning outcomes

By the end of this section you should have developed a knowledge and understanding of:

- One-off, batch and high-volume production systems
- Modular/cell production systems
- Just-in-time manufacture
- Bought-in parts and components, standardised parts
- The implications of these industrial production processes/procedures

Introduction

The main categories of manufacturing systems involve production processes that are one-off, batch and high volume. Factors that determine the selection of a manufacturing system are:

- The type of product
- Demand for the product
- Capital
- Premises
- Tooling
- Labour skills

One-off, batch and high-volume production systems

One-off production

One-off production refers to the manufacture of a single component or product. It is often referred to as jobbing production. One product is fully completed before the next is started. Examples include:

- Large-scale products such as ships, bridges and specialist stadium constructions

- Smaller-scale products such as jewellery, specialist furniture and bespoke clothing (suits and wedding dresses for example)

Key Points

One-off production

- Usually produced to a specific client specification
- Small level of demand
- Requires very high skill levels and high labour costs, resulting in high unit costs
- Generally low capital costs
- Worker satisfaction generally very high; often involved in every production stage from start to completion

Batch production

Batch production involves the production of batches of similar products. It refers to the scale of production – a few items to several thousand – and to the type of production, where components are processed together in a planned sequence. Cars used to be mass-produced with only one model manufactured. Modern car manufacturers produce batches with differing specifications, the quantities of each batch decided by customer demand. Examples include:

■ Small batches can be in low numbers, such as a batch of ten aeroplanes
■ Large batches can be in high numbers, such as several thousand pairs of training shoes

Key Points

Batch production
■ Flexible system, a wide range of products can be produced
■ Can react to demand, stop or increase production run
■ Workforce usually less skilled than one-off, will operate one or two processes
■ Medium investment needed for a range of machinery that can be set up for different operations
■ Workers sometimes have opportunity to work on more than one process

High-volume production

High-volume production is often referred to as mass production. High-volume production systems usually operate 24/7 and are used to manufacture high-demand items. Examples include:

■ Continuous-flow production systems take in the raw material at one end of the factory, and the finished product comes out the other end. These factories seem like a complex, continuous, fully automated machine. Glass, steel and paper are produced in this way
■ In-line production systems require the product to be moved from one process to another, usually using a conveyor system. Cars and domestic products such as televisions and kettles are produced in this way
■ Pens, clothes hangers, buttons and paper clips are produced in high volume using fully automated specialist machinery

Key Points

High-volume production
■ Minimal, if any, variation in product
■ Regular, high demand for product needed
■ Usually very high capital investment, low unit cost
■ Relatively low skilled workforce – repetitive task, mostly fully automatic production systems
■ Control/maintenance team required, stoppages costly
■ Often low job satisfaction; workforce usually only involved in small part of production cycle

Modular or cell production systems

These systems use a number of production cells or modules that are grouped together to manufacture a component or sub-assembly of a larger product. The cells or modules usually consist of production machines and include inspection and assembly units. Very often the cells are operated by a small multi-skilled workforce but can be fully automated.

Some large manufacturing systems process large batches in sequence through several dedicated process or manufacturing sections. This is often referred to as batch and queue production. Sections usually have large, expensive machines designed to minimise unit costs by mass-producing single identical components with minimal tool changes.

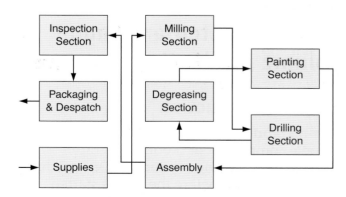

Fig 5.1 Layout of batch and queue manufacturing system – lots of 'down time' when one particular process causes a delay in production

This system largely requires advance orders and long production runs. It can be very wasteful as production can be held up if one section does not function correctly.

Storage space is required for batches between processes.

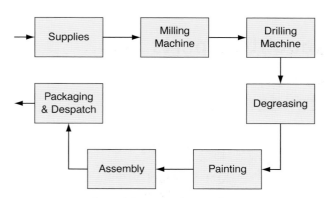

Fig 5.2 Layout of modular or cell manufacturing system – smooth flow of operations; flexible system enables variations in product

In modular or cell manufacturing the workstations are arranged in a logical manner to produce one complete item at a time, in a smooth and quick flow through the production process. The rate of production is decided by consumer demand. Production planning for the cells/modules must be accurately scheduled to ensure that the correct number of components/sub-assemblies is produced in time for the final assembly of the product. A hold up in one module is not as catastrophic as a hold up in an in-line production system.

Modular/cell production systems require careful positioning of workstations to enable minimal and quick movement of parts from one operation to another.

The workforce is often multi-skilled to offer maximum flexibility and enable rapid change of production.

The machines are often fitted with multiple tools and generally operate a rapid tool changing system. Powered clamping systems offering quick and easy location of the work-piece speed up production times. A standard size of manual

locking method, usually in the form of a chuck key or spanner, is used to avoid operators wasting time searching for the correct tool.

> **Key Points**
>
> **Modular or cell production systems**
> - Functional layout of modules/cells
> - Flexible system offering rapid change in production reacting to consumer demand
> - High level of job satisfaction; operators work in flexible teams carrying out different tasks, opportunities for training
> - All workforce responsible for quality control

Just-in-time manufacture

Many modern manufacturing companies such as Toyota, Dell and Rolls Royce operate a JIT system. The philosophy of JIT manufacturing is to meet consumer orders with a quality product with minimal delay and effective use of resources. The JIT system is sometimes referred to as 'lean manufacturing' as it focuses on giving customers value for money by reducing wastage.

The storage of materials and components needed to manufacture products requires space and, where appropriate, special conditions such as heating and ventilation. This adds to the overall manufacturing cost.

Lean manufacturing time line

1850	Eli Whitney used interchangeable parts when contracted by the US Army to manufacture 10,000 muskets. Engineering drawing conventions, tolerances and introduction of modern production machines.
1910	Henry Ford production line systems.
1950	Shigeo Shingo, Taiichi Ohno set up Toyota, just-in-time (JIT) production system.
1986	Toyota JIT system also referred to as 'stockless production' and 'world class manufacturing' (WCM)
2008	Many of the world's leading companies operate JIT systems: Dell's product cycle is four hours, supplies are delivered 90 minutes after an order is placed.

Companies set up detailed arrangements with reliable suppliers and distributors to ensure that advance orders are taken and regular deliveries of materials and components are made when required for manufacture. The products are distributed as soon as they are completed, removing the need for further storage.

Computerised stock control systems ensure that production is continuous.

Wastage is reduced in terms of:

- Storage space for materials, components and completed products
- Defective products – all of the workforce have a responsibility for quality
- Money invested in materials and components that will not be used and completed products that will not be sold
- Movement of the product through the factory is kept to a minimum
- Inefficient use of equipment – the system makes maximum use of production machinery and no waiting time between processing operations
- Labour misuse – appropriately skilled workers are used
- Product effectiveness – simplicity is a key feature of the system with the removal of product functions that are not necessary
- Downtime with new product run up – detailed plans are made to ensure seamless flow from completed product to new product

Key Points

JIT manufacture
- The partnerships between the manufacturer, suppliers and distributors is critical – if deliveries are late production stops
- Workforce relationships are also very important – staff absence or strike action can halt or delay production
- Workforce must be multi-skilled, flexible, have job satisfaction and be consulted in decision making
- A spirit of cooperation is vital; proposals for improvement are welcomed by management
- JIT is a very flexible manufacturing system and can react very quickly to changes in consumer demands
- The product is electronically tracked through the system to carefully monitor progress and ensure that the manufacturing schedule is efficient

Drawbacks of the system include:

- Delivery failure – Toyota production was halted as a result of a major fire that prevented supplies arriving
- Unsatisfactory workforce relationships may result in strikes or absenteeism, which holds up production
- Some environmental concerns over frequent road transport of supplies and despatched items
- Requires major suppliers and customers to be relatively close to the manufacturing plant

Bought-in parts and components, and standardised parts

Bought-in parts and components

Many products make use of similar parts or components. Different makes of computers often have the same make of hard drive or other internal components. Car manufacturers may set up sub-contractor arrangements for the supply of components such as headlamps and engine parts.

Key Points

Bought-in parts and components
- No need for production space for the components
- Speeds up overall production
- Quality assured by the component manufacturer, specified tolerances
- Specialist companies provide components, cost benefits through economy of scale
- Choice of suppliers if there are service/quality difficulties, cost benefits through price negotiations and loyalty contracts
- Reduces storage costs, components available when required

Standardised parts

Standardised parts are the common items that are required in the manufacture of a wide range of products such as screws, nuts and electronic components (batteries, resistors, capacitors etc.). They are usually small, simple items that are manufactured to guaranteed specifications and are of consistent quality. Other examples include:

- In the construction industry standardised components include doors, windows, sinks and other kitchen units
- Zips, buttons and other fastening devices are standard components used in the textiles industry

- In the manufacture of cars, a chassis may be used as a standard component and used in the production and development of several models

Key Points

Standardised parts

- The key points relating to bought-in components are relevant to standardised components
- Minimal interface and tolerance problems; standards usually generated by independent body, for example BSI
- Ease of maintenance; replacement parts for consumers

1 Discuss how manufacturers meet consumer needs in rapidly changing product markets.
2 Discuss the workforce issues to be considered when selecting an appropriate production system.
3 Discuss the importance of using standardised parts and components in the manufacture of either:
  - Domestic electronic products
  - Textile products
  - Cars
4 Discuss how tooling, assembly and labour skills can affect the scale of production of a product.

Section B — The use of digital technology in designing and manufacturing processes

Learning outcomes

By the end of this section you should have developed a knowledge and understanding of:

- CAD/CAM as used in industry/commerce
- Testing, modelling and rapid prototyping
- Stock control, monitoring and purchasing logistics in industry
- High-volume production and automation
- The implications of the use of digital technology

Introduction

Product designers throughout the years have adapted and embraced the developing technologies and changing methods of communication and processing in their designing and making activities. The last 40 years have seen a dramatic increase in the availability and use of digital systems in design and technology.

Computer aided design (CAD) and computer aided manufacture (CAM)

Global communication systems have greatly increased the flow of digital information in text, visual or audio format. Large, detailed documents can be transferred instantly between sites, regardless of distance. The internet has enabled online research opportunities and companies can make up-to-date checks on the competition and assess market trends. Marketing, advertising and sales opportunities have increased greatly with this worldwide phenomenon.

Table 5.1 CAD and CAM systems

CAD system could include	CAM systems could include
Computer Graphics tablet Scanner (flat and 3D) Internet access 2D/3D software	CNC lathe, miller, router Computerised embroidery Laser cutter 3D printer Plotter/cutter, vinyl cutter

CAD (basic 2D systems, although parametric systems where changing one minor element or dimension would automatically change the whole design were available in schools in the early 1980s) was introduced and initially used as a final output tool, to generate an accurate, detailed and dimensioned drawing of a proposal or solution.

The vector-based systems describe geometries that are able to be converted into machine code

Fig 5.3 CNC embroidery system

Fig 5.4 Laser cutter

and define tool paths for manufacture. The development and increase in sophistication and processing power of computers and the development of an increasing range of computer numerically controlled (CNC) machines signalled the single most significant contribution to the increase in manufacturing productivity.

Ideas could be shared and worked on in real time in different parts of the world, and could be sent to manufacturing units within the same facility or anywhere in the world. The development of 3D CAD tools enabled designers to quickly generate and explore ideas, not just create life-like presentation representations. Three-dimensional ideas can be realised using rapid prototyping technologies.

Other acronyms relating to the use of computers in designing and making include:

- CADD – computer aided design and drafting
- CAA – computer aided analysis
- CAAD – computer aided architectural design
- CAE – computer aided engineering
- CAPP – computer aided process planning
- CIM – computer integrated manufacturing

Testing, modelling and rapid prototyping

Testing and modelling

Computers can be used to test the feasibility of design ideas using a range of software. Various software including Finite Element Analysis programs can be used to predict the effects of loading and identify weaknesses and stress points. Simulations of functional aspects of components and assemblies can be used to ensure designs are feasible.

CAA software can be used to analyse the effect of external factors such as vibration, extreme temperature changes and variable loading on design proposals. The software will predict likely outcomes and give opportunities for improvements and modifications to proposals before production.

Fluid Dynamic software is used in the design of vehicles, analysing and measuring the effects of air flow, and in architectural design where the effectiveness of air conditioning and heating systems can be tested.

Computers are used extensively in 2D and 3D modelling of ideas. Images can be rotated, zoomed in and shown in an infinite variety of colours and textures. Modifications are carried out speedily and with relative ease.

Computer simulations are particularly useful when a real-life process is too dangerous for humans to carry out, takes a disproportionate amount of time to complete and is too expensive. Machine processes are simulated to test tool selection and settings: incorrect settings could result in very expensive damage to machinery and possibly lengthy delays in production.

Rapid prototyping

Several systems are available that translate 3D designs into solid, physical forms. Software is used to 'slice up' the 3D design into a series of adjacent horizontal layers which are then sent in sequence to a rapid prototyping system. They are then built up layer by layer in a number of different ways:

- **Laminated object manufacturing** – one of the earliest systems available, building up layers of adhesive-coated sheet material. This can be time consuming and the final object has a rough texture and may need additional finishing
- **Stereo lithography** – a laser traces the shape of a layer onto a bath of liquid resin. This cures the resin. The platform is lowered and another layer is traced until the whole object is created. Typical layer thickness is 0.1 mm
- **Laser sintering** – works in a similar way to stereo lithography. The laser traces the shape

onto fine heat-fusible powder (plastic, metal or ceramic). The powder becomes solid; another layer of powder is laid on top of the fused layer and the process is repeated until the object is completed
- **3D printing** – this system 'prints' a thermoplastic material (ABS) in successive layers on top of previous layers to build up a 3D shape. Complex shapes often require an additional support material to be printed to support the object while it sets. Objects can be painted or electroplated for a high-quality finish

Fig 5.5 3D printer

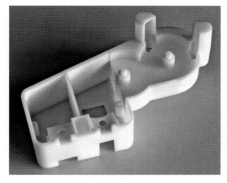

Fig 5.6 3D printed object: tape winding case

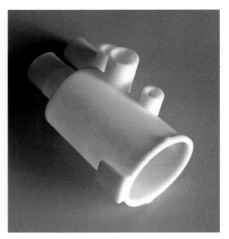

Fig 5.7 3D printed object

Stock control, monitoring and purchasing logistics

Stock is usually classified in three groups:

- **Materials/components** – bought in from suppliers to be used in the manufacture of products
- **Work in progress** – incomplete products currently being manufactured
- **Finished products** – assembled products of desired standard ready for distribution

Stock control enables production to flow without costly hold ups, ensures that sufficient raw materials and components of acceptable quality are purchased and customer demand is met.

Stock control systems used to rely on careful checks, regular stock takes and ordering, all carried out by individual employees. The system was time consuming and errors occurred.

Computerised systems, including the use of bar codes and other digital recognition processes to monitor stock, have speeded up the process considerably.

Links are easily made with the purchasing, marketing and sales departments. Accurate forecasts of predicted sales will ensure that sufficient orders are placed to meet demand.

The purchasing and logistics departments will make decisions on the suppliers who are reliable, competitively priced and are able to provide the

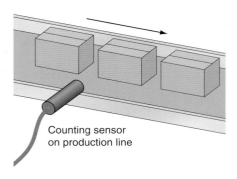

Fig 5.8 A counting sensor on a textiles production line

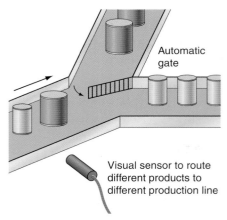

Fig 5.9 A visual sensor to route different food products to a different production line

materials and components of the required standard.

The system will include 'buffer' stock to cope with unforeseen problems or emergencies in the supply chain, or if there is a sudden and unexplained rise in demand.

Electronic Data Interchange

Electronic Data Interchange (EDI) is a way in which information can be exchanged between computers within a company or between companies. It is used extensively in stock control systems, enabling the speedy exchange of orders and invoices.

EDI is an automatic process, using an agreed standard for product codes, prices and location. It is an accurate and efficient system as it does not require extra human intervention, reducing

Table 5.2 Benefits and drawbacks of computerised stock control

Benefits of computerised stock control	Drawbacks of computerised stock control
■ Lower costs ■ Improve efficiency ■ Cater for fluctuating levels of demand ■ Efficiencies can lead to price reductions improving competitive edge ■ Very quick system ■ No manual checks required ■ Detailed, accurate, well-presented data available for print out or distribution in electronic format ■ Data can be selective for presentation to different groups ■ Data easily stored	■ Initial cost of set up and training ■ Sometimes only one person can operate system, so problems occur if absent or leaves firm ■ Possibility of manual input error ■ Software failure, virus attack ■ Computer breakdown can be costly; backup systems essential ■ Digital data may be accessed without security system

the possibility of errors through re-keying of data and costs are reduced.

Benefits of using EDI are:

■ Eliminating error
■ Eliminating paper
■ Reducing lead time through effective stock control
■ Confidence of trading relationships through overall increased efficiency and accuracy of system.

High–volume production and automation

Fig 5.10 An automated bottling plant

An introduction to high-volume production can be found in Section A of this chapter 'Manufacturing Systems'. This section covers high-volume production methods and principles in relation to digital applications and automation.

Computer Integrated Manufacture (CIM) is used by fully automated manufacturing plants. It is an integrated system including CAD, CAM, CAPP and CNC machinery, and also controls automatically guided vehicles (AGV) in an automatic storage and retrieval system (ASRS). CADCAM is at the centre of the system. CAD units can work independently or in collaboration to produce 2D and 3D images that can be analysed and tested before a prototype is built.

Process planning software is used to make maximum use of resources and ensure a rapid and consistent flow of operations with no down time due to bottlenecks.

Software is also used to predict and check for tool wear and plan-in appropriate maintenance tasks.

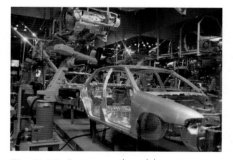

Fig 5.11 Automated welding

The final component designs are used to create tool paths for machining. CNC machines are used for a wide range of operations including cutting, pressing, stamping, embroidering, cropping, welding, forging, assembly and packaging.

Computer systems, using visual measurement and/or weighing sensors, are used for inspection

Table 5.3 Benefits and drawbacks of automated systems

Benefits of automated systems	Drawbacks of automated systems
■ Low labour cost ■ Low product cost through economies of scale ■ No production loss through disputes ■ Consistent quality ■ Very efficient, stock control system enables JIT manufacture ■ System checks and modifications easy to carry out	■ Expensive initial layout ■ System breakdown costly ■ Specialised workforce regular training ■ Protocols of all systems may not communicate ■ Require some flexibility in processing to enable manufacture of different products

at key processing points to ensure quality control; for example, in modern car manufacture the tools used to insert the screws that secure seat belts have angle and torque sensors to ensure that the screws are fitted correctly and at the precise angle.

Robots play a key role in automated systems. They are used for material and component handling and can be programmed and used in a wide range of applications. They carry out repetitive and boring tasks with consistency and precision, and can work in hazardous or hostile working environments. Most robots resemble mechanical arms and, in their simplest format, are used in 'pick and place' operations, for example positioning components on a circuit board for a computer. More sophisticated systems can be programmed to carry out multiple welding operations on the chassis of a car.

Section C — Commercial practice

By the end of this section, you should have developed a knowledge and understanding of:

- The role of marketing, including assessing consumer needs, product development, pricing, promotion and distribution
- Advertising

Introduction

Commercial practice refers to the range of activities by which companies linked to product design conduct their business. It includes marketing, product development, pricing, promotion, distribution and advertising.

Marketing

Marketing broadly refers to the identification and anticipation of customer needs and hopefully meeting those needs.

The manufacturing industry must take into account customer needs and wants in order to develop new and improved products that will give a competitive edge. New products are usually developed either because of market pull, where consumers demand a particular type of product, or by technology push, where new materials and/or technologies lead to innovative products that are released onto the market.

New product development can involve:

- **Investigation leading to idea/concept generation** – from research, consumers, employees, case studies, design consultancies

- Feasibility analysis – is it possible? Will customers want it? Can it be manufactured at a price that will generate a profit?
- Development – modelling and mock up designs
- Testing – working prototypes made to fully test function and safe usage
- User trialling – Beta testing product with selected consumers, feedback generated
- Modification – make necessary adjustments to the product
- Launch – possible small scale, gradual 'roll out' or full national launch of product

Market research

Sometimes referred to as 'in-bound marketing', market research involves finding out and analysing information about:

- Particular markets, the needs of consumers and target markets
- Who are the competitors and what are they doing
- Up-to-the-minute market trends
- Customer satisfaction relating to products and services

- How much customers might be prepared to pay
- How the product should be identified, naming and branding

Market research can be either primary or secondary or a mixture of both.

Once market research has been gathered and analysed, a marketing strategy can be put in place with the aim of maximising sales of the product.

This next stage in marketing is sometimes referred to as 'outbound marketing', including sales, advertising and promotions, customer service and satisfaction.

A company has to decide on its marketing mix: the best blend of the main elements depending upon the type of product and specific target market.

Key Term

Marketing mix – the basic components of a marketing plan or strategy, often referred to as the four Ps: product, price, promotion and place. Some include two more Ps: people and processes.

Table 5.4 Primary and secondary market research

Primary	Secondary
- Precise data, meets exact needs of company - Collected first hand using questionnaires, focus/user groups, surveys and field research, etc. - Can be costly	- Uses available data from magazines, reference books, government agencies - Provides information such as population trends and regional statistics - Can be out of date or incorrect

Table 5.5 Marketing mix

Product	Price
Product A product is examined on three levels: ■ The core product is the benefit of the product, for example the convenience of a car ■ The actual product is the tangible, physical product, for example the car ■ The augmented product is the customer service support offered, for example warranty, guarantee and after-sales service The quality of a product depends on factors such as its: ■ Aesthetics ■ Performance ■ Maintenance, ease of servicing ■ Durability ■ Range of features ■ Ease/effectiveness of use ■ Brand name	**Price** The price of a product may depend on: ■ Demand ■ Costs (need to cover all costs incurred) ■ Government taxes ■ Competition ■ Stage in the life cycle (price increases in growth and falls in decline) Methods of pricing include: ■ Penetration pricing – price set artificially low to gain market share; once achieved, the price is increased ■ Price skimming – if product has competitive edge, a high price can be set; this will fall with increased supply ■ Psychological pricing – charging £1.99 rather than £2 ■ Predatory pricing – undercutting competitors, creating price wars
Promotion Decisions have to be made on how best to promote the product and bring it to the attention of potential customers. The intention is to win new customers or persuade them to change brand loyalty. Methods include: ■ Short-term promotions such as Buy One Get One Free (BOGOF), competitions and coupons ■ Exhibitions and trade fairs ■ Publicity campaigns ■ Personal selling/sales representatives Can be expensive and risky, careful budgeting required. The acronym AIDA is used with promotion and advertising: draw attention, create interest, generate desire, invite action.	**Place** Or placement refers to the location where a customer can purchase a product. It is sometimes known as the distribution channel. It can include any physical store or shop as well as TV shopping channels and the internet. There are four main channels of distribution: ■ Manufacturer–Consumer e.g. mail order, farm shops ■ Manufacturer–Retailer–Consumer e.g. high-street stores ■ Manufacturer–Wholesaler–Consumer e.g. furniture ■ Manufacturer–Wholesaler–Retailer–Consumer e.g. medium-size convenience stores (large supermarkets often cut out the wholesaler) Functions of distribution channels include: *contacting* prospective buyers; *matching* the offer to the buyer's needs, *negotiating* agreement on price and terms, and *storing* and *transporting* the products.

Advertising

Advertising is used to bring a product to the attention of potential customers. When deciding on the appropriate way to advertise a product you should consider overall costs, the target market or audience and the most appropriate medium.

Table 5.6 Advantages, limitations and costs of advertising

Medium	Advantages	Limitations	Cost issues
Television	Very high, mass-market coverage; low cost per exposure; can generate powerful emotive response; can include images, sounds and special effects	Quick, fleeting exposure; target markets selected by scheduling, for example between 4pm and 6pm for children	Expensive: key viewing times very expensive
Radio	Very good local impact; high geographic selectivity; national radio effective for consumers on the move (for example in a car)	Audio only; fleeting exposure; fairly low attention, background 'noise'	Relatively low cost for local radio, higher rates for national commercial stations
Newspapers and magazines	Very good local markets, national or geographic selected coverage; broad acceptability; high believability; prestige; magazines often very high-quality images	Very short life; generally poor image quality in newspapers; no guarantees of positioning of advert	High cost in nationals and prestigious magazines; right-hand page often more expensive than left (readability)
Direct mail	Very high audience selection; can be personalised with mail-merge systems	Often discarded as junk mail; poor image	Relatively high cost per exposure
Billboards	Flexibility; high repeat exposure; local targeting; positioned in high traffic areas to catch mobile consumers; some are electronic with several adverts repeating	Little audience selectivity; image only; easily vandalised	Generally low cost; key sites (for example outside airports) can be very expensive
Online	High selectivity; instant; can be powerful, moving images and sound; can be as long as viewer wishes; can be interactive; direct access to supplier; increasingly popular	Small but rapidly growing audience; relatively low impact; anxiety over invasion of personal space, linked to spam and spyware	Low cost; purchases can be made directly from the advert.

Adverts are controlled by the Advertising Standards Agency, a voluntary body that is set up to ensure that adverts do not mislead or offend potential customers. Adverts are required to be legal, decent and honest, and to not cause offence to viewers or listeners.

The Independent Television Commission controls advertising on television and radio.

1 Discuss the appropriate media and method of advertising products for the following target markets:
  - Primary school-age children
  - Men and women aged over 65
  - Men aged 18–24 years
  - Women aged 45–55 years with significant disposable income
2 Discuss how manufacturers meet consumer needs in rapidly changing product markets.
3 Discuss the implications to the manufacturer of market competition.

Section D — Legislation

By the end of this section you should have developed a knowledge and understanding of:

- Trade description and sale of goods
- BSI standards applied to products/systems
- Labelling
- The implications of intellectual property – design rights and patents, registered designs, registered trade marks and copyright
- Regulations

Trade description and sale of goods

There are a number of Acts of Parliament that are designed to protect consumer rights. The Sale of Goods Act 1979 and the Supply of Goods and Services Act 1982 are concerned primarily with the sale of goods and services.

The Trades Description Act 1968 and the Consumer Protection Act 1987 are concerned with product liability, products incorrectly described or defective products.

The Consumer Credit Act 1974 is concerned with the purchase of products.

Key Points

Key features of the Sale of Goods Act:
- The goods must conform to the description given, for example a waterproof watch must not let in water
- Goods must be of satisfactory quality, based on what a reasonable person would accept as satisfactory considering factors such as age and price
- The goods must be of acceptable quality, fit for purpose and free from defects
- It is the seller, not the manufacturer, who is responsible if the goods do not meet acceptable standards
- Purchasers have the right to request their money back if the goods do not conform to contract. The claim must be made within a reasonable time period
- The purchaser, in some circumstances, can request a repair or replacement
- If repair or replacement are not possible or too costly, the consumer can request a partial refund, if they have had some benefit from the product, or a full refund if they have enjoyed no benefit from the product

Sale of Goods Act 1979

The Sale of Goods Act has been amended three times, the most recent amendment being in 1995. This law protects consumers and helps them to obtain redress when purchases go wrong.

Trade Descriptions Act 1968

The Trade Descriptions Act makes it an offence for a trader to apply, by any means, false or misleading statements, or to knowingly make such statements about services or goods.

Product labelling

Product labelling is covered under the Trades Descriptions Act. Labels must include accurate information to ensure that products can be used safely and correctly; for example, motorcycle helmets must have a label attached to them informing you not to paint them or apply any kind of solvent. These may damage and weaken the helmet, giving less protection in an accident. Aerosol cans must have a label warning the user to keep them away from heat for obvious safety reasons.

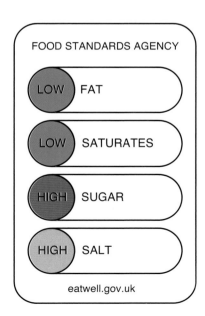

Fig 5.12 Food traffic light symbol

An increasing number of food manufacturers and supermarkets are using the traffic light system to give healthy diet information to consumers.

The Food Labelling Regulations 1996 relate to the labelling of food products. All ingredients of a food product must be listed in weight order.

Other information that must be included are: the 'use by' date, the 'best before' date and particular storage instructions.

For the labelling of sweets, the following information must be given:

- The true name of the sweets
- If the sweets contain any of the following types of additives, the category name for those additives must be stated: antioxidants, sweeteners, colours, flavourings and flavour enhancers, preservatives

Video films have to be labelled with a British Board of Film Classification symbol giving the age group for which they are suitable. It is an offence to hire or sell films to anybody under the age specified on the label.

The BS 2747 code of practice for textile care labelling recommends how information can be passed to the consumer on the washing, bleaching, ironing, dry cleaning and drying of textiles.

The symbols used on labels are consistent with those used on detergent packs, washing machines and irons.

Some further information on food labelling is given in Chapter 8, Section D. Further information on eco-labelling is given in Section F of this chapter.

Examples of quality and safety assurance labelling

Some products, such as tobacco, are required by law to include health warnings. The Tobacco Products (Manufacture, Presentation and Sale)(Safety) Regulations 2002 legislated for hard-hitting health warnings on tobacco packs

Fig 5.13 Machine wash at 40°C, reduced mechanical action

Fig 5.14 Cool iron (120°C) acrylic, nylon, acetate, triacetate, polyester

and also prohibited misleading terms such as 'low-tar', 'mild' and 'light' from tobacco packs.

On 29 August 2007, the introduction of picture warnings was announced. They will start appearing on cigarette packs from the end of 2008 and on other tobacco products from the end of 2009.

CE is not an abbreviation. The CE marking is a declaration by the manufacturer that the product meets all the appropriate provisions of the relevant legislation.

The CE mark refers to safety, not quality. It is mandatory and indicates conformity with

European safety requirements. It is the responsibility of the person placing the product on the market, to ensure that the product is correctly CE-marked.

Without the CE marking the product may not be placed in the market in the fifteen member states of the European Union and Norway, Iceland and Liechtenstein.

BSI British Standards

BSI British Standards is one of three divisions of the BSI group. The other divisions are BSI Management Systems and BSI Product Services.

BSI British Standards is the National Standards Body (NSB) for the UK and was the world's first NSB. It works with governments, industry, businesses and consumers to produce British, European and international standards.

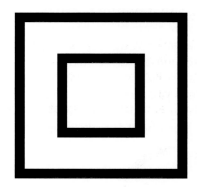

Fig 5.15 A safety label found on electrical products such as light fittings, fires and cookers, indicating that they have been double insulated (protected) throughout

Key Points

- BSI British Standards has 27,000 current standards — one for every 59 businesses in the UK
- The price of a standard reflects its complexity; the price range for BSI standards is £5 to £1150
- The most popular standard in the world — ISO 9001 Quality management systems (Requirements) — is used by over 670,000 organisations in 154 countries

A standard is an agreed, repeatable way of doing something. It is a published document that contains a technical specification or other

Fig 5.16 This BSI safety label can be found on electrical products such as light fittings, fires and cookers

precise information designed to be used consistently as a rule, guideline or definition.

There are five types of British Standard: specifications, methods, guides, vocabularies and codes of practice.

Standards help a company to:

- Attract and assure customers that products are safe and fit for purpose
- Demonstrate market leadership, creating a competitive advantage
- Continue to develop and maintain best practice

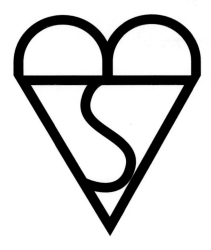

Fig 5.17 The Kitemark® is the world's best-known product and service certification mark

Many organisations have made it mandatory for their supplier's products to have been awarded the Kitemark® before they will place an order with them.

As consumers become increasingly informed about their choices, conformity to recognised standards becomes a key purchasing decision issue.

Some examples of British Standards include:

- BS EN 71-1:2005 + A4:2007 Safety of toys. Mechanical and physical properties.
- BS EN 71-2:2006 + A1:2007 Safety of toys. Flammability.

These standards would be essential for any individual or manufacturer involved in the design and manufacture of toys.

1 Discuss the role of the BSI in the design and making of products.
2 Discuss the importance to manufacturers of ensuring that their products are safe for the public to use.

Intellectual property

Intellectual property (IP) refers to all kinds of intangible (not physical) types of property that people can own, for example the creative outcomes from the mind such as design ideas, written material, artistic and musical composition.

It allows you to own the things that you create in a similar way to owning physical property. You can control the use of your IP, and use it to gain reward or protect from copying.

The main types of IP are: design right, registered designs, patents, trademarks and copyright.

Design rights

Design rights concern the rights of the creator of a design or designs, unless a third party commissions the work.

Registered designs

A registered design will give you ownership rights for the appearance of a product, protecting both the shape and the pattern or decoration.

Patents

A patent gives a designer protection against copying of the technical and functional aspects of his/her invention without permission. It covers details such as how it works, how it is made and what it is made of. A patent gives the creator rights for up to 20 years including:

- Allowing others to make copies of the invention
- Selling copies of the invention
- Offering copies of the invention for sale

A patent cannot be awarded for a range of IPs including:

- A scientific or mathematical discovery
- A literary, dramatic, musical or artistic work
- An animal or plant variety

Trademarks

A trademark is used by a company or an individual to identify and distinguish its products from those of others. It could be in the form of a word, name, song or a symbol.

Some familiar examples are the shape of a Coca-Cola bottle, the Apple symbol and the Nike swoosh.

Copyright

A copyright is a set of exclusive rights or protection given to creators of original ideas, information or other intellectual works. It is often seen as 'the right to copy' an original creation.

Copyrighted material can only be used or recreated with the owner's permission.

1 Use examples to discuss how legislation can protect purchasers of domestic electric products.
2 Discuss the benefits and drawbacks of the use of legislation to protect IP.

Section E — Health and safety of designers, makers and the public

By the end of this section you should have developed a knowledge and understanding of:

- The regulatory and legislative framework related to materials and equipment using Health and Safety at Work (HASAW)
- Control of Substances Hazardous to Health (COSHH) legislation
- Protection of the worker/operator
- Protection of the user/customer
- Protection of the environment
- Risk assessment

Health and Safety at Work Act

The Health and Safety at Work Act 1974 (HASAW) forms the basis of British health and safety law. It states that employers have a duty to ensure, so far as is reasonably practicable, that employees and other visitors are protected at work. Other visitors may include the self employed, sub-contractors and the general public.

'So far as is reasonably practicable' is a key phrase: employers do not have to take measures to avoid or reduce the risk if it is technically impossible, or the time, effort and cost is grossly disproportionate to the risk.

The Management of Health and Safety at Work Regulations 1999 give detailed guidance on what is expected of employers. Employers are also required to keep everyone involved informed of health and safety issues.

There are two organisations whose job it is to enforce existing health and safety law, and to act as a source of advice and information on health and safety matters:

- Health & Safety Commission (HSC), which considers and develops health and safety policy on behalf of the government
- Health & Safety Executive (HSE), which advises the HSC on the shaping of policy and is responsible for its implementation

Key Points

Key features of the Health and Safety at Work Act 1974:

Employers duties include:
- Make sure the workplace is safe and without risks to health by assessing risks
- Ensure plant and machinery are safe and that safe procedures of work are set and followed
- Ensure articles and substances are moved, stored and used safely by providing correct equipment and training
- Provide adequate welfare facilities including first aid arrangements
- Provide the information, instruction, training and supervision necessary for personal health and safety
- Make sure that work equipment is suitable for intended use, and that it is properly maintained and used
- Ensure that appropriate safety signs are provided and maintained

Employees duties include:
- Taking reasonable care for their own health and safety and that of others who may be affected by their actions
- Correctly using work items provided by their employer, including personal protective equipment (PPE), in accordance with training or instructions
- Using anything provided for health, safety or welfare correctly

Fig 5.18 Examples of personal protective equipment

Control of Substances Hazardous to Health Regulations 2002 (COSHH)

Using chemicals or other hazardous substances at work can put people's health at risk. Employers have a duty to control the exposure to hazardous substances and to protect both employees and others who may be exposed.

Failure to control exposure to hazardous materials can result in harm, from mild eye irritation, skin complaints and fainting as a result of fumes, to chronic lung disease or, on rare occasions, death. Consequences for the individuals harmed are obvious. For the employer they could be lost productivity and liability for legal action, including prosecution under the COSHH Regulations and civil claims.

Implementing the requirements of COSHH regulations can lead to:

- Improved productivity, using more effective controls (for example less use of a raw material or change of material or process)
- Improved morale, with a better employee understanding and compliance with health and safety requirements

Hazardous substances include:

- Substances that are used during work, for example adhesives, paints, cleaning materials and developing materials
- Substances that are created as a result of work activities, for example fumes from soldering
- Air-borne particles, for example dust

Asbestos, radioactive substances and lead are not included under this legislation as there is specific legislation relating to these materials.

For the majority of commercially available chemicals that we may use at work, for example soap and washing-up liquids, COSSH regulations are not applicable. COSHH regulations only apply to products if they have a warning label, for example bleach.

HSE guidance describes eight steps to comply with COSHH regulations:

1 **Assess the risks** to health from substances used in or created by your workplace activities.
2 **Decide what precautions are needed**. Do not expose employees to hazardous substances without considering the risks and taking the necessary precautions.
3 **Prevent or adequately control exposure**. Prevent employees from exposure to hazardous substances or adequately control it.
4 **Ensure that control measures are used** and that safety procedures are followed.
5 **Accurately monitor** the exposure of employees to hazardous substances.
6 **Carry out appropriate health checking** and monitoring where necessary.
7 **Prepare accident, incident and emergency procedures.** Have plans ready to implement when required.
8 **Ensure all employees are properly informed, trained and supervised** and that training is updated.

Other important health and safety regulations

- **Workplace (Health, Safety and Welfare) Regulations 1992** – covers issues including ventilation, heating, lighting, workstations, seating and welfare facilities
- **Personal Protective Equipment at Work Regulations 1992** – relating to protective clothing and equipment for employees
- **Provision and Use of Work Equipment Regulations 1998** – regarding the safe use of equipment and machinery provided for use at work
- **Reporting of Injuries, Diseases and Dangerous Occurrences Regulations 1995 (RIDDOR)** – employers must notify the Health and Safety Executive of certain occupational injuries, diseases and dangerous events

Risk assessment

Health and safety law requires employers to assess the risks to health and safety and make arrangements for implementing appropriate health and safety measures identified by those assessments.

In establishments where there are five or more employees, significant findings of the risk assessment and appropriate actions are recorded.

A risk assessment is a careful examination of what, in a workplace, could cause harm to people. Decisions have to be made as to whether enough precautions have been taken or if more could be done to prevent harm.

Workers and others have the right to be protected from harm. Failure to take reasonable control measures could result in injury or death, and lead to actions described earlier in this section.

We all already work to simple procedures that can control risks, for example keeping walkways clear so that people do not trip up. The law does not require all risks to be eliminated; it requires that all reasonably practicable measures have been taken to minimise the risk. In most cases an employer will ask a suitably qualified member of staff to identify hazards and quantify the risk.

A **hazard** is seen as anything that could cause harm, such as working at height, using machinery or working with electricity.

The **risk** is the chance that someone may be harmed by a particular hazard. The risk could be high, medium or low. An assessment would include an indication of the likely occurrence of the hazard together with an indication of how serious the harm could be.

HSE guidance describes five steps to carry out a risk assessment:

1 **Identify the hazards**. Inspect the workplace, carry out a tour with other suitably qualified personnel, ask employees their opinions, seek advice, use manufacturers' instructions or data sheets for chemicals and equipment, check accident and ill-health records – these often help to identify hazards.

2 **Identify who might be harmed and the nature of the harm.** Identify the groups or individuals who may be harmed, identify what type of injury or ill health might occur. Take into account special requirements, for example expectant mothers and new workers. Include others, for example cleaners, visitors etc. Consider the likely occurrence of the hazard: is it likely to be a very rare occurrence or high possibility.

3 **Evaluate the risks and decide on control measures.** The law requires that everything 'reasonably practicable' is done to protect people from harm. The easiest way is to compare what is being done with best practice. Seek advice on good or best practice. Aim to get rid of the risk; if this is not possible, ensure that the precautions you take control the risks so that harm is very unlikely. Control measures could include:

- Training – update regularly
- PPE (goggles, gloves, footwear)
- Guarding of machinery, fencing off areas

4 **Record findings and implement actions.** Write down the results of risk assessments, publish them for staff (not required for establishments with fewer than five employees). Do not produce lengthy statements: clear concise points are required. Accurately specify locations of hazards.

5 **Set a fixed period for review of risk assessment and update if necessary.** Risk assessments should be reviewed annually. A review date ought to be fixed when the assessment is completed. When changes are made in the workplace, new hazards may occur. A review should be made as soon as changes are introduced. Encourage a culture where employees will inform of particular hazards.

1 Use one example of legislation to discuss the protection of workers/operators in the workplace.
2 Discuss the issues to be considered by manufacturers when implementing risk assessment procedures.

Section F The impact of design and manufacturing on the environment

By the end of this section you should have developed a knowledge and understanding of:

- Issues relating to global sustainable development
- The energy needs during the life of a product or system – life-cycle assessment
- The terms 'availability', 'conservation' and 'pollution' relating to energy
- Recycling and green issues in product and systems design

Issues relating to sustainable development

Sustainable development is a socio-ecological process characterised by the fulfilment of human needs while maintaining the quality of the natural environment indefinitely. The Brundtland Commission coined what was to become the most-often quoted definition of sustainable development, as development that 'meets the needs of the present generation without compromising the ability of future generations to meet their own needs'.

The field of sustainable development can be conceptually broken into three parts: environmental sustainability, economic sustainability and social-political sustainability.

Why sustainable design is important

It is a commonly known fact that if everyone in the world lived as we do in the UK, we would need at least three planets to sustain us.

Fig 5.19

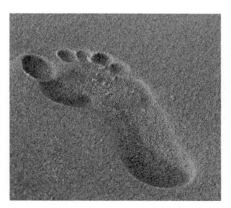

Fig 5.20

Some facts to think about

Have you ever thought about how much rubbish you and your family throw away every week? Or why we need to stop throwing so much of it away? Here are some waste facts.

- The UK produces more than 434 million tonnes of waste every year. Every year UK households throw away the equivalent of 3.5 million double-decker buses (almost 30 million tonnes), a queue of which would stretch from London to Sydney (Australia) and back
- On average, each person in the UK throws away seven times their body weight (about 500 kg) in rubbish every year. On average, every family in the UK consumes around 330 glass bottles and jars a year[1]. Recycling two bottles saves enough energy to boil water for five cups of tea
- Every year, an estimated 17.5 billion plastic bags are given away by supermarkets. We produce and use 20-times more plastic today than we did 50 years ago!
- Babies' nappies make up about two per cent of the average household rubbish. This is equivalent to the weight of nearly 70,000 double-decker buses every year. If lined up end to end, the buses would stretch from London to Edinburgh
- In Europe people keep a mobile phone for 18 months on average[2] Mobile phone chargers left plugged in cost £60 million per annum and produce 250,000 of CO_2[3]

We talk in terms of carbon footprints: a carbon footprint is made up of two parts: the primary footprint and the secondary footprint.

The primary footprint measures our direct emissions of CO_2 from the burning of fossil fuels, including domestic energy consumption and transportation.

The secondary footprint is a measure of the indirect CO_2 emissions from the products we use.

Calculate your own carbon footprint at www.carbonfootprint.com.

So what can we do about it?

As future designers, we need to consider the environmental, social and economic implications of our decisions and choices. These are all interlinked: all making involves materials; producing material consumes fuel; currently, most fuel processes cause pollution and affect climate change, damaging people's health. This can result in economic costs and social issues.

Many products are produced where workers are subject to poor working conditions and low wages, another social issue. Recycling has environmental benefits but also has economic costs and uses energy and can cause pollution, which can result in social issues. There are no right answers – all we can aim for is to be informed designers.

We can also try to:

- **Rethink** what materials and energy we use and the way we use it
- **Reduce** the materials we use and try to create products that are multifunctional
- **Reuse** materials or products
- **Repair** existing products rather than buying new ones; design products that can be repaired easily
- **Recycle** materials where possible; perhaps use recycled materials in new products
- **Refuse** to use certain materials or to buy certain products if they are not needed

More than 50 per cent of the world's energy is used by 15 per cent of the world's population[4].

- Choose an everyday activity and consider its environmental, social and economic issues, for example making a cup of tea or coffee. Think about the last cup of tea or coffee you made:
- Identify the stages involved from sourcing the ingredients to the end of their life.[5] Where did the tea or coffee come from and who was involved in its production? Had it travelled a long distance? Was it traded fairly? Who picked/harvested the tea/coffee? Where did the milk and sugar come from?
- Now think about the making phase. Did you measure the amount of water according to the number of cups that were to be made?

Fig 5.21

Did you leave the kettle and have to re-boil it again because the water had cooled?

- Now think about the end of the life-cycle. What did you do with any leftover water or with coffee filters or tea bags/leaves? What about the packages the tea or coffee came in? The dirty cups?
- Visit the Sustainable Design Award (SDA) website (www.sda-uk.org/lineups.html) for other ideas including: cleaning your teeth, disposing of waste at home, the last article of clothing you bought.
- Read articles in newspapers about issues relating to sustainability. There are many useful websites[6] and resources that will help you develop a greater awareness.

1 Discuss how the designer's responsibility extends beyond meeting the needs of the consumer and manufacturer.
2 Discuss the environmental implications of increased energy use.
3 Discuss ways in which consumers can help to conserve the environment.
4 Many UK-based companies manufacture their products in other countries. Discuss the moral and/or ethical considerations of the globalisation of product manufacture.

The energy needs during the life of a product or system

The table below outlines the energy needs in the life of a product or system and includes possible environmental implications. The stages described form the basis of a life-cycle assessment.

Life-cycle assessment

Life-cycle analysis, a raising of awareness of environmental issues through a detailed examination of the life cycle of products, was introduced in the late 1960s and early 1970s.

Current practice involves a life-cycle inventory (LCI) where detailed data is compiled followed by a life-cycle assessment (LCA), which interprets and evaluates the environmental impact of a product from 'cradle to grave' – from the extraction of materials required to manufacture the product to end of use and disposal.

Table 5.7 Energy needs in the life of a product or system

INPUTS		OUTPUTS
	Acquisition of raw materials	
	All products or systems are created from raw materials. Consider the energy needed to extract oil, ores and timber, and in livestock production systems. Look at the environmental impact of mining, deforestation and other issues related to the extraction of raw materials.	
	Transporting raw materials	
raw materials	Consider how raw materials are transported nationally and internationally, and examine the environmental impact, for example oil tanker disasters.	atmospheric pollutants
energy	**Processing raw materials**	solid waste
	Consider the energy requirements and environmental effects of transforming raw materials by chemical or physical processing methods, for example smelting and converting ores into usable materials	
	Manufacturing the product	
water	Most products require machine processing. The manufacturing industry requires energy for machines, lighting, heating, etc. Denim jeans are often dyed during manufacture: the chemicals used may have an environmental impact.\nTransporting components and completed products for distribution involves considerable energy use and impacts on the environment.	water pollution
	Using the product	
recycled products	Some products require no further energy in usage. Many products, such as cars, washing machines and dishwashers use significant amounts of energy. Some products, such as milk bottles, are reused; energy is used for cleaning before refilling. Detergents used may have an environmental impact.	recycled products
	Disposal	
	Collection of waste requires energy. Incineration centres use energy to dispose of waste, although many reclaim the energy created by incineration for useful purposes. Landfill systems may impact on the environment.	

LCA of a DVD

1 Acquire raw materials – bauxite (aluminium ore) mined, crude oil to make polycarbonate and dyes for printing

2 Transport – large container ships: significant energy use, possibility of oil tanker disaster (for example *Prestige, Exxon Valdez*)

3 Material processing – bauxite crushed, washed, mixed with chemicals, then smelted, rolled, machined or cast; very heavy power consumption; crude oil processed, with chemicals and natural damage

4 Manufacturing – polycarbonate injection moulded, disk coated with thin layer of aluminium, coat of lacquer applied, disks screen printed

5 Packaging – further material use, plastic cases, card boxes, printed using dye/ink, shrink wrapped; recycled card and plastic often used; energy needed during processing of packaging

6 Transportation/distribution – truck, rail, plane – all contribute to environmental damage

7 Usage – no additional energy needs during usage. If stored and used correctly will last for decades

8 Disposal – reuse, share, donate to others; recycled material can be reformatted – material can be cleaned, ground and turned into car parts, office equipment and cable insulation; 5.5 million unwanted software disks are discarded in landfill sites every year

Eco-labelling

Eco-labelling is a voluntary system, awarded by an independent third party, to identify and label products that have minimal environmental impact.

The flower is the symbol of the European eco-label scheme and identifies products and services that are a genuinely better choice for the environment. It is designed to help manufacturers, retailers and service providers to get recognition for good standards, and for purchasers to make reliable and informed choices.

The scheme is designed to:

- Help to achieve substantial environmental improvements
- Ensure that the award has credibility
- Encourage manufacturers, retailers and service providers to apply for the award
- Encourage purchasers to recognise the award and buy products and services that have the award
- Improve consumer environmental awareness

Fig 5.22 The European eco-label symbol

> **Produce a life-cycle assessment for the following products: denim jeans, disposable drinks cup (card), fizzy drinks bottle (plastic).**

Energy – availability, conservation and pollution

Availability

Fossil fuels account for 85 per cent of global energy use. Although estimates of available reserves vary, it is considered by some that, at current annual rates of production, we have left, worldwide, about:

- 155 years of coal
- 40 years of oil
- 65 years of natural gas[7]

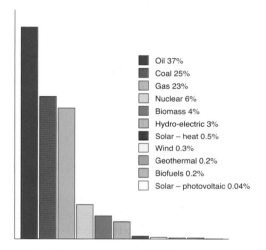

Fig 5.23 World energy usage 2005

Oil 37%
Coal 25%
Gas 23%
Nuclear 6%
Biomass 4%
Hydro-electric 3%
Solar – heat 0.5%
Wind 0.3%
Geothermal 0.2%
Biofuels 0.2%
Solar – photovoltaic 0.04%

Using fossil fuels to convert energy into a more useful form has a significant environmental impact. Burning fuels produces waste products due to impurities. Various gases such as sulphur dioxide, nitrogen oxide and other volatile organic compounds can have a harmful effect on the environment. The burning of fossil fuels creates carbon dioxide, which contributes towards global warming.

The USA, Canada, Russia and many European countries already have huge energy demands. India and China are emerging and rapidly growing industrial powers with plans for further expansion. Meeting these demands poses genuine concerns over the use of existing resources and the environmental effects of using existing technologies to generate energy.

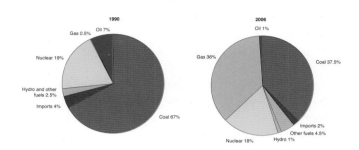

Fig 5.24 UK electricity production in 1990 and 2006

In the UK, electricity is the main source of power for industry, commercial and domestic use. Methods of electricity production have changed dramatically over the last 30 years.

The North Sea natural gas reserves, which helped to supply a significant percentage of our needs, are now dwindling, and importing gas from Europe and Russia has increased.

Global warming issues have emphasised the need to examine current practice and explore ways in which emissions can be reduced.

Renewable energy systems are being increasingly used to provide large and small scale, effective, environmentally acceptable power.

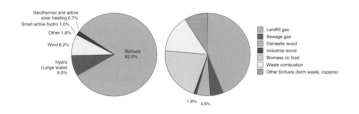

Fig 5.25 Renewable energy

Fig 5.26 Solar plant

All energy conversion systems have benefits, but also have some form of impact on the environment. The table below compares the main energy production systems.

Table 5.8 Energy production systems

Method	Description	Advantages	Disadvantages
Nuclear	Nuclear fission generates heat, heats water to generate steam, steam turns turbines, turbines turn generators, electricity distributed	Same cost as coal, no smoke or carbon dioxide emissions, lots of energy from small amount of fuel	Radioactive spent fuel, disposal of waste, serious potential effects of accidents
Gas/Coal/Oil	Fuel is burnt to generate heat, heats water to generate steam, steam turns turbines, turbines turn generators, electricity distributed	Readily available, ease of transport of fuel, gas-fired stations very efficient	Air pollution including carbon dioxide and sulphur dioxide, finite resources, visual impact of extraction, large stocks needed (coal)
Hydro-electric	Dam is used to trap water, water released turns turbines, turbines turn generators, electricity distributed	Very low cost once dam built, no air pollution, reliable, up to full power very quickly	Can impact on environment, (flooded area, reduced flow at base), initially expensive
Wind	Blades designed to catch wind, blades turn turbines using gears, turbines turn generators, electricity distributed	No fuel needed, no waste or greenhouse gases, can be used in remote areas	Unreliable, unsightly, old designs noisy, can harm flocks of birds
Solar photovoltaic	Photovoltaic cells convert light to electricity	Low cost after initial outlay, no pollutants or waste, used in small or large scale in remote areas	Expensive initial cost, unreliable, storage system needed
Tidal barrages	Barrage built across river estuary, turbines turn as tide enters (and when tide leaves), turbines turn generators, electricity distributed	Low cost after initial outlay, no pollutants or waste, predictable	Very expensive initial cost, environmental costs – can damage habitats, only generates power at set times during the day

Table 5.8 continued

Method	Description	Advantages	Disadvantages
Wave	Motion of waves forces air up cylinder to turn turbines, turbines turn generators, electricity distributed	Low cost after initial outlay, no pollutants or waste, OK for remote coastal areas	Unreliable, hostile environment, high maintenance, not a large power output
Geothermal	Cold water pumped underground through heated rocks, steam turns turbines, turbines turn generators, electricity distributed	No pollutants or waste, minor cost of pumping, resource 'free', very small stations, no negative visual impact	Only work in certain locations, can be unpredictable, possibility of gas emissions
Biomass	Fuel (wood, sugar cane, etc.) is burnt to generate heat, heats water to generate steam, steam turns turbines, turbines turn generators, electricity distributed	Readily available fuel, can use waste materials, low cost process	Air pollutants, requires large amounts of fuel, can be seasonal

Conservation and pollution

Society can make a contribution by:

- Conserving energy wherever possible: turn down central heating, share travel, walk or use a bike
- Making decisions on whether powered products and systems are necessary; are all gadgets useful?
- Using energy efficient products: refrigerators, freezers, washing machines, cars with good fuel consumption
- Using appliances efficiently: fill dishwashers, use just enough water in kettles
- Choosing reusable products over disposable ones
- Choosing products with minimal packaging, reducing the waste that you create

The generation of energy, particularly the production of electricity, has contributed greatly to atmospheric pollution. Coal, gas and oil-fired systems emit large amounts of carbon dioxide, which many scientists agree is a significant contributory factor to global warming.

The greenhouse effect

The earth is surrounded by a layer of gases including carbon dioxide (CO_2), methane (CH_4) and nitrogen oxide (NO). This layer allows the sun's rays to penetrate.

Around 30 per cent is deflected back by ice caps and clouds, but the majority of the rays are absorbed by the earth and oceans, and released back towards space as infrared radiation.

The layer of gases prevents all of the radiation from leaving the planet and traps enough to heat the lower atmosphere. Without this layer, temperatures would be at least 30°C cooler. Increased amounts of carbon dioxide, methane and nitrogen oxide are effectively making the blanket thicker and creating global warming.

Electricity production emits other pollutants. Over 80 tons of mercury, the most toxic heavy metal, is released into the atmosphere every year. Waste incineration plants add to this total.

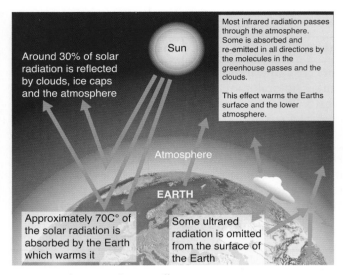

Fig 5.27 The greenhouse effect

Acid rain

Coal-fired power stations emit sulphur dioxide (SO_2), NO and polycyclic aromatic hydrocarbons (PHA) into the atmosphere. Regulations were introduced to limit emissions when the destruction of plant life and pollution of rivers was directly linked with power production plants.

A more accurate term is acid deposition, where SO_2 and NO are transformed into dry or moist secondary pollutants such as sulphuric acid (H_2SO_4), ammonium nitrate (NH_4NO_3) and nitric acid (HNO_3). The transformation of SO_2 and NO to acidic particles and vapours happens as they are transported in the atmosphere over great distances, up to thousands of kilometres. The acidic particles and vapours are deposited as either wet or dry deposition.

Wet deposition is acid rain, the process by which acids with a pH normally below 5.6 are removed from the atmosphere in rain, snow, sleet or hail.

Dry deposition occurs when particles of ash, sulphates, nitrates and gases are deposited on and absorbed by surfaces. They can then be converted into acids when they come into contact with water.

Flue gas desulphurisation (FGD) is one method used to remove pollutants. It is a 'wet scrubber' system used in many countries. Hot gasses are extracted and injected with limestone slurry. This combines with SO_2 to produce pH-neutral calcium sulphate, which can be extracted and used in the construction of roads.

Fine particle pollution is a real problem in many industrialised countries, causing a range of harmful effects including respiratory difficulties. The fine particles are a mixture of harmful pollutants (such as soot, acid droplets and metals) that are produced by combustion sources such as power plants and vehicles. These particles are either soot emitted directly from the sources or are formed in the atmosphere from SO_2 or NO emissions. The smallest (fine) combustion particles are of most concern because they can be inhaled deeply and be absorbed into the bloodstream, evading the human lung's natural defences.

Governments have reacted with legislation to impose targets for reduction of the key pollutants; companies are required to develop and use better technologies to filter and significantly reduce the emissions of SO_2, NO and mercury.

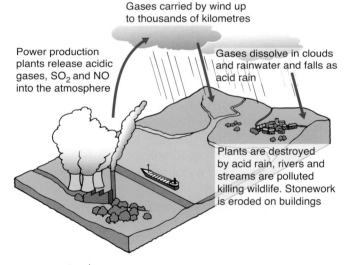

Fig 5.28 Acid rain

Recycling and green issues

Recycling prevents some environmental damage linked to the production of new materials. However the process of recycling does use energy for transportation and reprocessing.

We are better at recycling than we used to be: the household recycling and composting rate in

England increased to 31 per cent in 2006/7 from 27 per cent in 2005/6.

Key Term

The three Rs represent the **waste hierarchy** that lists the best ways of managing waste from the most to the least desirable. It is a central theme of EU waste policy:

- **Reduce** – not producing waste in the first place is the obvious solution and we can all play a part by thinking about how and why we produce waste
- **Reuse** – many of the things we currently throw away could be reused again and again with just a little thought and imagination
- **Recycle** – waste products can be turned back into raw materials and then used to make new products. This helps to conserve natural resources and energy

Fig 5.29 Recyclable aluminium

Aluminium and steel

Seventy-five per cent of all drinks cans are made from aluminium in the UK[8]. Recycling it uses only five per cent of the energy it takes to make new aluminium – and produces only five per cent of the CO_2 emissions. Just one recycled aluminium can saves enough energy to run a television set for three hours[9]

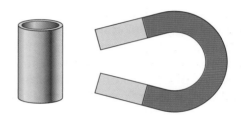

Fig 5.30 Recyclable steel

Steel recycling is common in industry – waste generated by the steel manufacturing process is re-melted and used over and over again. It never leaves the mill, refinery or foundry[10]. In the UK we use around 12.5 billion steel cans every year, or 600 per household, but nearly 10 billion of these still go to landfill. In 2005 we recycled around 50 per cent of steel packaging, including over 2.5 billion steel cans! The government target is to increase that to 54 per cent by 2008[11]. Producing steel from recycled material saves 75 per cent of the energy needed for steel made from virgin material[12]. Every steel can is 100 per cent recyclable, and it can be recycled over and over again.

Glass

Glass can be recycled again and again without losing its clarity or purity. In 2005 we recycled approximately 1.2 million tonnes of used glass (known as 'cullet')[13]. Making glass bottles and jars from recycled ones saves energy: the energy saved from recycling one bottle will power a 100-watt light bulb for almost an hour.

Paper and cardboard

We use 12.5 million tonnes of paper and cardboard every year in the UK. About one-fifth of the content of household dustbins consists of paper and card, of which nearly half is newspapers and magazines. This is equivalent to over 4 kg of waste paper and card per household in the UK each week.[14]

Recycled paper made up 80.6 per cent of the raw materials for the UK[15].

Textiles

Clothes recycling banks are currently only operating at about 25 per cent capacity[16].

Plastic

Recycling just one plastic bottle saves enough energy to power a 60-watt light bulb for six hours[17]. It takes just 25 two-litre pop bottles to make one adult-size fleece jacket[18]. Plastics for recycling are regularly bought in the UK and elsewhere for export to China. A tonne of plastic bottles for recycling can fetch around £200. It would make no economic sense to ship it half the way round the world to then simply dump it. In 2001, 66,813 tonnes of plastic were exported. This rose to 237,753 tonnes in 2005. Over half of the UK's plastics recycling is done via export.

Why are we sending plastic abroad to be recycled? China in particular has fast-growing manufacturing sectors; it's now a major manufacturer of plastic items. Therefore, if we want recycled plastics to be used again, it is inevitable that at least a proportion of our waste plastics will be exported to China to be reused. Don't forget, the UK has a huge economy and rate of consumption for its size, and we're still very much in the early stages of our plans for recycling.

We currently import goods from China: the ships that bring them would go back empty if they were not used to take secondary materials back to China. So, all recyclable materials that are exported to countries like India and China are shipped from the UK on otherwise empty container ships. The trade is robustly regulated by the Environment Agency.

Whilst steel and aluminium can easily be separated by magnets, we still need some way of identifying plastics quickly: 11 per cent of household waste is plastic, 40 per cent of which is plastic bottles. Up to 40 per cent less fuel is used to transport drinks in plastic bottles compared to glass bottles. Plastic packaging uses only around 2 per cent of all crude oil produced.

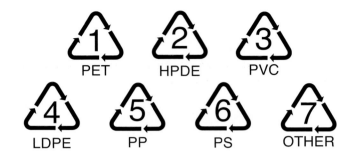

Fig 5.31

As designers we can aim to reduce, reuse and recycle. But the real message is perhaps to try to use less: less products, less materials, less energy, less resources. Less packaging results in less waste and less impact of the environment. By using less materials we can reduce our products' ecological impact.

Would it be possible to use materials that are easily recycled? If so, could you make it obvious to the consumer that it is recyclable?

How far are the materials you are using travelling to get to the factory? Could they be substituted for more locally sourced materials that will result in less impact?

Could your product and its packaging be designed so that it is easily recycled, with the materials easy to separate? Design for disassembly – with recycling in mind.

1 Discuss the environmental implications of using metals for food and drink cans.
2 Discuss the implications of using non-sustainable resources in disposable products.
3 Discuss the implications for the design of packaging to enable a reduction in the volume of disposable waste.
4 Discuss the implications of using recycled materials in the manufacture of products.
5 Many products are designed to have a limited life expectancy. Discuss the implications of the increased availability and use of 'throw-away products'.

By the end of this section you should have developed a knowledge and understanding of the ethical issues associated with:

- Environmental, moral, economic and social issues related to product design and manufacture
- The effect of fashion, trends, taste and style
- The effect of new technological developments
- Ethnic and cultural influences on the design and manufacture of products

All that glistens is not gold

A recent article from ActionAid (www.actionaid.org.uk) highlighted the plight of Nagina, a bangle maker from Pakistan.

Fig 5.32 Photo © Warrick Page/Pans/Action Aid

'Sometimes I work all night. I am always tired. Many women's fingers become badly affected by the chemical. We get allergies and our fingers become numb – when that happens we can't handle the bangles and we have to stop until the feeling comes back.'

Nagina is 28; she has five children and works almost 14 hours a day for approximately 60 rupees – the equivalent of one dollar, about 60p – to earn enough money to put four of her children through school. The youngest child, Dua, aged four, plays all day amongst the bangles and chemicals. Her house consists of one room 3 m × 4 m and is full of paint, thinners and other chemicals. If you buy a cheap bangle

from a market stall it is possible that Nagina's family made it.

Products are made by real people – there are moral implications for all of us when be buy things! The situation is complex. Organisations like ActionAid work along side their partners, such as the Labour Education Foundation, to slowly improve working conditions, health, education and childcare.

It's no good suddenly stopping buying non-ethical products – the families that make them might not be able to support themselves otherwise. You do, however, need to ask questions and challenge the ethics of those who design, manufacture and sell products, and support those organisations that are trying to change the lives of families for the better.

There may be questions on ethical issues in the exam. You will need to read and research round these issues and be able to quote actual examples of where individuals or groups are disadvantaged by unfair working conditions or trading practices.

Key Terms

Ethical – Correct, fitting, good, honest, honourable, just, moral, principled, proper, right, upright (Collins).

When you see the words 'moral implications', think of these things.

Ethnic – Cultural, folk, indigenous, national, native, racial, traditional (Collins).

You will also need to consider cultural values and make sure that no offensive symbols, images or materials are used in any products you design.

Make sure that you do not confuse 'ethnic' with 'ethical'.

Fig 5.33 ActionAid's logo

'If you think you are too small to be effective, you have never been in bed with a mosquito.'
Betty Reese

What can you do to make a difference?

Fig 5.34 'Fair trade and organic – the perfect combination. Certified fair trade organic cotton from Agrocel in India is used to make our popular range of tops. The tailoring at Craft Aid in Mauritius provides valuable employment for disadvantaged local people. Producing one top brings benefit to two producer groups.'
Traidcraft Catalogue, Winter 2007

Fig 5.35 'I like working with Agrocel. The cotton price is better. There are also credit facilities and we can get good quality seed.'
Satbir Singh, cotton farmer with Agrocel, India
Traidcraft photo: Shailan Parker

The thought-provoking statement – what can you do to make a difference – was adopted by Dame Anita Roddick (1942–2007), activist and founder of the Body Shop (www.thebodyshopinternational.com), one of the first major commercial success stories to be founded on ethical principles.

You can make a difference by being aware of sustainable issues and making informed choices when you design or buy products or source ethically produced materials.

Non-organic cotton contains only 73 per cent cotton – the rest is chemicals and pesticides.

Organisations such as Traidcraft only use ethically produced materials and ingredients, which help both the producers and manufacturers in developing countries. Through Traidcraft's work Dalbir Singh produces rice for a better price in India; Efiness Gwiriza, a basket weaver in Malawi, can now afford shoes for herself and her children; and Adriano Kalilii, a tea plucker from Kibena in Tanzania, can now afford the iron sheets to roof his house.

You need to be aware of the moral implications of fair trade and can help by buying fairly traded goods.

Fig 5.36 Rice in India (Traidcraft photo)

Fig 5.37 Basket weaving in Malawi (Traidcraft photo)

Fig 5.38 Tea in Tanzania (Traidcraft photo)

Use the internet to find out more:
- Traidcraft's website www.traidcraft.co.uk
- See the Flipside – Traidcraft's campaign to get young people involved in all things fair trade – www.seetheflipside.co.uk
- Need a speaker? www.traidcraft.co.uk/speakers

Many supermarkets now stock fair trade goods and ingredients. Get a taste for fair trade tea, sugar, coffee, rice, dried fruit and chocolate!

Fig 5.39 Traidcraft's logo

'I stand before you as a representative of an endangered people – as a result of global warming and sea level rise, my country may disappear from the face of the earth.'

Maumoon Abdul Gayoon, President of the Republic of the Maldives

What should you know about sustainable design?

The excellent resource *The Sustainability Handbook for D&T Teachers* (www.practicalaction.org) is also an essential read for all design and technology candidates. The chapter on the six Rs is abbreviated in the table below. It contains many fascinating facts and figures that will help with actual examples for your product study as well as the examination questions. There are more extracts from the book in Chapter 2, which deals with the product study; use the information in this section for 'Moral Implications'.

The six Rs

Table 5.9 The six Rs

RETHINK	How can it do the job better? Is it energy efficient? Has it been designed for disassembly?
REUSE	Which parts can I use again? Has it another valuable use without processing it?
RECYCLE	How easy is it to take apart? How can the parts be used again? How much energy to reprocess parts?
REPAIR	Which parts can be replaced? Which parts are going to fail? How easy is it to replace parts?
REDUCE	Which parts are not needed? Do we need as much material? Can we simplify the product?
REFUSE	Is it really necessary? Is it going to last? Is it fair trade? Is it unfashionable to be trendy and too costly to be stylish?

Unfair fashion

Are cheap clothes sustainable?

Clothing prices are now so low that shoppers are treating their clothes as practically disposable! But who pays the cost of cheap clothes?

One of the factors allowing UK retailers to sell clothes cheaply is low production costs. In many of the countries where value retailers source clothing, huge savings are made through low wages. Average hourly wages in 2000 were £0.46 in China, £0.38 in India, and £0.13 in Sri Lanka (calculated on 2005 exchange rate).
(www.corporatewatch.org.uk)

Fig 5.40 Extract from *The Sustainability Handbook*

Ethical electronics

Is being trendy always sustainable?

Advertisers are aware of the potential spending power of most age groups, especially young people and the so-called 'baby-boomers' born in the 1960s. We are encouraged to buy more electronic products and led to believe that a new item works better, or looks better, than a similar purchase made not that long ago.

Consumer electronics has followed the lead from fashion and music, and indeed these technologies have merged with an exponential synergy. The result is must-have products, which satisfy the drives of want and need.

The economic, environmental and ethical consequences are increasingly reported in newspaper headlines, such as:

'Dumped electrical goods: A giant problem'. In the same article, the sub-heading, **'Fashion beats functionality in a throwaway society'**, Martin Hickman, The *Independent*, 27 February 2006, news.independent.co.uk/environment/article347941.ece

'iPod City', The *Mail on Sunday*, 11 June 2006. The original report is unavailable on-line, but see a summary at:
'Inside Apple's iPod factories' (www.macworld.co.uk/news/index.cfm?RSS&NewsID=14915)

Fig 5.41 Extract from *The Sustainability Handbook*

Is there a seventh 'R'? Redesign – new technological developments can help to create a more sustainable environment. Watch out for the next one!

Product story

Belu now sell bottled water in a 'bio-bottle' made from corn starch that is composted back to soil in 12 weeks. They have not yet found a suitable material for the cap, so this needs to be recycled in the normal way. Innocent smoothies also use this type of bottle.

However, biodegradable tableware uses a greater weight of resources than ordinary disposable ones. (Vital Waste Graphics 2)

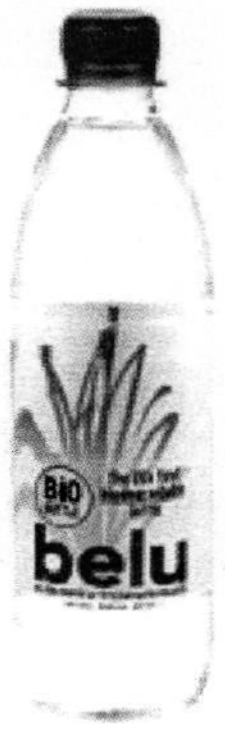

Fig 5.43

Fig 5.44 Practical Action's logo

Biodegradable packaging

Should food be packaged at all?

Most people associate recycling with food packaging: taking bottles to the bottle bank, washing out cans ready to be collected, saving milk cartons and aluminium cans for the recycling bin. We have all started to do something towards this sustainability issue, but is it enough?

More than half of all packaging is plastic, made from a non-renewable resource and the most difficult to recycle. Then there are mixed material packages such as Tetra Pak or Gualapack where recycling is even more difficult.

Product story

Tetra Pak is made up of 70% paperboard, 24% LDPE, a plastic, and 6% aluminium. It has been suggested that Tetra Paks can stop light depleting the vitamins and minerals in drinks, and that they are good because the paperboard is made from a renewable resource and is easy to recycle and transport.

At the moment, however, there is only one recycling plant in the UK for Tetra Pak cartons. On average, each household in the UK uses 2.3 kg of Tetra Pak cartons a year.

Fig 5.42 Extract from *The Sustainability Handbook*

Section H — Aesthetics and function, shape, form, colour and taste

By the end of this section you should have developed a knowledge and understanding of aesthetics and be able to:

- Develop a critical awareness of designed objects/products in terms such as colour, form, shape, taste, texture and surface finish
- Consider the way aesthetic aspects influence appearance, contrast, composition, harmony/disharmony

The study of aesthetics should not be considered in isolation, but should pervade the whole course. It is particularly relevant in the coursework units, F522 Product Study (AS), and F523 Design, Make and Evaluate (A2).

Questions relating to aesthetics will be set in Session 3, the written paper of unit F521 Advanced Innovation Challenge (AS) and in unit F524 Product Design, the Advanced GCE written examination.

The introduction of new and innovative products is essential to any economy. Products have to be designed to suit the users' needs. They must be affordable, ecologically sound, work well and be aesthetically pleasing. Very often, customers are drawn to products because of their visual impact.

Aesthetics, in its widest interpretation, is involved with our senses – vision, hearing, taste, touch, smell – and our emotional responses to objects and things.

The Bauhaus, a design school founded by Walter Gropius in 1919, was very influential in shaping an understanding of design and taste. Design was considered crucial and integral to the production process rather than merely a visual 'add on'.

'Form follows function' was a phrase often used to counteract the historically prevalent view that beauty was achieved by including additional features, not necessarily useful features. Architects and industrial designers in the 20th century were beginning to show that the form or shape of an object should be based on its intended purpose or function.

There are many examples of products that demonstrate a sensitive consideration of both aesthetics and function, some examples of which are shown in Figures 5.54 to 5.57.

Fig 5.45 Apple iPod, Jonathan Ive (2001)

Key Points

Aesthetic failure

In *Design for the Real World*[19] Victor Papanek describes how automotive designers in Detroit, looking at making car dashboards more symmetrical and aesthetically pleasing by relocating ash trays, controls and switches, could have resulted in 20,000 fatalities and 80,000 serious injuries over a five-year period. This happened as a result of drivers having to over-reach for controls, diverting attention from driving for split seconds longer.

Shape

Shapes are formed as a result of closed lines. Shapes can be visible without lines when a designer establishes a colour area. They may be composed from parts of different objects in an arrangement; they can be gaps, or negative shapes between the objects. Basic shapes include circles, squares, triangles and polygons, all of which appear in nature in some form or another.

Form

Form refers to the three-dimensional quality of an object. When light from a single direction hits an object, part of the object is in shadow. Light and dark areas within an image provide contrast that can suggest volume.

Colour

Colour is very important in product design. It creates responses by stimulating emotions and can excite, impress, entertain and persuade. Colour can also create negative associations.

Research suggests that 73 per cent of consumer purchasing decisions are made in-store. Catching the consumer's attention and conveying information effectively are critical to successful sales.

A designer must be aware of how people respond to colour and colour combinations.

Apple broke with tradition by introducing iMac computers in a wide range of colours. They realised that home computers did not need to look like the usual office machine and that customers wanted a more visually interesting and appealing design. The success of the iMac rescued a brand that had suffered $1.8 billion of losses in two years.

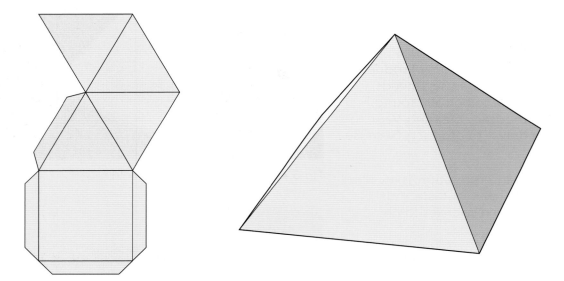

Fig 5.46 A 2D shape produces a 3D form

Define the following aesthetic features. Give examples of where they apply to products.

Visual	Hearing	Taste	Touch	Smell
Pattern Balance/ composition Rhythm Harmony/contrast	Loudness Pitch Melody	Sweetness Sourness Texture	Texture Comfort Temperature	Strength Pleasant/unpleasant

Table 5.10 Colour associations in product design

Red	Aggressive, passion, strong and heavy, danger, socialism, heat
Blue	Comfort, loyalty and security, for boys, sea, sky, peace and tranquillity, cold
Yellow	Caution, spring and brightness, joy, cowardice, sunlight
Green	Money, health, jealousy, greed, food and nature, inexperience
Brown	Nature, aged and eccentric, rustic, soil and earth, heaviness
Orange	Warmth, excitement and energy, religion, fire, gaudiness
Pink	Soft, healthy, childlike and feminine, gratitude, sympathy
Purple	Royalty, sophistication and religion, creativity, wisdom
Black	Dramatic, classy and serious, modern, evil, mourning
Grey	Business, cold and distinctive, humility, neutrality
White	Clean, pure and simple, innocence, elegance

Taste

Taste depends upon the individual. In product design we often refer to 'good taste' or 'bad taste'. There are accepted standards, often legally defined, but we all have our own views and preferences regarding what is 'aesthetically pleasing', and what combinations of shapes, colours and aromas work.

Taste is often linked to aesthetics and the personal appreciation of beauty. Product designers will research thoroughly to ensure that their creations will appeal to targeted groups and will acknowledge that taste changes quickly, depending on factors such as peer pressure and celebrity endorsement.

Section J — Ergonomics and anthropometrics

By the end of this section you should have developed a knowledge and understanding to be able to:

- Demonstrate an understanding of ergonomics when designing products
- Interpret and apply anthropometric data when designing

Key Term

'If an object, an environment or a system is intended for human use, then its design should take into account the characteristics of its human users.' Stephen Pheasant[20]

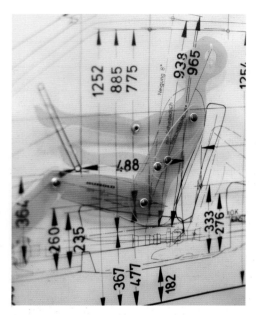

Fig 5.47

Ergonomics

Ergonomics is about 'fit' between people and the things they do, the objects they use, and the environments they work, travel and play in.

If a good fit is achieved, the stresses on people are reduced. They are more comfortable, they can do things more quickly and easily, and they make fewer mistakes. So when we talk about 'fit', we don't just mean the physical fit of a person, we are concerned with psychological and other aspects too.

Ergonomics can be thought of as usability. In the example of a mobile phone, the layout of the buttons are important; they should be the correct size and distance apart. The size of the text on them should be clear to read. Colour can play a big part in this. The edges of a product and its weight affect how comfortable it is to hold and use. The volume of ringtones, and the way in which it vibrates could also affect its usability. All these are ergonomic factors.

Looking at examples of 'bad' design is one way to learn about ergonomics. Some websites with illustrated examples of products that are hard to use because they do not follow human factors principles are shown in the following box.

See the following websites:
www.baddesigns.com
www.hfes-europe.org/badergo/ec_bad.htm
http://ergonomics.about.com/od/
ergonomicbasics/ss/bad_designs.htm

Anthropometrics

Anthropometrics are people measurements. Anthropometric data comes in the form of charts and tables. They may provide specific sizes, such as finger lengths and hand spans, but they also offer average group sizes for people of different age ranges. Other sizes to consider are heights, reach, grip and sight lines.

There are three general principles for applying anthropometric data:

- Design for the extreme
- Design for adjustability
- Design for the average

Percentiles

The use of percentiles is an important aspect of anthropometrics. The sizes of the human body given in anthropometric data are usually presented in tables, and normally include the 5th percentile, the 50th percentile (the average) and the 95th percentile. Look at some anthropometric data and notice the different percentile measurements.

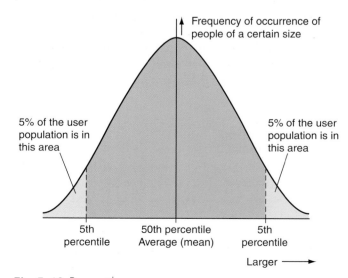

Fig 5.48 Percentiles

The 50th percentile is the most common size, the average.

The 5th percentile indicates that five per cent of people (or one person in 20) is smaller than this size.

The 95th percentile indicates that five per cent of people (or one person in 20) is larger than this size.

Very few people are extremely large or very small.

Which percentile, and therefore which size to be used from anthropometric data tables, will depend on what is being designed and who will be using it. Is the product for all potential users or just the ones of above or below average dimensions? If you pick the right percentile, 95 per cent of people will be able to use your design.

Consider which percentile you would need to use for the following:
- The height of a doorway
- The size of a handle for an upright vacuum cleaner
- Ventilation slots in the casing of an electric motor on a lawn mower
- The diameter of a screw-top water bottle

EXAMINER'S TIPS

Students commonly mix up ergonomics and anthropometrics. Ergonomics is the study of how we use and interact with an object/tool, etc. Anthropometrics is the study of the sizes of the human body. Anthro in Greek means man; metric should help you remember measurement.

Using human factors in designing

Hand tools are anything that can be manipulated by the hand. Bad design may cause result in slower work and more error, and possibly injury or accidents. The grip and the level of muscle exertion needed must be considered. The wrist angle is also important. Slight contouring of the grip or flared handles can increase comfort and reduce slippage in sweaty hands. Handles should be at least

100–125 mm long for power grip. Longer handles distribute forces on fingers.

Fig 5.49 Hand tools

Smooth handles for tools requiring wrist rotation should be avoided because of the increased risk of slippage and rotational wrist damage. Padding handles reduces the force needed to grip the tool. The effect of tool weight is also a factor. Use of rests, supports, two hand grips, etc. can all help to decrease the effort required to use heavy tools.

When designing, try to design tools and hand-held items for operation with both hands. When only right-handed users are considered, left-handers may be at an increased risk of injury. A simple experiment can demonstrate how the wrong angle can affect strength and grip in your hand.

- Stand with your arms at your side. Bend your right elbow at a 90-degree angle. Hold your hand out straight with your palm open; your wrist should be straight. Place two fingers from your left hand in your palm; keeping your right wrist straight, make a fist around these fingers and squeeze hard as a test of your grip strength.

- Now, repeat the procedure, but this time bend your wrist at a sharp angle so your palm faces in towards you. Was your grip affected at this position? Have you ever tried to complete a task that was much more stressful than normal because your body was twisted or extended in an odd way?

More activities of this type are available[21].

Fig 5.50

Fig 5.51

Design of seating

An estimated 50 per cent of people in the developed world suffer some form of back complaint and many of these are related to poor seat design. How we sit and what we sit on affects the health of the spine. There is no single ideal sitting posture. We can't design a chair for the best way to sit. We need a variety of chairs that allow us to sit in a variety of postures.

Seat design criteria

Optimum seat height is controversial, but a minimum height should be 380 mm based on the 5th percentile in women. Approximately half of a person's body weight is supported by less than 10 per cent of the body when seated.

Seat contouring and cushioning can help to distribute weight over a larger area causing the pelvis to rotate forward and promoting better posture.

A good seat angle helps users to maintain good contact with the backrest. For most purposes a five to ten degree angle is recommended. From a study of students, the preferred seat back angle for comfort is 15 degrees. This is in keeping with other studies by Etienne Grandjean where VDT operators have preferred 13–15 degrees backward incline. Higher backrests give better support; an optimal angle seems to be between 100 and 110 degrees from research.

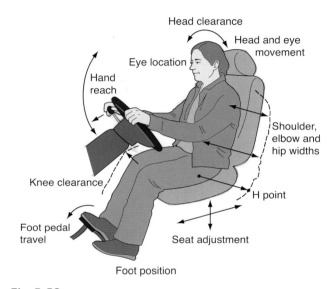

Fig 5.52

Dynamic dimensions are more difficult to measure but can be captured using mock-ups. Using real people is always recommended. Car designers build full-scale mock-ups and test them with real people.

Using ergonomics and anthropometrics in your project work

Ergonomics is an important part of research. Whatever you design it must fit the person who is to use it. For example, if you are to design a desk storage unit you may need to collect statistics regarding the size of hands. If you are designing a wrist band – you would need to know the size of a wrist; a hand-held snack should take into consideration the size of a hand.

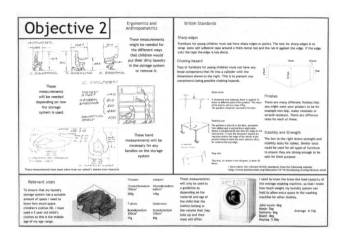

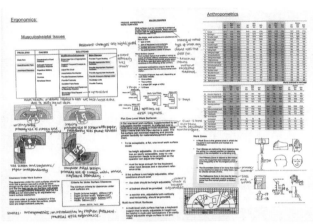

Fig 5.53

The same guidance applies to collecting data about animals. For instance, if you are designing an automatic animal feeder it may be necessary to collect statistics regarding hamsters or gerbils.

EXAMINER'S TIPS

When using anthropometric data or making ergonomic recommendations, add notes and labels to the diagrams – an explanation is necessary. State clearly how you intend to use the data you have collected (no comments = no marks).

Using and obtaining ergonomic and anthropometric data

Some useful reference books and websites are given at the end of this chapter.

Be careful with online data. It often uses measurements from military personnel (a sample of 'healthy' people). The data may be dated (the population may be changing year over year). Remember that measurements are taken with nude subjects (normally, people wear clothes and shoes at work).

Use the data to help design a rough prototype model, then find some testers (perhaps friends or family). Match your users' age if possible. Have your testers mimic real-life use of your design and record what difficulties, clearances, reaches etc. they have.

Record what you observe, not what they say, and take formal measurements. Make adjustments or tailor the design.

Record what percentiles are accommodated by the improved design and, explicitly, consider the consequences for users who cannot be accommodated: safety, discomfort, extra instructions – can you stop them buying/using?

By adding data about people into the design process, a product or environment can be designed so that *all* users are accommodated, not just those who resemble the designer!

Aim to design for all – inclusive design

This does not mean you have to design for all six billion people on the earth, but it means to aim to exclude as few as possible during the whole design process. Designing for all is also known as inclusive design, designing for the widest possible audience.

Design has traditionally catered only for the perceived majority of so called 'normal' users. But what is 'normal'? The vast majority of us will suffer at least temporary disability at some point in our lives, and we will grow old, we are tall, short, fat, male and female. The majority of products are designed for 'normal' users – who are fit, non-disabled, male, single, young, of average weight and height – whereas people fitting these categories are often the minority.

If we are to realise a future in which everyone participates fully and enjoys equality, we need to break down the traditional barriers between 'regular' and 'specialised' design and create a new inclusive design language. Why should any design be difficult for anyone to use?

The principle of inclusive design is that 'all design – landscape, engineering, architectural, product or process – should respond to the greatest diversity of human need possible. While it is acknowledged that it is impossible to design any environment to the exact specifications of every group, much less every individual, it is perfectly possible to design-out known barriers and design-in a high degree of flexibility.'

The UK Institute of Inclusive Design actively promotes inclusive design and, most recently, the Helen Hamlyn Research Centre and the Design Council have created a new inclusive design education resource website that illustrates inclusive design case studies (www.designcouncil.org.uk/inclusivedesign and www.ukiid.org/index.html).

Why should we take note?

'By 2020, close to half the adult population of the UK will be over 50 years old. In the UK, only the 50+ age groups have increased significantly in size over the past 100 years. With age, people change physically, mentally and psychologically, [with] minor impairments in eyesight, hearing, dexterity, mobility and memory. At present, such changes have a significant impact on older people's independence due to an unnecessary mismatch between the designed world and their changed capabilities.'

Roger Coleman, Professor of Inclusive Design and co-Director of the Helen Hamlyn Centre at the Royal College of Art

EXAMINER'S TIPS

Observation is a design method that can be used to identify the problems that can arise when people interact with products, services and environments. Design with the user in mind!

1 Describe, using sketches and notes, three examples where anthropometric data has been taken into account in the design of a product of your choice.

2 Describe, using sketches and notes, three examples where ergonomics have been taken into account in the design of a product of your choice.

Section K Technical data

By the end of this section you should have developed a knowledge and understanding of:

- How to use and interpret technical data

Throughout the process of designing and manufacture of a product it will be necessary to gather and interpret information and data. This could be in the initial stages of design to gather, analyse and relate anthropometric data; in the development of a product to find suitable materials, components or ingredients for the product; or during the planning of the manufacturing process to organise suitable tooling and machinery for successful production.

The precise source and nature of the data, and its significance to each stage of the work of a designer, will depend enormously on the field of work. Whatever the field, information is often available from:

- Official independent (non-commercial, often government) sources, for examples the Food Standards Agency, British Standards Institution (BSI), Building Regulations
- Trade organisations, for example the Timber Research and Development Association (TRADA), International Snack Foods Association, Society of Motor Manufacturers and Traders (SMMT)
- Manufacturers' data sheets
- Other published sources, such as directories, books and magazines

Although traditional printed media should never be forgotten as a source, information is increasingly available electronically. This brings huge benefits in terms of:

- Speed of access
- Continually updated information
- Ease of processing (for example the use of spreadsheets for sorting or calculation, utilising numerical data)
- Ease of storage and presentation

There are many specialist internet search engines that support the design process by collating information sources in particular fields of design, for example the engineering search engine www.globalspec.com.

Some of these search engines operate as free sources, while others operate a charging policy as commercial sources for business users.

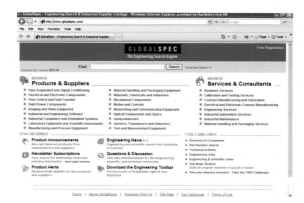

Fig 5.54

Fig 5.55

Fig 5.56

Section L — Principles and techniques of testing applied to product design

By the end of this section you should be:

- Aware of a range of tests to identify characteristics/properties of materials

Introduction

The ultimate aim for a well-designed product is to function efficiently in the situation for which it was designed. In the case of a car seat, it is essential that the finished product fulfils its design specification. Ultimately this means that the product should protect a baby or small child during extreme circumstances in a car crash. The impact could be head on or from the side, and any final test on the product needs to take account of this. Manufacturers go to great lengths to test products, often exceeding recommended test criteria; the details below show this. Before a finished product is tested a whole range of tests will have been carried out on the individual materials used to make it to make sure that they are suitable to withstand the conditions in which they will be expected to function.

Some common mechanical tests applied to materials:

- Tensile
- Hardness
- Impact
- Abrasion

These are known as 'destructive tests' as the samples used are damaged during the test process.

Some large samples of material are either too large to move or may have hidden defects within them: 'non-destructive tests' are then used:

- X-ray
- Ultrasonics
- Shore scleroscope

Sensory testing of food samples are undertaken during development or on a final product.

Fig 5.57

Destructive testing

Tensile testing: using a 'tensometer'

A turned test piece is subjected to tension through a worm drive gear system, which applies force through a spring beam. A force/extension graph is plotted using a stylus – this follows a mercury column, which magnifies the deflection of the beam.

Information gained: elastic limit (A), yield point (Y), maximum load (M), final breaking force after 'necking' at B.

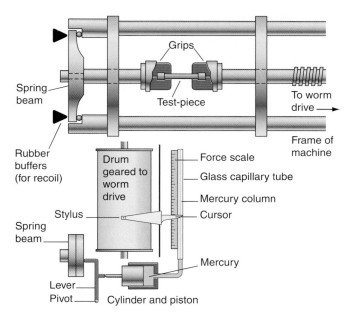

Fig 5.58

Note that Figure 5.59 shows a force/extension graph. This is different from a stress/strain graph. When necking occurs the cross-sectional area reduces (s/s = B1).

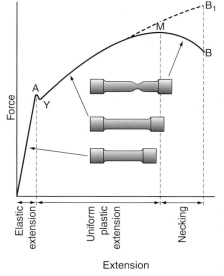

Fig 5.59

Brinell hardness testing

To carry out a Brinell hardness test a 'compression cage' is fitted to the tensometer. The sample to be tested is placed between the anvil on the left and the Brinell ball on the right. For steel, a 5 m diameter (D) is used with a force of 750 kg. The diameter (d) is measured and the Brinell number is found by reference to a table.

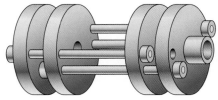

Fig 5.60

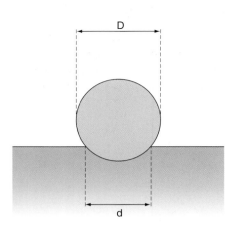

Fig 5.61

Fig 5.64

Vickers pyramid hardness test

For very hard materials a pyramid-shaped diamond is used instead of a ball bearing. Results are very accurate – a diamond won't deform. A microscope scale is used to measure the results of Brinell/Vickers indents.

Izod impact test for toughness/brittleness

In an Izod impact test, a 'notched' test piece is fixed in the test vice. A heavy pendulum is released from a set position. As the pendulum breaks the notched test piece it absorbs energy.

As the pendulum swings past, it drags a pointer with it. The pointer stops at the highest point of the swing. This indicates the amount of mechanical energy used. Comparative toughness/brittleness can then be ascertained.

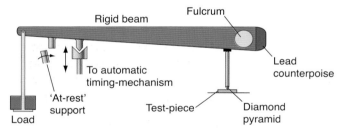

Fig 5.62

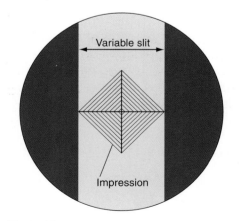

Fig 5.63

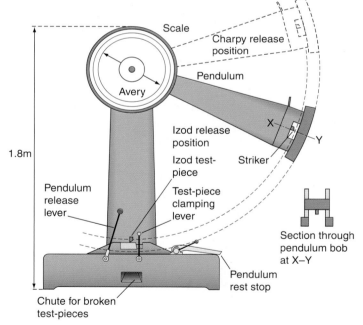

Fig 5.65

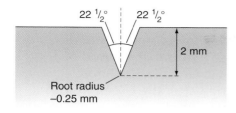

Fig 5.66

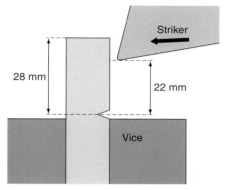

Fig 5.67

British standards test for abrasion

This may look less complicated by comparison but it is the actual British standard test for 'wear resistance'. This is ideal to test textiles, which are fixed to one block and abrasive paper to the other. For comparative tests, pressure/frequency is constant.

Textiles can also be tested in tension with special grips fitted on to the tensometer: these spread the load and prevent tearing.

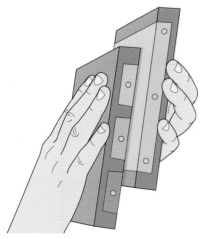

Fig 5.68

Sensory testing of food

Food-focus candidates will need to carry out appropriate sensory tests both in product development and final product stages. Sensory testing should ideally be carried out in special stimulus-free booths.

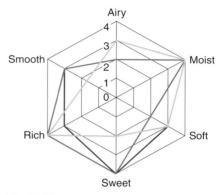

Fig 5.69

Non–destructive testing

X–rays and ultrasonics

Some very large castings could have hidden internal defects inside them that manufacturers need to check before any expensive processing occurs on imperfect raw material. There are two ways of achieving this: X-rays and ultrasonics.

X-rays are dangerous and a large shield is essential for health and safety purposes. Any defect within a casting is clearly shown on an X-ray film.

Ultrasonics works on the same principle as ultrasound used in maternity wards to give images of unborn babies. Any defect within a metallic structure will be picked up and recorded as an ultrasonic vibration.

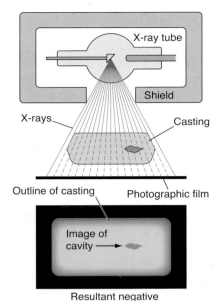

Fig 5.70

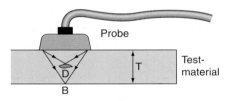

Fig 5.71

The Shore scleroscope

It is inconvenient to move some very large samples of metal so a portable method of testing hardness is invaluable. The Shore scleroscope is named after the Greek word 'skleros', which means hard.

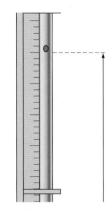

Fig 5.72

A glass tube can be held firmly above a large ingot or a very large casting, then a diamond-tipped 'hammer' is released from a standard height. The height of the rebound measured against a scale gives a comparative hardness index.

Figure 5.72 shows a simple version of this which could be easily made in a workshop – the diamond hammer is replaced by a ball bearing, making it a very useful portable device.

| Section M | Quality control and quality assurance |

By the end of this section you should have developed a knowledge and understanding of:

- Quality control
- Quality assurance
- Total quality management

Quality counts

Every product is designed and manufactured with a particular customer in mind. It is vital that the end product fulfils the requirements and expectations of the customer. These may include appearance, performance, availability, delivery, reliability, safety, maintainability, value for money and price. It is important that an organisation knows and understands their customer's needs and expectations, and then puts in place the procedures and systems to ensure it meets them.

Failure to meet the customer's specification will result in a product that does not sell as well as expected and such a response will result in wasted materials, wasted resources during the manufacturing process, and damage to the reputation of the manufacturing company. This could have a drastic effect on the future success of the company.

Total quality management

An important consideration for the organisation of the company is how can they guarantee the manufacture of a quality product. It is because of a company's desire to gain customer satisfaction that total quality management (TQM) procedures are set up. Companies who implement TQM are constantly seeking improvement, trying to continually improve the performance of its organisation and its products and services. TQM emphasises the importance of the whole manufacturing process, reviewing and monitoring every stage of management and manufacture. Checks are made at every stage from the delivery of resources through to the final delivery of the product to the customer. In order to be effective TQM relies upon every employee within the workplace to be responsible for their quality standards. If any faults are found then these need to be corrected immediately; by doing so this allows repairs to be carried out or changes to the production methods made. By applying such procedures faults are identified at an early stage and the amount of rejected items can be reduced.

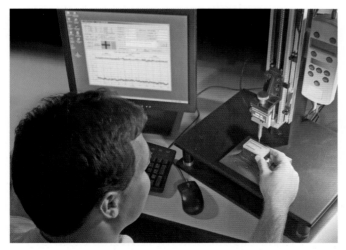

Fig 5.73 Testing being carried out as part of the quality control process

Quality assurance

Quality assurance is carried out by the company to see that the product meets the quality standards set. A series of planned actions and procedures are set up to check the product before, during and after manufacturing operations have taken place. The aim of the process is to prevent failure and to make sure that quality of the product is right first time and every time. Quality assurance is the responsibility of everyone and it should be built in to the process from the beginning of day one until the end of the production process.

Quality control

Quality control supports the quality assurance process and is used to set up ways of checking quality against the required customer standards or to see that items have been made within set tolerances. It involves using an inspection team who are looking for items that are not up to the specific standards. Inspections take place at identified stages during manufacturing as well as after the final item has been assembled. Inspections can be carried out on all, or a sample, of the products, depending upon the number of items being produced.

Inspection checks can be carried out in a variety of ways, including:

- Simple visual checks
- Detailed data comparisons
- Checking accuracy of dimensions
- Flammability tests
- Tasting
- Checking weight
- Electric circuit checks
- Safety checks

It is the ultimate aim of every company to produce products that quality control procedures identify as having zero faults. However, when mass producing certain products it may be impossible for every product to be identical and therefore a tolerance has to be applied. Such procedures allow slight agreed differences to be approved and for the product to be passed as meeting the product specification when inspections are carried out.

Very few products and components need to be manufactured to a zero tolerance. Manufacturing to high levels of accuracy will be costly. The specifying of tolerances enables an adequate level of accuracy to be provided and keeps costs to a minimum.

Section N — Smart and modern materials

By the end of this section you should have developed a knowledge and understanding of:

- Up-to-date development of materials and their application in product design

Scientists and technologists have been developing and creating new materials with useful properties since the 19th century. Plastics are often regarded as 'modern' materials, yet cellulose nitrate was introduced at the Great International Exhibition in London in 1862. Further technological advances in product design will depend increasingly upon the development of new materials with valuable properties.

Smart materials respond to changes in temperature or light and adapt in some way. Some smart materials have a 'memory' and can revert back to their original state. An example of this type would be medical threads, which knot themselves. Temperature and/or light could be used to activate the 'memory material' to revert to its original shape.

Smart or modern materials may be high-performance materials, such as genetically engineered dragline spider silk, which is used to produce super-strong, super-light military uniforms. The materials that will have the greatest impact are those that sense conditions in their environment and respond to those conditions. These materials function both as sensors and actuators.

Modern materials are developed to perform particular functions and have specific properties; they are intentionally developed, rather than being naturally occurring changes.

Many smart and modern materials are developed for specialised applications, though some eventually become available for general use. In the last ten years a range of smart materials has been produced for personal, domestic, medical, transportation and telecommunication applications.

Smart goggles are made using a low-cost electrochromic sheet that changes colour and shade using minimal power.

Materials are being developed that repair themselves, for example a bridge could reinforce itself and seal cracks during an earthquake. Aerospace engineers are developing smart materials, which can automatically seal cracks in airplane wings. Cars are being designed with 'intelligent crumple zones', using smart materials to regain their original shape after an accident.

There are many uses of smart materials in medicine. In order to prevent the collapsing of arteries, small tubes are inserted. The smart material tubes are injected directly into the vein, without the need for a complicated operation, and take on the required form in the affected artery, opening out and improving blood circulation. The changes in the smart material are triggered by body temperature. Smart bandages and plasters can use bacteria sensors to warn of infection using colour changes.

Smart materials allow complex items to be disassembled easily and in a cost-effective manner. Components that may have been difficult to remove and thrown away in the past can now be recycled or reused. Screws can be made from shape polymers that, under heat, return to their original shape, retract their threads and are easily removed from their holes.

Smart material – Textiles examples

Medical plasters can be made using smart fabrics; they can be encapsulated with antiseptic substances to encourage healing. Some plasters or bandages are made from smart fabrics that change colour in reaction to body temperatures, indicating whether a wound is healing or not.

Smart fabrics can change colour when exposed to UV light and revert back when out of the light source: a useful warning for over exposure to sunlight. Smartex, an Italian firm and Textronics, from the US, have developed vests, which can remotely monitor heart rates and measure respiration.

Biomimetics is the study of how good design can be extracted from observing nature. The controversial 'Fastskin', used in the manufacture of swimming costumes, was developed as a result of careful observation of the shark, its skin texture, and how it swims through water. Performance increases significantly when using the material. Some manufacturers claim that it gives swimmers an unfair advantage.

Different fibre profiles can be made to contract or expand, loosening and tightening clothing to make the wearer feel warmer or cooler.

Smart fibres can function as conductive 'wires' and react to signals from electricity, heat or pressure. These fibres can be used to monitor body temperatures or can be combined with electronic devices such as mp3 players, communication systems etc, giving additional functionality to fashion items.

Food makes up a large percentage of waste in the UK. When food becomes less fresh, chemical reactions take place and bacteria build up. Smart labels have been developed that react to chemical/bacterial action and change colour, informing the consumer.

Smart material – Food examples

Examples of modern and smart materials in Food include:

- Probiotic yoghurts and drinks
- Genetically modified foods
- Modified starch – it responds to changes in temperature so that it is used in products which become thicker at a higher temperature, and runny at lower temperature. An application of this might be a pizza topping.

Intelligent packaging materials for food are being developed which can regulate air and moisture movement. Photochromic materials can be used on the packaging to clearly indicate 'use by' dates.

Polymorph

Polymorph (polycapralactone) is often used as a modelling material. It is heated by water or a hair dryer and becomes easily mouldable at 62°C and takes on a solid form, very similar in performance to nylon. It is also available in liquid form: it is liquid at room temperature but solidifies at approximately 2°C.

Shape memory alloys

Early use of nickel-titanium alloys (Nitinol) established that they had a 'memory'. When heated they retained their original shape. Gold-cadmium and some alloys of brass all have a 'memory'. Available in wire or sheet form, shape memory alloys are used in robotics (copying muscle function) and are being developed to operate the wing flaps on aeroplanes. They are used in modern buildings in automatic air vents.

Piezoelectric materials

When a piezoelectric material is deformed it gives off a small electrical discharge. Also, when an electrical current is passed through a piezoelectric material, it increases in size by up to a four per cent change in volume. A piezoelectric material is used as the airbag sensor in a car. The material senses the force of an impact on the car and sends an electric charge that sets off the airbag.

Chromic materials

Chromic materials refer to materials that radiate a colour, erase a colour or change it as a result of external stimuli:

■ Photochromic materials change colour with changes in light intensity. Usually, they are colourless in the dark; when sunlight or ultraviolet radiation is applied the molecular structure of the material changes and it exhibits colour

■ Thermochromic materials change colour reversibly with changes in temperature; they are usually in the form of semi-conductor compounds, liquid crystals or metal compounds

■ Electrochromic materials change when electricity is the external stimuli

■ Piezochromic materials change when pressure is the external stimuli

■ Solvatochromic materials change when liquid is the external stimuli

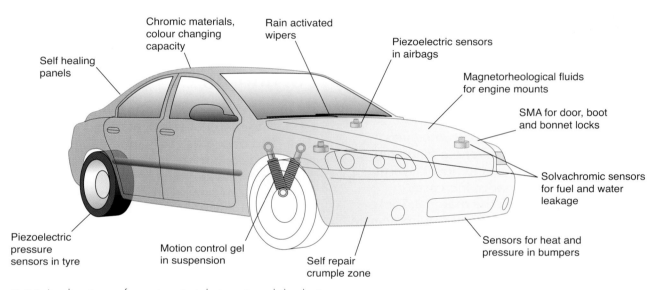

Self healing panels

Chromic materials, colour changing capacity

Rain activated wipers

Piezoelectric sensors in airbags

Magnetorheological fluids for engine mounts

SMA for door, boot and bonnet locks

Solvachromic sensors for fuel and water leakage

Piezoelectric pressure sensors in tyre

Motion control gel in suspension

Self repair crumple zone

Sensors for heat and pressure in bumpers

Fig 5.74 Applications of smart materials in automobile design

REFERENCES

[1]SOURCE: British Glass

[2]SOURCE: UNEP Vital Graphics

[3]SOURCE: www.guardian.co.uk www.wasteonline.org.uk

[4]SOURCE: World Bank

[5]SOURCE: adapted from Sustainable Design Award (SDA) www.sda-uk.org

[6]FURTHER READING: www.myfootprint.org
www.stepin.org
www.informationinspiration.org.uk
www.climatechallenge.gov.uk/communicate.html
www.climatechange.wmnet.org.uk
www.icount.org.uk
www.peopleandplanet.org
www.sda-uk.org
www.woodcraft.org.uk/projects/
www.climatechange.org

[7]SOURCE: BP plc, Statistical Review of World Energy 2006

[8]SOURCE: Waste Watch

[9]SOURCE: www.alupro.org.uk

[10]SOURCE: Waste Watch

[11]SOURCE: www.cspr.co.uk

[12]SOURCE: www.scrib.org

[13]SOURCE: www.britglass.co.uk

[14]SOURCE: Waste Watch

[15]SOURCE: NNIEAG

[16]SOURCE: www.e4s.org.uk

[17]SOURCE: Recoup

[18]SOURCE: WRAP

[19]'SOURCE: Design for the Real World' Victor Papanek

[20]SOURCE: Stephen Pheasant. Author of a number of Ergonomics/Anthropometrics texts, published by BSi (British Standards Institute in 1990s[I have more details if needed JFG]

[21]Ball State University: http://www.bsu.edu/web/jcflowers1/rlo/anthropometrics.htm

FURTHER READING
http://www.ergonomics4schools.com/
http://www.openerg.com/index.htm
http://www.roymech.co.uk/Useful_Tables/Human/Human_sizes.html
http://www.openerg.com/studentzone/studdata.htm
http://www.designcouncil.org.uk/en/About-Design/Design-Techniques/Ergonomics/
http://www.ergonomics.org.uk/
http://www.dti.gov.uk/files/file21811.pdf
http://www.dti.gov.uk/files/file21830.pdf
http://www.dti.gov.uk/files/file21827.pdf
http://stinet.dtic.mil/cgi-bin/GetTRDoc?AD=ADA244533&Location=U2&doc=GetTRDoc.pdf
Anthropometrics—Stephen Pheasant

Ergonomics—Stephen Pheasant
The measure of man and woman—Henry Dreyfuss
Ergonomics for beginners— Dul and Weerdmeester
DTi Anthropometric Data for Chidren
DTi Anthropometric Data for Adults
The Measure of Man and Woman: Human Factors in Design by Alvin R Tilley
Bodyspace: Anthropometry, Ergonomics and the Design of Work by Stephen Pheasant et.al.

CHAPTER 6

Built environment and construction

Learning outcomes

By the end of this section you should have developed a knowledge and understanding of:

- **Procedures for site preparation** – walk-over, desk-top and soil investigations
- **Functional requirements of a foundation** – strength, stability, soil types and ground movement
- **The selection and uses of foundation types** – deep narrow strip, wide strip, raft, pad and short bored pile
- **Temporary supports for foundation trenches** – timbering, trench boxes and vertical steel sheeting

Introduction

Site preparation is an essential procedure to undertake if a proposed project is to be successful. The information collected and recorded from a preliminary investigation is used to help in the design and construction of a building project. The extent of the investigation will depend on the type and size of the project, its purpose and the loadings to be supported by the ground.

The main purpose of a foundation is to safely transmit the building's dead loads (the walls, floors and roof, etc.), together with the imposed loads of people and furniture and wind loads, safely to the ground on which it rests. The foundation also provides a level base on which the building can be constructed. The subsoil in contact with the foundation is compressed and reacts by exerting an upward pressure to resist the loads.

The choice of foundation will depend on the soil conditions, type of building, the building's loadings, any economic considerations and time factors.

Temporary support of a foundation trench is required when a risk assessment identifies that the sides are likely to collapse. The main factors to be considered if temporary support is required are the safety of operatives working in the trench, whether any collapse may affect the foundations of adjacent buildings, and whether it will allow the construction work to proceed in an orderly and safe manner.

The government has set a target for house builders to produce 240,000 homes a year by 2016, 60 per cent of which have to be on sites that have been previously used, that is brownfield sites.[1] This is due to:

- The public's growing intolerance of derelict areas, abandoned buildings and rundown places caused by the decline and closure of many traditional industries in the 1980s and 1990s
- A shortage of greenfield sites in some areas of the UK, particularly in and around towns and cities

Key Terms

Brownfield – usually urban land that has previously had a structure built on it. The site will often require remedial work before the new building can commence due to the ground being contaminated by chemicals and so on. Government grants are often available to encourage companies to build on such sites.

Greenfield – an area of open land that has not had a previous use, or may have been used for agricultural purposes. Often situated in rural areas, this type of site is usually more expensive to buy if planning permission has been granted.

Fig 6.1 Brownfield site

Fig 6.2 Greenfield site

Site preparation

Site preparation consists of three main stages:

- Desk-top study
- Walk-over survey
- Sub-surface investigation

Desk-top study

The purpose of a desk-top study is to provide a useful guide about the site prior to a site visit. This information is available from a number of sources including local authority records, local libraries, Ordnance Survey and British Geological Survey maps, mining records and previous site investigation reports.

Walk-over survey

The aim of a walk-over survey is to confirm and supplement the information collected during the desk-top study. Points to be noted:

- Signs of landslip, erosion of slopes or local subsidence
- Evidence of highly compressible soil, for example peat or silt
- Species, height, spread and condition of trees on or near the site
- Are there any ponds, streams, wells, etc., and signs of flooding?
- Signs of existing foundations, basements, sewers or mains services
- A high water table is indicated by bounciness under foot

Sub-surface investigation[2]

There are various methods of sub-surface investigation depending on the size of the project and its cost. Some specialist methods include aerial photography using digital and thermal imaging cameras. However, the usual methods are:

- **Trial pits** are holes dug by a mechanical excavator in the subsoil up to 4 m deep, which allows a detailed visual examination to take place. Trial pits are the best method of investigation for the majority of small structures, for example houses
- **Boreholes** are appropriate for larger structures where the subsoil may need to be examined to a depth of up to 20 m because the site has been a landfill, or other soft strata is expected. Typically, a 150 mm diameter hole will be bored in the ground using a mechanical

drilling rig; the resulting soil samples are analysed off-site in a specialist laboratory

Foundation types and uses

There are two main types of strip foundation: traditional wide strip and deep narrow strip.

Traditional wide strip foundation

This is used for low-rise domestic dwellings, or similar buildings.

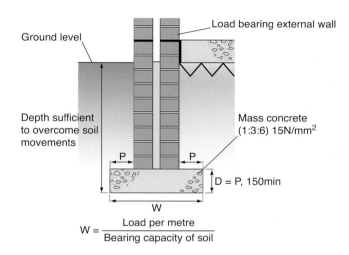

Ground level

Load bearing external wall

Depth sufficient to overcome soil movements

Mass concrete (1:3:6) 15N/mm²

P   P

D = P, 150min

W

$$W = \frac{\text{Load per metre}}{\text{Bearing capacity of soil}}$$

NB. In all cases W must give adequate working space which is usually 450 to 600mm minimum depending on depth of excavation.

Fig 6.3 Traditional wide strip foundation

Deep narrow strip foundation

This is used as a cheaper alternative to the traditional wide strip due to it being less labour intensive. The use of ready-mixed concrete means that less material is stored on site, making the process clearer and easier to manage. Also, no working space for bricklayers in the excavation is required.

Fig 6.4 Excavation of deep narrow strip foundation

Fig 6.5 Deep narrow strip foundation with trench blocks built to ground level

Raft foundation

A raft foundation is a continuous slab of reinforced concrete covering an area equal to or greater than the base of the building. Raft foundations may be used for lightly loaded buildings on soils with poor load-bearing capacity, where variations in soil conditions require considerable spread of the loads, or where differential settlement is likely. Raft foundations are often used for buildings on sites previously used for landfill waste.

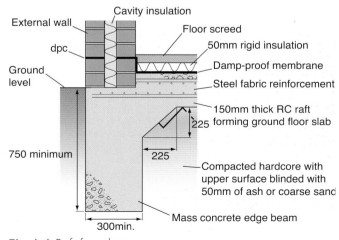

Cavity insulation

External wall

Floor screed

dpc

50mm rigid insulation

Ground level

Damp-proof membrane

Steel fabric reinforcement

150mm thick RC raft forming ground floor slab

225

225

750 minimum

Compacted hardcore with upper surface blinded with 50mm of ash or coarse sand

300min.

Mass concrete edge beam

Fig 6.6 Raft foundation

Short bored pile foundation[3]

Piled foundations will be required where the soil at the usual foundation depth – up to 1.5 m – will not support the building's loads. Holes of 250–300 mm diameter spaced at 1.8–2.5 m are bored into the ground and are filled with concrete, the top part being reinforced with steel bars. Reinforced concrete ground beams are then cast over the pile heads from which the load-bearing walls are built.

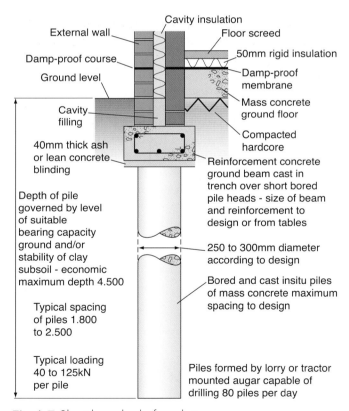

Fig 6.7 Short bored pile foundation

Temporary support for foundation trenches

All types of subsoil have individual characteristics that affect their ability to remain stable during excavation: they will form a natural angle of repose unless given temporary support. A high water table, the time of year, the period of time the trench will remain open and surcharges from the proximity of buildings or vehicles can also adversely affect the stability of the trench, causing the sides to collapse. Under the Construction (Design and Management) Regulations 2007, a risk assessment must be undertaken by a competent person to ensure the excavation work is planned, managed and supervised, and so preventing accidents.[4]

Methods of temporary trench support

Timber is the traditional material used for support although, with the advent of modern techniques such as steel sheet shoring systems and trench boxes, its use is declining.

Open timbering

Used mainly for supporting excavations in hard or firm soils; as the name implies there are spaces left between the vertical timbers or poling boards with horizontal struts fixed tightly across the width of the trench.

Close timbering

In this form of support there are no spaces left between the vertical timbers, therefore it is ideal for granular or wet soils.

Steel sheet shoring systems

Used as an alternative to close timbering, the vertical boards are replaced by profiled steel sheeting held in place by either manual or hydraulically adjustable struts. The hydraulic version is the preferred method as it allows the horizontal struts to be adjusted by a person operating the pump outside the trench.

Fig 6.8 Steel sheet shoring system

Trench box

In very poor soil that will not support itself, steel trench boxes are the best option because operatives do not need to be in the trench to construct the support. After aligning the box with the trench, it is pressed into the ground using an excavator's back actor arm; the soil is then removed from within the box in stages as it is lowered.

Fig 6.9 Trench box

1 a) Identify the objectives of a site investigation.
   b) Describe the site investigation procedure.
2 Explain the functional requirements of a foundation.
3 Recommend a suitable foundation for a small detached house for the following ground conditions and, in each case, give reasons for your choice:
   a) A level site with clay subsoil. The water table is 2.5 m below ground level and the site used to be used for agriculture.
   b) A gently sloping site on firm sand. The water table is a minimum of 4 m below ground level.
   c) A level site with large trees close to the proposed house. The ground is clay.
   d) A level site that is mostly filled ground. The fill has an uneven depth that ranges between 4 m and 9 m but is relatively stable.
4 Use annotated sketches to show the temporary earthwork support required in a foundation trench that is 1.2 m wide by 2 m deep in moderately firm soil.

Section 2   Floors

Learning outcomes

By the end of this section you should have developed a knowledge and understanding of:

- **Functional requirements of a floor** – strength, stability, resistance to ground moisture, durability, fire safety, and the resistance to the passage of sound and heat
- **The selection and use of ground floor types** – ground supported concrete slab, suspended concrete slab, suspended pre-cast concrete floor slab and suspended timber
- **The selection and use of upper floor types** – reinforced concrete, suspended timber, and double floor with a steel beam
- **The selection and use of floor finishes** – timber strips, tongue and groove boarding, wood blocks, particle board and plywood panels

Introduction

The main purpose of a floor is to span between its supports and carry all imposed loads of people and furniture, and the dead loads of flooring and ceiling finishes, without failure, unacceptable deflection or settlement.

Other considerations include that the choice of material used should be economical for the required span, sound transmitted from storey to storey should be kept to acceptable levels, fire resistance is essential to restrict fire spreading upwards or downwards, resistance to moisture and water vapour, and reduce heat loss from the lower floor. The successful selection of both ground and upper floor types will depend on the chosen method satisfying these criteria.

The floor finish is a vital element in the design of a building. The following factors should be considered: resistance to wear and, in particular, water, impact and chemicals; comfort criteria (sound control, resilience, slipperiness, warmth and overall appearance); cleaning; cost and fire resistance.

Ground floor types

Ground-supported concrete slab

The main use for this type of floor is for buildings on level ground that is resistant to movement. The main advantage of this type of floor is that any building shape or area can be accommodated. Prior to the floor being constructed it is usual to build the external and internal load-bearing walls to the height of the damp-proof course; this allows the inner face of the wall to be used as permanent formwork. The floor slab is comprised of a sub-base of 150 mm consolidated hardcore (i.e. stone or rock). To fill in any voids between the stones and to prevent the horizontal damp-proof membrane from being punctured, a compacted 50 mm layer of sand or ash blinding is laid directly on the hardcore. To comply with the Building Regulations Approved Document L Conservation of Fuel and Power, a layer of rigid thermal insulation will be incorporated prior to the 100–150 mm thick concrete floor slab being laid.

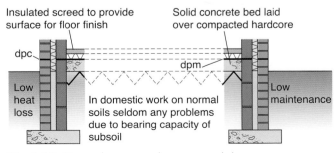

Fig 6.10 Ground-supported concrete slab

Suspended concrete slab

Where the ground is sloping, has poor load-bearing capacity or is liable to clay heave or shrinkage, it is advisable to use a suspended floor slab as it does not rely on the ground for support. There are two main types: pre-stressed hollow core or solid concrete planks and pre-stressed concrete beams and blocks.

Due to the components of both types of floor being manufactured in a factory and then installed on site, they can be constructed in all weather conditions; backfill to the foundations is eliminated; an immediate safe working platform is provided and the installation time is quicker through the absence of in-situ concrete and the corresponding curing time. Each type of floor is also appropriate for the construction of upper floors.

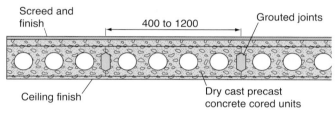

Fig 6.11 Suspended concrete slab

Key Term

Shrinkage and heave – these can cause damage to foundations and are linked indirectly to the roots of adjacent trees. **Shrinkage** in some clay soils can occur during prolonged periods of dry weather and is compounded by thirsty trees such as willow, oak and poplar taking up water from the subsoil. **Heave** is the opposite as it occurs in wet weather and is compounded by the removal of moisture-dependent trees, which causes the excess water to raise the ground level.

Pre-stressed concrete planks

The 1200 mm wide and 110 mm deep pre-stressed concrete planks can span up to 6 m between load-bearing walls, eliminating any sleeper or intermediate supporting walls. It is usual to complete the construction with a damp-proof membrane and rigid thermal insulation sandwiched between a 65 mm concrete screed. However, some manufacturers recommend a particle board finish laid on a damp-proof membrane as an alternative to a concrete screed.

Pre-stressed concrete beams and blocks

A series of inverted T beams are positioned on supporting walls with concrete infill blocks placed between the beams. The spacing of the supporting walls is governed by the economical span of the inverted T beam. A concrete screed or particle board finish can be applied (similar to the concrete plank method). To eliminate the build up of unwanted gases the void beneath the floor must be ventilated.

Fig 6.12 Pre-stressed concrete beams and blocks

Suspended timber

Timber ground floors consist of boarding supported by a framework of joists. The boarding can be of timber tongue and groove boards or sheet material such as plywood or particle board. The most important features of this type of floor are to ensure adequate ventilation to the void below the floor finish and to prevent rising moisture reaching the floor. Therefore, the timber joists rest across brick sleeper or honeycombed walls that are capped with a horizontal damp-proof course. To allow a cross flow of air beneath the floor, provision must be made in the external walls for appropriate air vents. To achieve the required thermal insulation for the floor quilt insulation is held in plastic netting between the joists. It is good practice to treat the joists with a preservative prior to assembly.

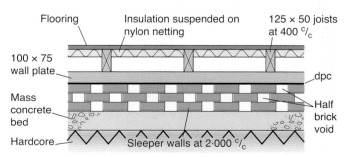

Fig 6.13 Elevation of a sleeper wall at mid span of the floor joists

> In a small group, look at examples of the use of thermal insulation to prevent heat being lost through the ground floor of a house.

The main difference between ground and upper-floor construction is that larger unobstructed areas are required under upper floors, resulting in wider floor spans and deeper joists. Due to the tendency for the deeper joists to twist and consequently cause the ceiling finish to crack, a system of strutting is used between the joists. A further function of an upper floor is to provide lateral restraint to the external walls by means of galvanised steel straps and joist hangers.

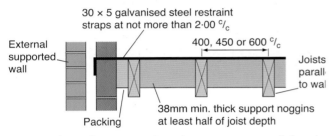

Fig 6.14 Lateral restraint when the joists are parallel to the wall

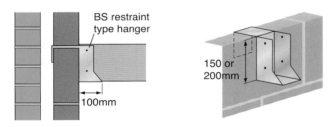

Fig 6.15 Lateral restraint when the joists are at right angles to the wall

Where the span of the floor is greater than 4.9 m the depth of the joists required would be uneconomic, therefore a steel or laminated timber cross beam is introduced at mid span to carry the ends of the joists.

Fig 6.16 Suspended timber upper floor

Floor finishes

When specifying a floor finish it is important that the use of the building is assessed: every effort should be made to match the flooring with the intended use. The floor finish is usually applied to a structural base, but can also provide another function as part of the floor structure, for example tongue and groove boards. The following factors should be considered when specifying a floor finish:

- **Wear** – resistance to wear is a major factor, especially the concentration of foot traffic and its movement around the floor
- **Appearance** – the floor finish can determine the scale of a room and dictate how it is perceived by those using it, for example warm or cold, informal or formal
- **Sound control** – a soft flooring that is not masked by furniture can contribute to the absorption of airborne sound, for example soft flooring can absorb up to 60 per cent of that absorbed by acoustic ceiling panels
- **Resilience** – a 'dead' floor finish is tiring to walk on so concrete would be an unsuitable finish for a sports hall floor. A finish such as wooden strips laid on top of battens would be more suitable
- **Slipperiness** – the degree of slipperiness is determined by measuring the frictional resistance of the floor finish. The occurrence of slips is increased by a frequent number of joints in the floor finish, for example quarry tiles
- **Warmth** – the floor finish should not allow too great a heat loss through heat transfer. Conversely, the floor should not retain too much heat so it becomes too warm
- **Cost** – low initial cost is often associated with high maintenance cost and rapid deterioration. The total cost of the flooring finish should be related to its life expectancy

Traditionally in domestic construction softwood boards manufactured from redwood or whitewood in long lengths and narrow widths with a tongue and groove on opposite edges were the favoured floor finish.

In modern construction softwood tongue and groove floorboards have largely been superseded by particle board and plywood sheets. The size of a typical sheet is 2400 × 600 × 18 mm thick, therefore approximately four to five times the width of a typical floorboard. Having a tongue and groove on all edges means that the sheets can be laid across the joists without the ends needing the support of a joist, reducing time and waste.

Generally, both particle board and plywood have good wear resistance and should be maintenance free as long as they are not allowed to be wet for long periods. The main advantage is their relative cheapness in comparison to softwood boards.

In more prestigious buildings, hardwood strip flooring (less than 100 mm face width) is often the preferred floor finish. Similarly, 225 × 75 × 25 mm blocks of hardwood can be laid in a variety of different timbers and patterns to create colour and grain effects. Typical hardwoods used for the floor finish include beech, oak or mahogany.

1 Use sketches to show the constructional details of how a suspended timber ground floor is supported at mid span. Label three features.
2 Describe two methods of incorporating thermal insulation into solid concrete ground floor construction.
3 a) Use annotated sketches to show details of a domestic ground floor using pre-stressed concrete T beams and blocks.
   b) State three benefits of using this form of construction.
4 Explain with diagrams how lateral restraint of a wall is achieved at upper floor level when:
   a) The timber joists are at right angles to the wall.
   b) The timber joists are parallel to the wall.
5 Identify and describe the features of four types of floor finish.

Nailing T and G boarding

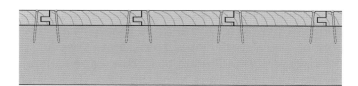

Fig 6.17 Tongue and groove boarding

Section 3 Walls

Learning outcomes

By the end of this section you should have developed a knowledge and understanding of:

- **Functional requirements of a wall** – strength, stability, resistance to weather, durability, fire safety, resistance to the passage of heat and sound, security and aesthetics
- **The selection and use of external wall types** – cavity walls in masonry, solid walls, timber and light-gauge steel framed
- **The selection and use of supports over openings** – arches, concrete and pressed-steel lintels
- **The selection and use of internal non-load bearing partition walls** – timber, masonry and proprietary types

Introduction

Although walls are often thought of as a means of providing shelter from the elements, they fulfil a number of other functions including supporting and offering mutual stability to the upper floors and roof, giving privacy to the occupants from outside the building and between rooms inside the building, keeping the inside cooler or warmer than the outside ambient temperature, and modifying the sound entering or leaving the building.

Depending on the structural form of the building, the external wall can be load bearing or non-load bearing. The traditional method of forming load-bearing external walls is a brick and block cavity wall. Alternatively, framed construction with posts of timber or light-gauge steel carry the imposed loads, whereas the external skin provides a barrier to the external elements. Therefore, the choice of wall materials has a direct relationship to the functions of the wall, in that a load-bearing wall needs the good compressive qualities found in bricks and blocks. Conversely, a non-load bearing wall may provide protection from the elements by means of a glass curtain wall.

Openings in walls are needed to provide natural light, access and ventilation. Support to the area of wall above an opening is provided by either a horizontal beam, such as a lintel, or some form of arch.

The main purpose of a non-load bearing or partition wall is to divide the space within a building into rooms. The other requirements of a normal partition wall are strength, stability, sound insulation and fire resistance.

Functional requirements of a wall

Strength and stability

A wall's strength is determined by its resistance to the stresses set up in it by its own weight, the imposed loads and lateral wind loads. Stability is measured by a wall's resistance to overturning by lateral forces and buckling caused by being too slender in relation to its height. The Building Regulation's Approved Document A Structure gives guidance for the stability of low-rise buildings.

Resistance to weather

The external walls of a building need to provide resistance to rain and wind penetration. The amount of resistance required on a particular wall will depend on its height, the locality, and degree of exposure to the elements. The locality of the wall is especially important as wind force and rainfall can vary considerably: a wall built on a site near the coast is more likely to present greater problems than one built further inland. Wind has an influence on rain penetration because it can force the water through pores or cracks that it might not otherwise penetrate.

Fire resistance

Walls have an important role to play in the resistance to fire: they are used to compartmentalise buildings so that a fire is confined to a certain area, to separate specific fire risks within buildings, to form escape routes for the building's users, and stop a fire spreading to adjacent buildings.

Keep in mind that fire resistance applies to the whole element of the building, i.e. in this case the wall and not the individual materials that make up the wall. Building Regulations Approved Document B Fire Safety provides guidance for the minimum acceptable length of time a specific element must resist failure due to fire.

Resistance to the passage of heat

The external walls, together with the roof and ground floor, must reduce the speed of heat passed through them to the outside air in order to maintain a satisfactory internal environment. Conversely, the walls, roof and ground floor should also prevent the inside of the building from becoming too warm in hot weather.

The wall is comprised of different materials, each conducting heat at a different rate largely depending on its density. Generally, the denser the material the greater its resistance to heat passing through it. The Building Regulation's Approved Document L Conservation of Fuel and Power offers practical guidance to designers on how to satisfy the required standard.

Resistance to the passage of sound

The sound insulation quality of a wall is an important design consideration, particularly for internal and party walls (that is, walls shared between two separate houses). If an external wall's other functional requirements are satisfied, the passage of sound to the inside of the building should be excluded to an acceptable level. Sounds are classified as either airborne or impact sounds, the determination being the source producing the sound. Approved Document E Resistance to the Passage of Sound provides guidance for the maximum values of sound transmittance.

Aesthetics

An external wall, apart from making a visual impact, makes a contribution to the character of the building. The aesthetic appeal will involve the colour, texture and shape of the materials used and the style of the building. The choice of materials will be made largely to satisfy the functional requirements identified previously in this section. However, the materials used for the external walls of adjacent buildings and the requirements of the local planning authority will also need to be considered.

External wall types

Cavity walls in masonry

A cavity wall comprises of two leaves or skins tied together with wall ties. The outer leaf is usually of facing brick built in stretcher bond while the load-bearing inner leaf is constructed of lightweight or aerated concrete blocks. The functions of the cavity are to prevent any water that has penetrated the outer leaf from reaching the inner leaf and to improve the thermal efficiency of the wall as the air in the cavity is a good insulator.

The stainless steel or plastic ties that connect the leaves should be positioned at 900 mm centres horizontally and 450 mm vertically at 450 mm offsets. Where an opening occurs the ties should be spaced at no more than 300 mm vertically.

To achieve the standards of thermal insulation required by Approved Document L Conservation of Fuel and Power of the Building Regulations it is necessary to add some form of insulation to the wall. This can be in the form of partial cavity fill, full cavity fill or the insulation applied to the outer face of the inner leaf as part of the dry lining process.

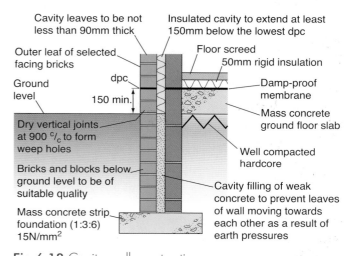

Fig 6.18 Cavity wall construction

> **Key Term**
>
> **Bond** – the arrangement of bricks or blocks in a wall to a set pattern to maintain an adequate lap. The most common type is stretcher bond, which can be used for half brick walls and also for walls of one brick thickness. The amount of lap is half a brick. This bond is quite attractive and provides adequate strength for normal forms of construction, such as cavity walls and internal blockwork partitions.

Solid walls

Although the use of cavity wall construction for dwellings has largely replaced solid walls due to the thickness of wall required to resist the penetration of rain, they do have a use for the restoration of Victorian or Edwardian buildings, garden or boundary walls, decorative brickwork, retaining walls and drainage inspection chambers.

Invariably, solid walls are at least one brick thick; this necessitates the bricks to be arranged in a regular pattern or bond to distribute any intended load throughout the length and thickness of the wall. The two most commonly used bonds for solid walls are:

- **English bond** normally used for walls that require the maximum amount of strength, it consists of alternate courses of headers and stretchers
- **Flemish bond**, which has considerable lateral strength through a bonding arrangement of alternate headers and stretchers on the same course

Fig 6.19 A solid wall built in English bond

Fig 6.20 A solid wall built in Flemish bond

- In a small group investigate a range of methods of incorporating thermal insulation into a cavity wall and, where applicable, how it is held in place.
- Within your group produce a list of the types of horizontal damp-proof course used in wall construction to prevent rising dampness.
- Choose a building you are familiar (such as your school, college or home) and identify the different types of brick/block bonding arrangements that are used to build the walls.

Timber-framed walls

Timber framed is a form of wall construction whereby the structural frame members are fabricated from strength-graded (formerly known as stress-graded) timber and connected together to transmit the building's loads to a suitable foundation onto which they are fixed.

The framework forms the shell of the building and a suitable cladding is attached externally to provide weather protection. The framework can be constructed up to a height of six storeys. Because the timber frames are manufactured in a factory, the site operations become merely off loading the frames and final assembly onto the prepared foundation base. This creates a number of advantages over the traditional in-situ masonry wall construction. The advantages of timber-framed construction are:

- A reduction in the site erection time, so ideal for fast-track construction
- Improved thermal insulation over a masonry wall of the same thickness
- A wide variety of external finishes are available
- Dry construction methods allow quicker occupation of the building (no drying out time is required)
- Adverse weather conditions do not greatly affect the progress of the construction work
- The prefabricated components are manufactured under ideal factory conditions so quality should be assured and waste should be minimal
- The timber frames have a low dead weight, which leads to a reduced foundation size resulting in a lower cost

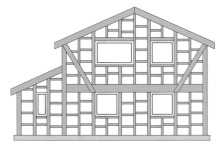

Fig 6.21 A timber-framed house

The main types of timber frame are **platform** and **balloon**.

Platform-framed walls

The most widely used method, whereby the walls are prefabricated in single storey height panels. Once the ground-floor wall panels have been erected, the first-floor joists and floor covering are fixed to the head of the walls. This creates a platform for the upper-storey walls to be erected. Further storeys can be added by continuing this process. Finally, the roof is erected on top of the upper-wall panels.

A variation of this method that utilises factory construction to a much greater extent is volumetric or modular housing units. Using this method complete box or room-sized units are delivered to the site and stacked together to form the complete structure. The units are normally fully decorated with the mechanical and electrical services incorporated, ready to be connected to the mains supply. Many of the budget 'travel' hotels and drive-in fast food outlets that are popular in the UK today are constructed in this way.

Fig 6.22 Exterior face of platform-framed wall construction

Fig 6.23 Interior face of platform-framed wall construction

Balloon-framed walls

This method uses timber frames that are the full height of the building. The studs or vertical members are continuous from damp-proof course level to the eaves. The wall panels are erected to their full height in one operation, with intermediate floors being supported on a horizontal timber bearer fastened to the studs.

> ### Key Points
>
> Although timber-framed houses comply with the same building regulations as those constructed from masonry, there are some points that need attention:
> - Severe damage to the frame will occur if the house is flooded
> - The frames are vulnerable to insect attack if not properly treated with a preservative
> - If not stored in dry conditions before erection, the moisture content of the timber will increase causing shrinkage cracks in wall and ceiling finishes on drying
> - To prevent fire spreading within the wall's cavities, fire stopping is required around all openings, horizontally at joist and eaves levels, and vertically to break up large areas of cavity

Light-gauge steel-framed walls

Although steel is usually associated with large office and industrial buildings, the use of light-gauge steel frames for houses is becoming more common. The galvanised cold-rolled sections are used in a similar way to the timber framed method (platform framed, including volumetric or modular and balloon framed).

The advantages of factory prefabrication are similar to those identified for timber framed construction, although providers of the steel frames suggest that their frames are easier to assemble than timber due to the reduced weight, greater dimensional accuracy, and no moisture content, making the individual members more stable and future extensions easier to accommodate.

Fig 6.24 Light-gauge steel-framed wall construction

Supports over openings

There are a number of ways in which openings can be made in external walls to accommodate windows and doors. However, the chosen method must safely support a combination of wall, floor and roof loads from above and prevent lateral damp penetration. Historically, the head of a window opening was in the form of an arch; in modern construction a horizontal beam or lintel is the usual method of ensuring no loads are transmitted directly to the window or door frames.

Arches

An arch is a decorative but expensive method of spanning an opening. Arches are usually classified by their shape (segmental, semi-circular, semi-elliptical and so on) and are formed using tapered bricks or stones, for example voussoirs, or regular-shaped bricks with tapered joints.

Traditionally, the arch is constructed over a wooden temporary support or centre until the bricks are self-supporting. Modern arches are usually built using a standard profile stainless or galvanised steel cavity tray and arch support that prevents dampness passing to the inner wall and also removes the need for temporary support.

Fig 6.25 Segmental brick arch over a window opening

Concrete and pressed-steel lintels

A lintel should have an adequate bearing of at least 100 mm on the sides or jambs of the opening so the loads are transferred to the wall without undue stress being created in either.

Concrete lintels can be pre-cast or cast in situ. Because tension is developed in the bottom of the lintel, steel reinforcement is needed. Pre-cast lintels are available in modular lengths, usually in increments of 150 mm, to suit a range of standard door and window openings.

When the cavity insulation is continued down to the head of the door or window frame, a

stainless or galvanised steel and polyester-coated lintel is used to support the masonry above the opening. The advantages of this type of lintel are that it is lightweight, easy to handle and, in cavity walls, can eliminate the need for an integral damp-proof tray (the purpose of the damp-proof tray is to collect any water that has penetrated the external leaf and to direct it to the weep holes in the wall).

Fig 6.26 Concrete lintels over window openings, with brick walls built in stretcher bond

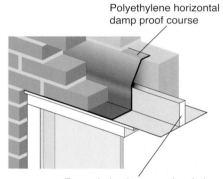

Polyethylene horizontal damp proof course

Expanded polystyrene insulation in the hollow core of the lintel

Fig 6.27 Pressed-steel lintel

Internal non-load bearing partition walls

The main purpose of partition walls is to divide a large floor space in a building into smaller compartments. Other functions include the provision of a measure of fire resistance, sound and thermal insulation.

Timber stud partition wall

The stud partition takes its name from the vertical members or studs that are spaced at horizontal centres to accommodate the sheet lining material, such as plasterboard or plywood. The timber framework should be capable of supporting its own weight and resist impact damage from doors closing, people and furniture.

The advantages of a stud partition are that it is quicker and easier to erect than the masonry types and it is much lighter in weight. Due to its open framework, services can be easily incorporated into the wall by cutting notches or drilling holes in the studs. Thermal and sound transmission can be improved by placing insulating material in the spaces between the studs after one side of the partition has been boarded.

To provide additional strength and prevent the studs from distorting, which can cause cracking of the finish, one or two rows of noggings are fixed horizontally. The vertical spacing of the noggings will depend on the size and thickness of the lining material.

Although timber is the usual material for the construction of the partition, a similar framework arrangement using metal studs, although more expensive, is becoming popular.

Fig 6.28 Timber stud partition walls

Non-load bearing masonry partition wall

Lightweight concrete block partitions are built directly off the floor surface if it is concrete; however, if it is built off a timber floor, a timber sole plate at the base and head at the ceiling will be required to improve lateral stability. To further aid stability the door frames should be storey height and fixed to the floor and ceiling. To prevent shrinkage cracking and provide continuity as a sound barrier at an abutment, the wall should be adequately tied into the structure.

Proprietary partition walls

Paramount partition wall system

This is a lightweight non-load bearing, low-cost partition suitable for domestic dwellings. Paramount partitioning consists of two layers of plasterboard factory bonded to a cellular cardboard core. It can be obtained with tapered edges and ivory-coloured faces for decoration, or with square edges and grey faces for skimming with gypsum plaster. The panels are available in a range of sizes but, for most domestic applications, 2400 mm high by 1200 mm wide and 50 mm thick panels are used.

The panels are held in position by timber battens running round the perimeter of each partition and fixed vertically between the junction of two panels. The panels can be erected quickly and the shrinkage cracking associated with timber stud and block partitions is also avoided.

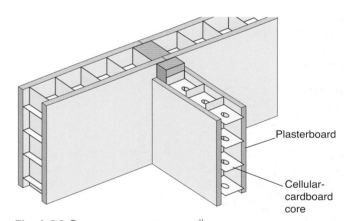

Plasterboard

Cellular-cardboard core

Fig 6.29 Paramount partition wall system

Laminated partition wall system

This type of non-load bearing partition consists of a laminate of three layers of plasterboard bonded on site. It has adequate sound and fire protection and is quickly erected. The outside face can be finished with skimming plaster, but it is usual to use ivory-coloured face boards that are intended to receive the decoration directly.

The thickness of the wall is usually 50 mm but, where extra sound or fire insulation is required, thicker boards or extra layers can be used.

The construction procedure begins with a 38 × 25 mm batten being screwed to the floor, adjacent wall(s) and ceiling to form a perimeter frame. The first board is cut and nailed to one side of the timber frame; a special bonding compound is then applied to the inside face of the board and a central layer of 19 mm plasterboard is cut between the battens and pressed into position. The bonding compound is applied to the face of the central board and the outer layer placed in position and nailed to the perimeter frame. The partition should achieve full strength after four days.

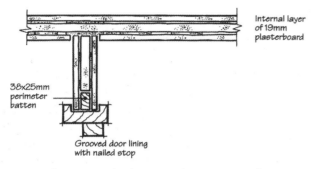

Fig 6.30 Plan view of a laminated partition wall system

1 A building developer cannot decide whether to use a timber frame or traditional brick-and-block cavity wall construction. Discuss what may influence the choice of the developer.

2 a) Using annotated sketches, explain two methods of providing cavity wall insulation to be incorporated during construction. You must include cavity wall ties and their spacing.

   b) Identify three types of damp-proof course used in traditional house construction.

3 a) Use annotated sketches to show the type of brickwork bond that could be used in a 225 mm thick garden wall.

   b) Use annotated sketches to show the type of brickwork bond that could be used in a domestic cavity wall.

4 Describe two methods of spanning window and door openings in brick-and-block cavity walls.

5 A housing developer is planning a new estate of three- and four-bedroom houses. The developer is considering lightweight concrete block or timber stud partitions for all non-load bearing walls. Identify and evaluate the advantages for each type of construction.

6 Using notes and sketches, describe three types of construction to form non-load bearing internal partition walls.

Section 4 Roofs

Learning outcomes

By the end of this section you should have developed a knowledge and understanding of:

- **The functional requirements of a roof** – strength, stability, resistance to weather, durability, fire safety, resistance to the passage of heat, sound and air leakage, security and aesthetics

- **The selection and use of roof types** – double, pre-fabricated trussed rafter, and flat

Introduction

Traditional thinking has a building's roof as having one main function: to provide shelter from the elements. However, significant changes in roof design and roofing materials, and more onerous requirements of the Building Regulations over the last few decades, have influenced the functional requirements of the roof structure, for example to stabilise the external walls, reduce the spread of fire to other adjacent buildings, provide thermal and sound insulation, and so on.

Despite the changes it is important to have a good understanding of older, or traditional, roof construction and the materials used, because there are many roofs that will need to be repaired or replaced in the forthcoming years. Also, with people recognising the value of having a habitable space within the roof, some elements of traditional roof construction may return.

Functional requirements of a roof

Strength and stability

A roof is constructed to support the dead loads of the roof members and its coverings, together with any imposed loads such as snow and wind, without undue deflection. A roof's strength mainly depends on the individual strength of the materials used and the format in which they are assembled.

A flat roof relies on adequate support from the walls on which it bears and the depth and thickness of the joists, whereas a pitched roof relies on the triangulation of the roof members, for example ties and struts.

The improvement in modern technology has enabled lighter roof construction; however roofs are now more vulnerable to wind pressure. Therefore, galvanised steel retaining straps fixed to the internal leaf must be used to prevent uplift and timber bracing fixed to resist racking of the trussed rafters (the domino effect).

Resistance to weather

The exclusion of rain and snow largely depends on the roof covering, which may be impermeable layers of 3G (a polymer-modified bitumen membrane with a reinforced polyester fibre base and bituminous felt) or a series of individual units of clay or concrete tiles and slates that are laid so they overlap or interlock and the rain runs off the roof into guttering.

Durability

The durability of a roof largely depends on the roof covering's ability to exclude rain because regular penetration will cause the roof structure to decay or corrode. The tiles and slates used for pitched roofs are generally maintenance free, however most flat roof coverings have a limited lifespan owing to the extremes of weather.

Fire safety

The main consideration of the Building Regulations Approved Document B Fire Safety is the means of escape for occupants to a place of safety. Realistically, as the roof structure is likely to take some time to collapse after ignition, evacuation by the occupants should not present a problem. The fire resistance requirement for roofs is to limit the spread of flame across the surface of the roof covering to adjacent buildings.

Resistance to the passage of heat

A roof structure's materials and coverings are invariably poor insulators against the transfer of heat. Therefore, to comply with the requirements of the Building Regulations Approved Document L Conservation of Fuel and Power, some form of thermal insulation will be required. This usually takes the form of laying insulating material between or across the ceiling joists; this has the effect of reducing the heat loss to the roof space and heat gain in warmer weather conditions.

Resistance to the passage of sound

A roof's resistance to sound is not usually a consideration unless the building is close to an airport, railway or busy road. The mass of the roof's covering has the most effect on airborne sound (the greater the mass, the better the sound reduction).

Air leakage

Efficient ventilation of the roof space is essential in order to prevent harmful condensation that can cause timber to rot or metal to corrode. In modern houses, large quantities of water vapour are produced by everyday activities, for example cooking, taking a shower, and so on. The roof space is a problem area because the air is cool due to the level of insulation: once the warm moist air finds its way into the void condensation is likely to occur.

The usual method of providing cross ventilation is the use of continuous ventilation at the eaves along opposite sides of the building. However, care should be taken not to over ventilate the roof space when attempts are made to make the rest of the building airtight.

Aesthetics

The appearance of a roof will largely be influenced by the context and locality of the building. Decisions will be made on the visibility of the roof and its covering from the ground or the roof covering used for adjacent buildings.

The selection and use of roof types

It is worth noting that roofs are classified as flat when the pitch is 10 degrees or less, and pitched when greater than 10 degrees.

> **Key Terms**
>
> - **Clear span** – the clear horizontal distance between the supporting walls
> - **Effective span** – the horizontal distance between the outer faces of the supporting timber wall plates
> - **Pitch** – the slope of the rafters; given in degrees or as a fraction (rise/span)
> - **Rise** – the vertical height to which the rafters rise above the wall plate
> - **Eaves** – the overhanging portion at the feet of the rafters
> - **Gable-end roof** – the triangular part of the end wall that is built up to the underside of the roof slopes
> - **Hipped-end roof** – the hipped end is formed by the intersection of two, usually similar, slopes and at right angles to the main roof

Double roofs

When the length of the rafters is more than the economic limits of the timber section size, a cross beam or purlin is introduced at mid span. In the case of a traditional gable roof, the purlin is a solid beam with each end being built into the inner leaf of the cavity wall. When a purlin is required in a hipped roof, it is supported by timber struts bearing on load-bearing walls. In modern roof construction the purlin takes the form of a pre-fabricated deep truss with open triangulated members, or with a plywood web that spans from gable to gable that is either bolted to the inner leaf or supported on steel brackets.

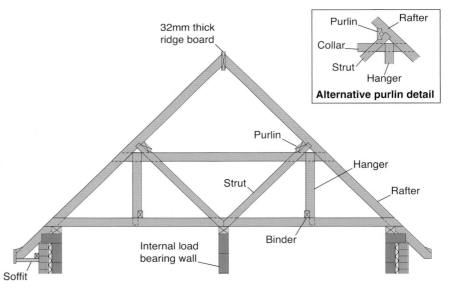

Fig 6.31 Double or purlin roof

Pre-fabricated trussed rafter roofs

Trussed rafters are triangulated roof frames that are designed to give a clear span between the external walls that support them. The trusses are designed for each specific building and will vary in relation to:

- The location of the building with reference to unusual wind conditions
- The overall span
- The pitch
- The type and weight of the roof covering
- The size and position of any water tanks
- The amount of overhang at the eaves, and the spacing of the trusses

During the factory pre-fabrication of trussed rafters a number of quality control checks should be made to ensure that they will meet their intended purpose, that is, the timber used is strength graded and conforms to BS4978:1973; the joints between members are closely fitting with connecting plates that are of the correct size and are fixed properly; the span, rise and pitch meet the design specification; the size of each member is correct; and there are no significant defects in the timber, for example bowing, twist, etc.

After pre-fabrication the trusses are delivered to the site, lifted into position by a crane and fastened to the wall plate by means of a metal truss clip. To link the trusses together longitudinal binders are fixed over the ceiling ties and under internal ties near to the roof apex. To complete the bracing, diagonal boards are fixed to the underside of the top chord from the gable-end eaves to the apex. Lateral restraint to the gable walls at top chord and ceiling tie levels is achieved by using steel straps at 2 m centres connected to the two trusses on each side of the wall. A similar arrangement is used to strap the wall plate down to the inner leaf of the external wall, preventing the roof from lifting in high winds.

> ### Key Term
>
> **Strength-graded timber** – strength grading is a visual or mechanical procedure for determining the strength of a piece of structural timber. Each piece of strength-graded timber is clearly marked to include its moisture content at the time of grading, the strength grade, the timber species, the British/European Standard and the mark of the certifying body.

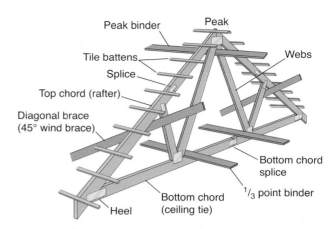

Fig 6.32 Elevation of a trussed rafter

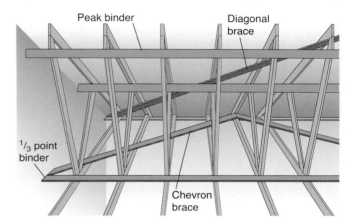

Fig 6.33 Wind bracing of the roof structure

The advantages of trussed rafters over traditional roof construction are:

- No internal support is needed from load-bearing partitions
- Spans of up to 12 m can be easily achieved
- Forty per cent less timber is used than in a traditional roof

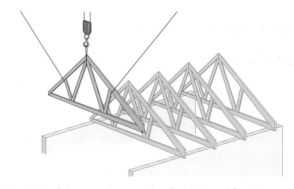

Fig 6.34 Pre-fabricated trussed rafter being lifted into position

- Reduced labour costs on site due to the amount of pre-fabrication
- Quick erection of the structure and space saving on site, with no need for timber storage

> Investigate how ventilation can be provided at the eaves to prevent condensation occurring in the roof space.

Flat roofs

The structure of a flat roof is similar to that of an upper floor, the exception being that a roof will require a small slope to allow rainwater to run off into a gutter. Also, the loadings will usually be less than those of a floor, allowing smaller joists to be used. Apart from the dead load of the materials used, typical imposed loadings on a flat roof that need to be considered are foot traffic during maintenance operations and occasional snow loads.

Spacing of the joists is similar to a suspended floor and will depend on the thickness of the deck to be supported. To achieve the desired roof slope or fall, a tapered firring piece is fixed to the top of each joist; this allows the ceiling to be kept level.

> - In a small group, identify a range of methods used to achieve the slope that would allow rain to fall from a flat roof.
> - Research a range of galvanised steel fixings used in the manufacture and installation of timber-trussed rafters and timber-framed walls. Make a comparison with traditional nailing and jointing methods for the construction of these building components and list their advantages and disadvantages.

The two methods of providing thermal insulation are: cold deck and warm deck. The cold deck method has the insulation supported by the ceiling boards. A minimum 50 mm air circulation space between the insulation and decking, together with matching eaves vents, allows moisture vapour to be removed. The warm deck incorporates rigid insulation sandwiched between the waterproof covering and the roof decking. It is common practice to use a 1000 g polythene vapour barrier beneath the insulation to maintain control of any vapour and so prevent the timber from rotting.

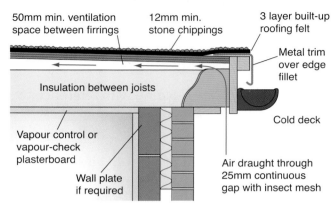

Fig 6.35 Cold deck flat roof construction

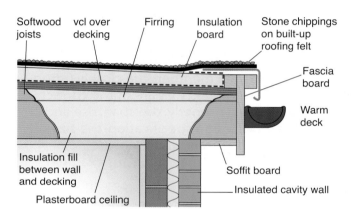

Fig 6.36 Warm deck flat roof construction

1 a) Identify the design criteria that are required before the manufacture of a trussed rafter can commence.

   b) Describe three quality control checks that may be carried out during the manufacture of a trussed rafter.

   c) Explain the benefits to the builder of using standardised trussed rafters for the production of a roof.

2 a) In a double roof what is the function of a purlin and how are the ends of the purlin supported?

   b) Describe the various ways in which a purlin can be supported in mid span.

3 a) Use notes and sketches to identify where bracing and restraints are essential for the stability of the whole roof structure.

   b) Explain why the timber used in trussed rafters is treated with preservatives.

4 a) Use an annotated sketch to show the elevation of a conventional trussed rafter.

   b) Draw and label an eaves detail for a typical trussed rafter roof of a domestic dwelling. Choose three of the components you have labelled and describe their function.

5 a) Explain the two main methods of providing thermal insulation to a flat roof and include a summary of their respective advantages and disadvantages.

   b) Describe with sketches how the fall can be provided for a flat roof.

Section 5   Internal surface finishes

Learning outcomes

By the end of this section you should have developed a knowledge and understanding of:

- **The functional requirements of an internal surface finish** – durability, aesthetics, ease of cleaning and prevention of mould growth or insect attack
- **The selection and use of internal surface finishes** – wet finishes, dry linings, joinery in softwood and hardwood, self finishes and paint finishes

Introduction

The choice of surface finishes for a building is important because they form the interface between the building and its users and, subsequently, how we perceive the building.

Surface finishes can affect our senses because they are seen and touched. The choice of colour has a psychological effect on how we feel in that environment, for example walls painted red or blue have a contrasting effect. Some surface finishes are required to be touched in their use, like that of a stair handrail. Although there is a variety of surface finishes available, they fall into two categories: self or natural finishes and applied finishes.

Functional requirements to be considered prior to specifying an internal surface finish

- The durability of the finish – this will not only depend on the properties of the finish but also on the properties of the material it is applied to and the bond between the two materials
- Is the finish hardwearing and does it have good impact resistance?
- The costs to install and the implications for on-going maintenance and replacement costs
- Does the finish have to be waterproof or moisture resistant, for example in kitchens and bathrooms where there will be condensation?
- Should the finish be easy to clean, hygienic and resist mould growth?
- Is the finish required to have a low surface spread of flame rating or absorb sound?
- Does the finish have to be attractive or decorative?
- Is a visual finish required to produce high levels of contrast, for example stair nosings (the projecting part of the stair tread), for safety reasons?
- Would a touch-sensitive or tactile finish be an aid to people with visual difficulties?
- Is the applied finish ecologically sound, in that it minimises pollution and does not contain pigments and other substances that may be harmful to people, animals or plants?

The selection and use of internal surface finishes

Wet finishes

Plaster

Plaster is a material that is spread over irregular walls and ceilings to provide a smooth, vertical and level finish. Once dry it can be painted or wallpaper applied. Other characteristics of a plaster finish include the resistance to fire and surface abrasion; the provision of sound and thermal insulation; and a hygienic surface.

When the plaster is mixed with water a hydration reaction occurs and a set or hardening is produced. For most walls a 13 mm thick backing or undercoat is applied followed by a 3 mm finishing coat. Ceilings will only require a 3 mm skimming coat, but joints in the plasterboards should be first covered with a paper tape to prevent cracking of the hardened plaster.

Sand-and-cement screed

Although most floor finishes have been considered earlier in this chapter, a 30–50 mm layer of sand and cement applied to the surface of a concrete slab provides a suitable finish to receive the floor covering. Other functions of a screed include services, for example central heating, water and gas pipes; under floor heating cables/pipes can be embedded; and protection of the damp-proof membrane and thermal insulation materials.

Dry finishes

Dry lining

A common alternative to using wet plaster for the finish of internal walls is to dry line the walls with plasterboard. The boards can be fixed to the backing wall by means of a recommended adhesive applied to both contact surfaces, nailed or screwed to vertical and horizontal timber battens, or pressed into dots and dabs of wet finishing plaster. A flat surface is obtained by filling the tapered edge joint with a special plaster, applying a joint tape and a final coat of the filling plaster. The advantages of dry lining are:

- It is quicker than using wet plaster because no drying out period is required
- The small gap between the back of the board and the wall provides improved thermal insulation
- It requires less skill than traditional plastering
- It is less prone to material failure and labour error
- Paint, wallpaper and other finishes, such as tiles, are applied directly to the plasterboard surface

There are some disadvantages regarding this method, mainly that it is more expensive, and the gap between the board and the backing wall creates an additional air passage. However, airtightness can be achieved by providing a continuous strip of wet plaster around the perimeter of the board prior to fixing to the backing wall.

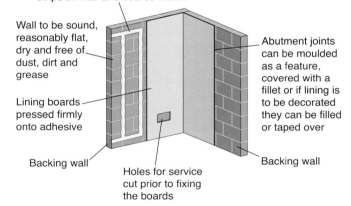

Strips of contact adhesive applied to wall and backs of boards as recommended by manufacturer so that strips on wall and boards match

Wall to be sound, reasonably flat, dry and free of dust, dirt and grease

Abutment joints can be moulded as a feature, covered with a fillet or if lining is to be decorated they can be filled or taped over

Lining boards pressed firmly onto adhesive

Backing wall

Backing wall

Holes for service cut prior to fixing the boards

Fig 6.37 Dry lining detail

Wall tiles

Wall tiles are classified by the material from which they are made, the finish, shape and size. They are used to enhance a wall's appearance and to provide resistance to moisture and abrasion. The most common form is the glazed ceramic tile, although stainless steel, plastic, glass and mosaic can be used.

Tiles can be fixed to a range of backing materials, for example plaster, sand cement rendering, sheet materials, etc., but it must be strong enough to support the weight of the tiles without losing adhesion. The backing should also be dry, smooth and rigid. After a day of being fixed with a thin bed of special adhesive, the tiles are grouted by pressing a grout paste into the joints.

Self finishes

Some materials such as facing bricks, stone, and some timbers and metals, do not require an applied finish to the surface. However, the quality of the work and materials is critical and, in particular, any joints in brick or blockwork, because they will not be covered up. If the self-finished materials are chosen carefully it can help to reduce construction time and cost. Also, at the end of the building's life the materials used can be recycled more easily because they have not been compromised by the application of a finish.

1 Recommend a type of internal surface finish for the following situations and, in each case, give reasons for your choice.
   a) A modern house with a brick-and-block cavity wall.
   b) A wet finish for an internal load-bearing wall constructed from 100 mm dense concrete blocks.
   c) A solid brick external wall, where the inside face of the wall is very uneven and a wet finish is required.
2 What are the advantages of dry lining compared with wet plastering?
3 a) Identify the types of wall tiles in common use.
   b) Explain which backgrounds are suitable to receive wall tiles.
   c) Describe the methods of fixing and grouting wall tiles.
4 Explain the function of a sand-and-cement floor screed.

REFERENCES

[1] Regarding building on brownfield land: www.englishpartnerships.co.uk

[2] BS5930:1999, Code of Practice for Site Investigations HMSO.

[3] Chudley, R and Greeno, R. (2006) Building Construction Handbook, 7th Edition, Butterworth-Heinemann.

[4] The Health and Safety Executive's Information Sheet: Construction Sheet No 8 Safety in Excavations. www.hse.gov.uk/pubns/cis08.pdf

FURTHER READING (GENERAL)

Emmitt, S. and Gorse, C. (2005) Barry's Introduction to Construction of Buildings, Blackwell Publishing

Foster, J.S. and Greeno, R. (2006) Mitchell's Structure and Fabric 7th Edition Parts 1 and 2, Prentice Hall.

Chudley, R. and Greeno, R. (2006) Building Construction Handbook, 6th Edition, Butterworth-Heinemann.

www.british-gypsum.com

Engineering

Learning outcomes

By the end of this section you should have developed a knowledge and understanding of:

- Common processes for working with engineering materials – drilling, sawing, shaping, abrading (see Chapter 10 Manufacturing)
- Processes used to manufacture products from metal – milling, turning, casting, modifying characteristics using heat, pressing and stamping
- Processes used to manufacture products from plastic – compression moulding, injection moulding, vacuum forming, rotational moulding, extrusion and blow moulding (see Chapter 10 Manufacturing)
- The design of simple jigs, presses and moulds
- Joining methods using fittings, adhesives, heat and common joints (see Chapter 10 Manufacturing)

Introduction

The aim of this section is to enable you to recognise and select a range of engineering processes suitable for the manufacture of a typical electromechanical product. You should be able to evaluate a specific process and to describe specific techniques and possible sequences required for the manufacture of a product. The ability to select suitable processes and machine tools to manufacture a component or product based upon a given product specification is important.

Common processes for working with engineering materials – drilling, sawing, shaping, abrading

This material is covered in Chapter 10 Manufacturing

Processes used to manufacture products from metal – milling, turning, casting, modifying characteristics using heat, pressing and stamping

The following pages are intended to give you a deeper understanding of what is required in order to manufacture a typical metal-based product to a required specification, giving consideration to the future maintenance of the product and its performance throughout its working life.

The manufacture of any given product will often require a great deal of knowledge and understanding of the materials being used, and skill in manipulating those materials to produce the required end result.

Initially it is vitally important to evaluate which process is most suitable for the manufacture of a specific product.

These considerations may take several forms and may well depend on the criteria laid out in the product specification, for example:

- Material used
- Quality of the final product
- Costs, both manufacturing and final selling price
- The quantity to be manufactured
- Tolerances (accuracy of the engineering process)

Secondly, the manufacturing team must consider which of the following engineering processes (and associated techniques) are the best to be used and to what extent:

- Material removal – includes turning, milling, sawing, shearing and drilling; how much material can safely be removed in one cut?
- Basic shaping – includes forging, bending, moulding, pressing and stamping; can the final product be produced in one piece?
- Joining and assembly – includes nuts and bolts, screws rivets, soft soldering, welding, brazing; how are the component parts of the finished product to be connected together, and will it be a permanent connection?
- Surface finishing – what will the finished product look like, under what conditions will it be used, and how can consistent quality be maintained throughout the manufacturing process?
- Heat treatment – can the basic properties of the material being used be enhanced? Can the hardness and toughness be modified by the application of heat?
- Chemical treatment – as for surface finish, will the final product require a special corrosion-resistant finish?

Designers and engineers have to consider some or all of the above factors prior to commencing manufacture.

1 Study the following list of common products. State briefly which of the above processes and associated techniques you think are used to make them. Some may require more than one process/technique:
- An open-ended spanner
- A bicycle frame
- A motor car gearbox casing (excluding the gear train)
- A gear cog
- A 13 amp electric plug (all components)
- A hammer (the head)

Material removal

When manufacturing any sort of metal product it is inevitable that some amount of material will have to be removed from the metal blank in order to produce the required shape.

Three of the most common procedures for removing excess material include:

- Turning
- Milling
- Drilling

It can sometimes be quite difficult for engineers to decide which of the three techniques listed above would best be used in the manufacturing process since, often, more than one of the three could be used. In cases such as this, the production team would have to consider a further set of criteria. These would include:

- What is the basic shape of the product?
- What material is being used and what is the (possible) best process for this material?
- What are the required tolerances (the level of accuracy and precision required)?
- What is the size of the batch; how many are to be made?
- Is one of the available processes/techniques quicker than another?
- What are the cost implications of using one process when compared to another?

In order to make an informed choice in the manufacture of a particular product, there is a requirement to be familiar with the basic material removal process as well as the more specific techniques within that process, for example:

- Turning – typically used to produce cylindrical shapes, tapers, holes and screw threads (both male and female)
- Milling – this process is generally used to produce smooth, flat surfaces, slots and sometimes curved surfaces
- Drilling – used typically to produce holes all the way through a material, as well as to produce blind holes, counterbores and countersinking

1 Complete the table below to show which of the three processes listed above is used in the manufactured products shown. One has been done for you as an example.

Turning

Turning is done on a lathe (see Figure 7.1); the most common type found in engineering workshops and manufacturing industries is called a centre lathe. The main function of the turning process is to produce parallel cylindrical shapes or tapered cylindrical shapes to a very high degree of accuracy.

Fig 7.1 A centre lathe

The basic turning process involves the use of a single point cutting tool in which material is removed from the outside diameter of the workpiece, or a twist drill used to bore a hole into (or through) the centre of the horizontal axis of the workpiece.

There are many types of lathe cutting tools that are available to the engineer. Some of those are shown in Figure 7.2.

	Centre punch	Brake disc	Bolts	Printed circuit boards	Car engine cylinder head
Turning	√				
Milling					
Drilling					

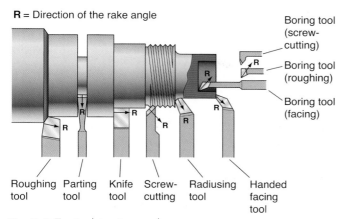

R = Direction of the rake angle

Roughing tool | Parting tool | Knife tool | Screw-cutting | Radiusing tool | Handed facing tool

Boring tool (screw-cutting)
Boring tool (roughing)
Boring tool (facing)

Fig 7.2 Typical turning tools

The removal of material is achieved by clamping the material that is to be worked on (the workpiece) firmly into a work-holding device.

There are two main forms of chucks. One is a three jaw self-centring chuck and the other is a four jaw independent chuck. A third work-holding device is called a faceplate. Their uses depend on the initial form of the workpiece.

Three jaw self-centring chuck

This type of chuck can be used to hold a wide variety of both cylindrical and hexagonal workpieces. The three jaws of the clamping facility can move inwards or outwards together. They are driven by the scroll plate which in turn is rotated by a chuck key. This ensures that the centre line of the workpiece always lies on the centre axis of the lathe spindle.

This type of chuck is used when end facing, turning external surfaces and boring. It should not be used when working on material that is

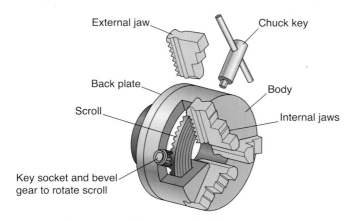

External jaw
Chuck key
Back plate
Scroll
Body
Key socket and bevel gear to rotate scroll
Internal jaws

Fig 7.3 Three jaw self-centring chuck

not truly cylindrical or hexagonal as this would put undue loading on the jaws of the chuck, which may mean that the workpiece is not securely fastened.

Four jaw independent chuck

In this type of chuck, each of the four jaws moves independently on its own screw thread and has its own chuck key socket. It is used primarily for holding square bar, hot rolled black bar and irregular-shaped castings and forgings. Because each jaw operates independently of each other, it has greater gripping strength than the three-jaw chuck, although it is more time consuming to set up prior to the cutting operation.

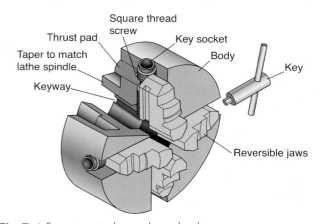

Thrust pad
Square thread screw
Taper to match lathe spindle
Key socket
Body
Key
Keyway
Reversible jaws

Fig 7.4 Four jaw independent chuck

Face plate

This chuck allows irregularly-shaped casting or forgings that are too big for a standard four-jaw chuck to be securely held in place. It also enables diameters and faces to be machined on workpieces that are parallel or perpendicular to a surface which is pre-machined to a flat finish. The pre-machined face is placed against the face plate allowing the parallel surface to be machined.

If a hole has to be bored directly through the workpiece, the workpiece is separated from the faceplate by the use of parallel bars. Also, if a workpiece is of a very irregular shape or is eccentrically mounted in the face plate, then a considerable out-of-balance force will be present. This causes excessive vibration, which may lead to premature failure of the machine bearings and also a poor surface finish. In order to prevent this balance weights attached to the face plate are used as shown in Figure 7.5b.

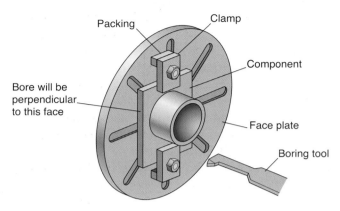

Fig 7.5 Face plate

Milling

This process allows for the rapid removal of material using multi-tooth cutting devices. The workpiece is secured to the machine work table and is fed under the cutter. There are two basic types of milling machines:

- **Horizontal millers** – the tool spindle axis is in the horizontal plane
- **Vertical miller** – the tool spindle axis is in the vertical plane

Additionally, there is a third type called a universal miller. It is similar to the horizontal miller but has a table that can be swivelled through a prescribed angle.

All three types of machines are used to machine flat surfaces, slots and steps.

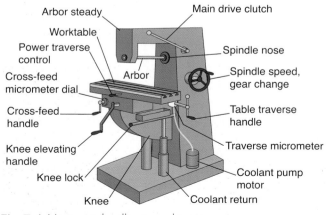

Fig 7.6 Horizontal milling machine

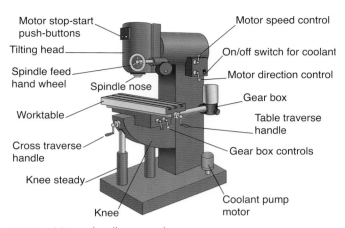

Fig 7.7 Vertical milling machine

Both types of milling machines have a work table that can be raised or lowered and also moved horizontally in two perpendicular directions. On most milling machines it can also be set to traverse automatically beneath the cutter.

The spindle, which holds the cutting device, can be driven either by a variable speed drive motor or through a gearbox so that the correct cutting speed can be selected depending on the nature of the material being worked.

For milling operations, there is a wide variety of cutting tools, all of which comprise of a series of wedge-shaped teeth that have been ground with suitable rake and clearance angles.

Horizontal milling

Four of the most common cutting tools used in horizontal milling operations are:

- **Slab cutters** – sometimes known as slab mills or roller mills; they are used to produce wide, flat surfaces
- **Side and face cutters** – these have their cutting teeth around the periphery and the side faces of the tool; they are generally used for light facing operations and for cutting slots and steps in a workpiece
- **Slotting cutters** – these are somewhat thinner than either of the two above cutters and have cutting teeth on the periphery only; they are used for cutting narrow slots and keyways in shafts

- **Slitting saws** – the thinnest of all milling cutting tools; they are used to cut very narrow slots and also to cut material to size (parting off)

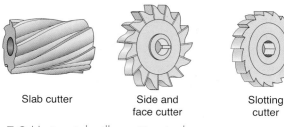

Slab cutter   Side and   Slotting
                   face cutter   cutter

Fig 7.8 Horizontal miller cutting tools

Vertical milling

Four of the most common cutting tools used in horizontal milling operations are:

- **Face mills** – used to machine wide, flat surfaces and for cutting steps; they have cutting teeth on the periphery and also on the end face of the tool. These tools produce flat surfaces more accurately than slab mills as they 'generate' the surface: every part of each of the teeth on the end face passes over the whole surface of the workpiece
- **Shell end mills** – also used for generating flat surfaces; they are smaller than face mills and are mounted on a stub arbour
- **End mills** – these cutting tools also have teeth on the periphery and the end face; they are used for relatively light facing operations and for milling slots in a workpiece
- **Slot mills** – basically end mills with two cutting lips; they are used primarily for the accurate milling of slots and keyways

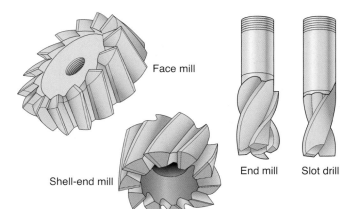

Face mill

Shell-end mill   End mill   Slot drill

Fig 7.9 Vertical miller cutting tools

1 What are the two different surface shapes that can be machined on a centre lathe?
2 Name three different ways in which a workpiece can be secured in a centre lathe.
3 Which kind of lathe cutting tool would you use for taking large initial cuts from a workpiece?
4 Name the two common types of milling machine.
5 What type of cutter would you use to machine a wide, flat surface when using a horizontal milling machine?
6 What type of cutter would be used if accurate, narrow slots and keyways were required when using a vertical milling machine?

Casting

There may be a need to produce quite complex shapes during the making of an engineered product; this is not always possible using standard material removal techniques so the product is made in one piece, usually by using a casting process. There are several methods by which a component can be produced by casting including:

- Sand casting
- Die casting
- Gravity die casting
- Pressure die casting
- Investment casting

Sand casting

Some form of crude metal casting was practiced by early civilisations over 6000 years ago. Its popularity reached a peak during the industrial revolution in the early part of the 19th century where the casting of base iron ('cast iron') was used extensively.

Fig 7.10 A cast iron bridge

Although the majority of today's products are now made from steel, cast iron is still used in the manufacture of machine tool beds and some engine blocks due to its inherent compressive strength.

Sand casting is a relatively straightforward process in which an impression is made in damp sand using a pattern of the required product. The pattern is slightly oversized to compensate for the shrinkage of the final product during the cooling stage.

Wood is quite often used to make the pattern as it is a relatively cheap material that can be easily worked into the desired final shape. Sand is used, again, because it is cheap, strong enough to withstand the pressures of the molten metal and permeable enough that it can allow the escape of the hot gasses produced during the casting process. Green sand is generally used as it contains small amounts of clay which help to bind the fine particles of sand together.

Castings can vary in size from small components that would fit in your hand up to very complex products weighing many tonnes. Although the casting method varies according to the product being made, the basic process remains the same.

Depending on the complexity of the final product, the pattern may have to be made in two halves, which are joined together using locating dowels as shown in Figure 7.11. This is called a split pattern.

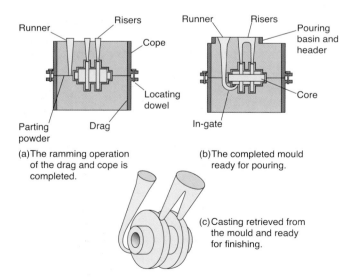

Fig 7.11 The sand casting process

One half of the pattern is placed face down on a 'turnover board' and the lower half of the two-part steel mould, called a drag, is placed around it. A fine sand called parting powder is then sifted over the pattern until it is completely covered (this is to assist the final removal from the mould) and the whole is then packed tightly with green sand. The drag is then inverted, the turnover board removed and the second half of the pattern located onto the first using the locating dowels.

The top half of the mould, known as the cope, is then placed in position and the process is repeated. The only difference is that two extra items are inserted before packing with sand: these are known as a runner, which allows for the pouring of the molten metal into the cavity of the mould, and the riser (which is always placed in the highest part of the cavity), which allows the gases to escape and also shows when the mould is full. These two pieces also act as reservoirs for the molten metal so that the casting can draw down additional metal as it cools.

To prevent cohesion of the two parts of the mould a parting material such as graphite powder is dusted onto the sand in the drag before the cope is placed in position. Once the mould has been packed, it is carefully separated and the pattern removed to leave a mould cavity. At this stage, if required, ready-formed sand 'cores' can be placed into the mould cavity to produce holes in or through the finished casting. The molten metal is then poured into the assembled mould.

During the solidification stage, the metal in the mould contracts slightly, hence the need for a slightly larger pattern at the outset.

This process allows for the manufacture of complex-shaped engineered products from virtually any metal that can melted. Although the sand can be reused, the moulds cannot and have to be remade each time. This makes it a very time-consuming process.

Die casting

The major disadvantages of the basic sand casting process are:

- The mould has to be remade for every component manufactured
- The overall accuracy of the finished product is poor

This is overcome to a great degree by the use of **die casting**. In die casting, molten metal is poured into reusable steel moulds called 'dies' and the resultant casting is, in general, left in its finished state with little or no further machining required. This is a much faster process and subsequently leads to a reduction in production costs.

However, there is one drawback of this process in that it can really only be used for casting non-ferrous materials, mostly aluminium and zinc, as the higher temperatures needed for casting steel and iron damage the expensive dies and lead to premature loss of accuracy and finish of the product being cast.

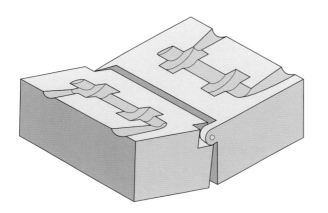

Fig 7.12 Typical steel mould used in die casting

Gravity die casting

This process also uses a permanent metal die; it closely resembles the processes of basic sand casting in that the die is filled with molten metal by the natural force of gravity. If cores are

needed, they are often made so that they can collapse in order to allow them to be withdrawn from the casting. Where this proves to be a problem then, just like the sand casting method, sand cores are used. This process is not suitable for zinc-based alloys as it tends to promote a fairly coarse grain in the finished product.

Pressure die casting

As the title suggests, this process utilises pressure. The pressure is applied to the molten metal as it is fed into the dies (moulds) using a simple plunger. This technique ensures that the molten metal makes good contact with the walls of the casting dies and, by maintaining the pressure throughout the cooling stage, gives a sharp and well-defined casting that requires little or no additional machining.

If the dies are cooled when filled the components will solidify quicker and the casting can be removed while solid but still hot, enabling the process to be repeated again giving a much faster turnaround of components.

One disadvantage of this type of process is that the cost of producing the dies is quite high. It only becomes economical if a large number of components are being made.

Investment casting

This process is also known as lost wax casting. It is used where a very complex component that requires a high degree of accuracy is needed, and one that is difficult to machine after casting.

The moulds used in this process are made from fine refractory materials that can withstand very wide heat ranges and which can provide fine dimensional accuracy. The process involves the following procedures:

- A wax pattern is made of the component being manufactured; if there are a large number of these components being made, then a mould would be made to produce the wax pattern
- The wax pattern is then coated with a refractory slurry (either by dipping or spraying); as the slurry dries it forms a hard, brittle shell which now forms the 'die'

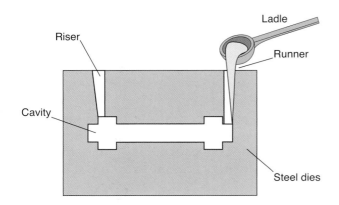

Fig 7.13 Gravity die casting

Labels on figure: Riser, Cavity, Ladle, Runner, Steel dies

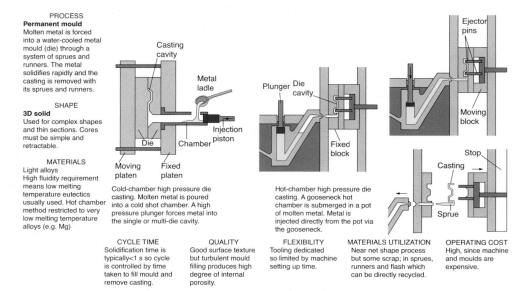

PROCESS
Permanent mould
Molten metal is forced into a water-cooled metal mould (die) through a system of sprues and runners. The metal solidifies rapidly and the casting is removed with its sprues and runners.

SHAPE
3D solid
Used for complex shapes and thin sections. Cores must be simple and retractable.

MATERIALS
Light alloys
High fluidity requirement means low melting temperature eutectics usually used. Hot chamber method restricted to very low melting temperature alloys (e.g. Mg)

Cold-chamber high pressure die casting. Molten metal is poured into a cold shot chamber. A high pressure plunger forces metal into the single or multi-die cavity.

Hot-chamber high pressure die casting. A gooseneck hot chamber is submerged in a pot of molten metal. Metal is injected directly from the pot via the gooseneck.

CYCLE TIME
Solidification time is typically<1 s so cycle is controlled by time taken to fill mould and remove casting.

QUALITY
Good surface texture but turbulent mould filling produces high degree of internal porosity.

FLEXIBILITY
Tooling dedicated so limited by machine setting up time.

MATERIALS UTILIZATION
Near net shape process but some scrap; in sprues, runners and flash which can be directly recycled.

OPERATING COST
High, since machine and moulds are expensive.

Fig 7.14 Pressure die casting

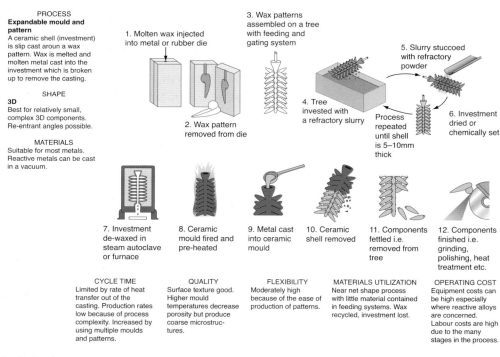

PROCESS
Expandable mould and pattern
A ceramic shell (investment) is slip cast aroun a wax pattern. Wax is melted and molten metal cast into the investment which is broken up to remove the casting.

SHAPE
3D
Best for relatively small, complex 3D components. Re-entrant angles possible.

MATERIALS
Suitable for most metals. Reactive metals can be cast in a vacuum.

1. Molten wax injected into metal or rubber die

2. Wax pattern removed from die

3. Wax patterns assembled on a tree with feeding and gating system

4. Tree invested with a refractory slurry

5. Slurry stuccoed with refractory powder

6. Investment dried or chemically set

Process repeated until shell is 5–10mm thick

7. Investment de-waxed in steam autoclave or furnace

8. Ceramic mould fired and pre-heated

9. Metal cast into ceramic mould

10. Ceramic shell removed

11. Components fettled i.e. removed from tree

12. Components finished i.e. grinding, polishing, heat treatment etc.

CYCLE TIME
Limited by rate of heat transfer out of the casting. Production rates low because of process complexity. Increased by using multiple moulds and patterns.

QUALITY
Surface texture good. Higher mould temperatures decrease porosity but produce coarse microstructures.

FLEXIBILITY
Moderately high because of the ease of production of patterns.

MATERIALS UTILIZATION
Near net shape process with little material contained in feeding systems. Wax recycled, investment lost.

OPERATING COST
Equipment costs can be high especially where reactive alloys are concerned. Labour costs are high due to the many stages in the process.

Fig 7.15 Investment casting

- The die is now placed in a preheated furnace or autoclave, and the wax is melted out; this process leaves a cavity and also assists in the setting of the refractory mould
- Molten material is then poured into the cavity and allowed to solidify; when solid, the refractory lining of the mould is broken off

This process allows castings to be made from virtually any material that can be melted and gives a high degree of accuracy and complexity. It is an expensive process however as it is very slow.

1 For the manufacture of the following components, state which casting process is the most suitable?

- A motor vehicle engine block
- A model toy car
- A turbine blade for a jet engine
- The housing for a small electric motor
- A water pump for a motor vehicle

Modifying materials or component characteristics using heat

During the manufacture of metal components, especially where a stamping, bending, forming or pressing process has been employed, the materials often become 'stressed' and the natural properties inherent in the materials become modified.

In order to relieve these induced stresses and to further enhance the function of the component or the material used, these materials are subjected to a heat treatment. The most common forms of heat-treatment processes are:

- Annealing
- Normalising
- Hardening
- Tempering
- Case hardening

Annealing

This process is carried out on materials that have been 'cold worked', by drawing, extruding, rolling or pressing. These process cause materials to be deformed beyond their elastic limit and, in doing so, deform the crystal or grain structure of the material.

As the material is further worked, the grain structure becomes more distorted and the material becomes 'work hardened', making it much more difficult to deform. If further cold working is necessary the material must first undergo a softening process in order to restore its malleability and ductility.

This process is known as annealing. The material is heated up to its re-crystallisation temperature, normally in an oven or a furnace, and held there (or 'soaked') for a given period of time. This allows new crystals or grains to start to form at the points where the original distorted grains are most stressed (bends or reduced-thickness areas).

The time during which the material is 'soaked' is determined by the amount of cold working and the use to which the component will be put.

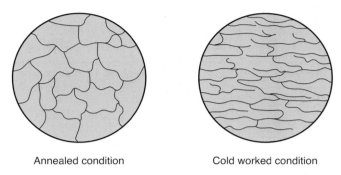

| Annealed condition | Cold worked condition |

Fig 7.16 Grain structure – the effects of annealing

Plain carbon steels, copper, aluminium and some brasses can be softened by annealing.

Care should be taken however as, if the material is held at the re-crystallisation temperature for too long, it can become too soft for further working.

For plain carbon steels, the material is heated in the oven/furnace until it achieves a 'cherry-red' colour, corresponding to the top of the graph shown in Figure 7.17. The oven or furnace is then turned off and the material left in it to cool.

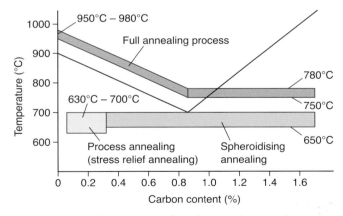

Fig 7.17 Annealing process for plain carbon steels

Steels with a carbon content above about 0.4 per cent are not naturally very ductile and seldom, if ever, become work hardened. However, they may have been 'quench hardened' and require softening before further work can be carried out on them. In this case, a process called 'spheroidising' annealing can be employed in preference to a full annealing process. This is carried out at slightly lower temperatures to enable finer grain refinement.

For aluminium, copper and cold working brasses, they may be quenched when re-crystallisation is complete. Annealing temperatures for these materials are shown in Table 7.1.

Table 7.1 Annealing temperatures for aluminium, copper and cold working brass

Material	Annealing temperature
Pure aluminium	500–550°C
Pure copper	650–750°C
Cold working brass	600–650°C

Normalising

Steel components that have been 'hot worked' – hot formed to shape by forging or hot stamping – often contain quite appreciable internal stresses as a result of uneven cooling. This often causes distortion in the component when it is subsequently machined. In this case a 'normalising' process is used to remove the internal stresses and refine the grain structure.

In this case, the materials are heated to the same temperatures as for a full annealing process but are then removed from the oven/furnace and allowed to cool in still air. This provides a faster rate of cooling, leaving a finer grain structure and a stronger material.

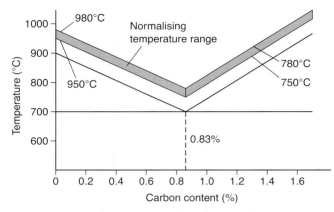

Fig 7.18 Normalising process for plain carbon steels

Hardening

This process is carried out on plain carbon steels with a carbon content above 0.3 per cent, again by heating them to the temperature of the annealing process. They are then quenched in cold water or oil depending on the carbon content and the degree of final hardness required.

Quenching in cold water gives the material maximum hardness while cooling in oil, being a less violent action, leaves the steel with a little less hardness and makes it less brittle.

Steels with a high-carbon content (above 0.9 per cent) are oil quenched as water quenching could cause the materials to crack.

Table 7.2 Examples of tempering temperatures depending on the components' final use

Component	Tempering colour	Temperature
Edge tools, lathe turning tools	Pale straw	220°C
Turning tools	Medium straw	230°C
Twist drills	Dark straw	240°C
Taps	Brown	250°C
Press tools, knives, scissors	Brownish purple	260°C
Cold chisels, screwdrivers	Purple	280°C
Springs, needles	Blue	300°C

Tempering

Components are often required to be both hard and strong at the same time. Quench-hardened plain carbon steels are too hard and brittle for further machining or forming (for example scribers, screwdriver blades, and so on). Tools such as drills and chisels would chip and crack very easily.

A tempering process is one in which some of the brittleness is removed while still retaining the hardness of the steel. The hardened components/materials are reheated to a specific temperature depending on their final use (see Table 7.2) and then quenched in either oil or water.

Where these processes are used on a regular basis in industry 'tempering furnaces' are set up and maintained at the required temperatures. In small workshops, tempering is often carried out on single or very small batches of components using a gas flame to heat them up. The components are polished before heating: the oxide-coloured film that then spreads over the surface of the polished component when it is heated gives a good indication of the temperature and the point at which they should be quenched.

Case hardening

Generally, low-carbon steels (normally referred to as mild steel) have too low a carbon content to allow them to be quenched hardened. Instead they can undergo a case hardening process in which the surface hardness is enhanced while the core of the component remains relatively soft yet tough. The process is suitable for gears and cams, which require hard-wearing surfaces.

Components that are to be case hardened are first 'carburised'. This involves prolonged 'soaking' of the carbon-bearing material at the same temperature used for annealing. Over a period of time the carbon soaks into the steel and allows the carbon content to rise on the surface of the component. The depth of penetration, and hence the thickness of the case hardening, depends on the length of time of soaking. This carburised surface can then be hardened by heating and quenching.

Pressing

This term describes any press-related operations, which can be categorised as follows:

- Piercing – where a suitable punch shears a hole in a piece of metal
- Blanking – where the punch shears the required shape from the metal
- Notching – where the punch shears an open-sided hole in the metal
- Cropping – where the punch shears a plain or shaped length from the metal
- Bending – where a suitable punch shapes the metal by a folding process
- Drawing – a process by which a punch produces a cup or dish shape from a piece of metal
- Forming – where the metal is forced into the shape of the surface contours of a die

The above terms cover all of the shaping and manipulative techniques used in sheet metal work and, to a lesser extent, on plate metal. (Metal up to about 3 mm in thickness is classified as sheet metal and above 3 mm as plate metal).

Virtually all presswork operations are carried out on sheet metal as the vast majority of engineered and manufactured components are made from metal that is less than 2 mm thick.

The main users of these manufacturing techniques are the automobile, aeronautical, heating and ventilation industries, along with canning and container making, and domestic appliance industries. Examples include:

- Car body panels
- Aircraft panels
- Air conditioning units
- Washing machine and tumble dryer carcases

As a result, more than half of the production of metal products in the western world involves the use of sheet metals. It should be noted though that large presses can be used to fold and form materials up to about 50 mm in thickness, and presses exist that can cater for capacities exceeding 45,000 tonnes (these are generally used in large forging operations however).

The majority of press-related operations are confined to large batch production work due to the high costs and lengthy tool-setting involved.

The tool costs for producing a motor vehicle body part, for example, can exceed several thousands of pounds. Production runs need to be high so as not to add significantly to the retail price of the finished product.

Presswork operations on this scale use up vast amounts of sheet steel, so it is usually supplied in large diameter rolls of a pre-determined width to suit the product being made. Operations of this type are virtually all automated now (or, at the very least, semi-automated) with only very small batches of products being suited to manual press operations.

As mentioned above, presses can be manual or power operated to suit the working conditions. Manually operated presses are obviously limited in the magnitude of the forces that can be applied to the metal, with large fly presses being capable of operating at just a few tonnes at best.

The arbor press

This type of press is usually operated by a handle working through a rack and pinion system. There are also foot-operated versions but these are not quite as powerful.

They are used for punching, bending and cropping operations on relatively thin gauge material or for light assembly work. Where they are used today, they are more than likely operated pneumatically, with the air supply operating the ram directly.

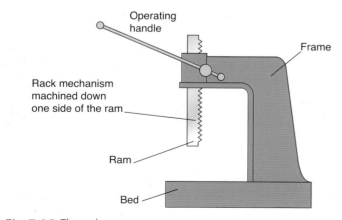

Fig 7.19 The arbor press

The fly press

This type of press is ideally suited to the operations outlined above and, despite its limitations, it is relatively cheap to install and its tooling costs are low. It consists of a heavy cast iron frame with a large diameter square thread operating the ram as show in Figure 7.20.

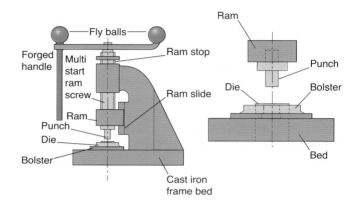

Fig 7.20 The fly press

The ram is usually constrained by a dove-tail slide so that it moves accurately in a straight, vertical direction. The operating handle is attached to a cross bar that is fixed to the top of the square thread. A large cast iron ball is fixed to each end of the cross bar (smaller, lighter models may have only one).

As the operator swings the handle, the rotational kinetic energy imparted to the cross bar (and the balls) is stored with the downward movement of the ram, and is expended as the punch makes contact with the metal workpiece.

The punch spigot fits directly into a hole in the ram and is locked into position by means of a simple clamp. The die is then located in a bolster that is clamped to the press bed. A hole lines up with the holes in the die and the bolster so that the punchings can drop through into a container.

An adjustable stop, in the form of a collar on the upper part of the ram thread, can be set to limit further movement of the ram once the punch has entered the die.

Fly presses are often used to prove and check the alignment of larger tool sets for use on large power presses

Power press

There are many types of power presses, all of which are designed for different types of presswork operations, such as blanking or piercing, bending and drawing.

In each case the operating energy is derived from an electric drive motor, stored in a large flywheel and delivered to the workpiece by means of a mechanism that changes the rotary motion of the flywheel into linear motion at the tool, in a similar manner to the way in which the screw on the fly press changes the rotary motion of the handle and the fly balls to linear motion at the tool.

In the following example, the open-framed power press, this is generally achieved by means of a connecting rod and a linkage mechanism.

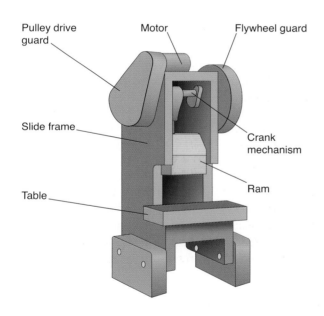

Fig 7.21 Open-framed power press

The flywheel is driven by an electric motor through a series of 'vee' belts; energy is then stored in it. The crank is driven from the flywheel by a clutch mechanism that allows one revolution of the flywheel when a foot pedal is pressed. During the operation of the press tool on the metal workpiece, energy is expended and the flywheel slows down. Speed is then built up again and the energy restored before the next downward stroke.

The only difference between the fly press and the power press is one of available power: the power transmitted to a fly press is limited by a human operator.

In the case of a power press, the operator only has to press a foot pedal and the motor power is transmitted to the machine. The power press, of course, has a much more rapid action than the manually operated fly or arbor press and, with an automatic feed of work to the tool head, the pedal can be locked down to give continuous operation.

In general fly presses are much cheaper to buy, set up and run, and lend themselves to the use of fairly simple tooling and low volumes of production. Power presses on the other hand are more expensive to buy, set up and run, but are quicker and lend themselves to high rates of production.

Stamping (blanking)

Pressing and stamping is often associated with high-volume production. A great deal of time is spent in planning the blanking layout for a particular component: the orientation of the component on the metal strip is important as it affects the economic use of the material. The direction of the grain of the material must also be considered as it may affect the final product, especially if a subsequent bending process is necessary.

Some examples of how a saving in the economy of a material can be achieved are shown in Figure 7.22.

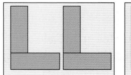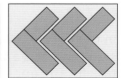

Fig 7.22

From the examples in Figure 7.22 it can be seen that for maximum efficiency of material, it is beneficial to have more than one punch carrying out the stamping operation and that the production rate is increased by the number of punches being used. The increased cost of the punches is more than offset by the increase in the production rate and associated economy of material. The logical conclusion is to arrange for multiple blanking layouts, the limit being governed only by the size of the component being produced and the width of the steel sheet supplied.

Most steel producers today will accurately slit the steel coils to any desired width, so multiple layouts should not present a problem. The only limiting factor is that, when planning for multiple stamping layouts, one has to bear in mind that the minimum 'land' (space) between adjacent components and the outer components; the edge of the strip cannot be less than the thickness of the material being used.

Figure 7.23 shows a typical layout for the stamping of blanks to make coins. Here, six punches are used and the arrangement usually has to be staggered to allow for the outside diameter of the die. The production rates for this arrangement can be in excess of 10,000 per hour per machine.

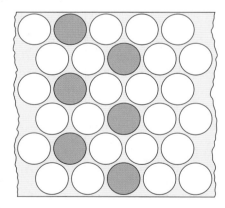

Fig 7.23

In the past arranging the layout for multiple stamping was a very skilled and time-consuming task for the engineer or manufacturer. With today's modern technology of course, a computer program is used to ascertain the maximum usage of the metal strip (referred to as a nesting process).

Processes used to manufacture products from plastic – compression moulding, injection moulding, vacuum forming, rotational moulding, extrusion and blow moulding

This material is covered in Chapter 10 Manufacturing.

Design of simple jigs, presses and moulds

Jigs and fixtures

The terms jigs and fixtures are closely related and are often interchangeable. The only real difference between the two is in the way in which the tool is guided into the workpiece.

Jigs and fixtures are production tools that are used when a number of duplicate components parts are to be made accurately and the correct alignment between a tool and the workpiece must be maintained. In order to provide this means of repetitive accuracy, a jig or fixture would be designed and made to hold, support and locate each part of a component to ensure that each part is drilled or machined in precisely the same way, within the limits or tolerances of the product specification.

A jig is a special device that holds, supports or is placed onto a part to be machined. It is a production tool that it not only locates and holds the workpiece but also guides the cutting tool as the engineering operation is performed. They are usually fitted with hardened steel bushings for guiding drills or other such cutting tools (see Figure 7.24a). Small jigs are not necessarily fastened to the drill table: they can be hand held. However, if the holes being drilled are larger than 6 mm then they are generally clamped securely to the drill table.

A fixture on the other hand is a production tool that locates, holds and supports the workpiece securely during a machining operation. Set blocks and feeler gauges are used in conjunction with fixtures to reference the cutter in relation to the workpiece (see Figure 7.24b).

Fixtures are always securely fastened to the table of the machine on which the work is being carried out. Although largely used for milling operations, they can be used on a wide variety of machine tools. They vary in design from very simple and inexpensive tools to quite complex, expensive devices.

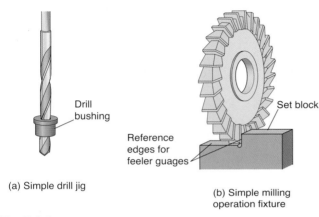

(a) Simple drill jig

(b) Simple milling operation fixture

Fig 7.24

Jigs are divided into two distinct classes: boring jigs (Figure 7.25), which are used to bore holes that are either too large to drill conventionally or are of an odd size, and drill jigs (Figure 7.26), which are used for a whole range of drilling operations including:

- Reaming
- Tapping
- Counterboring and reverse counterboring

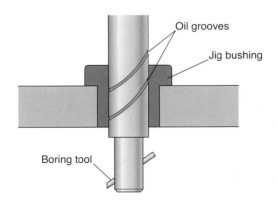

Fig 7.25 Simple boring jig

- Chamfering
- Countersinking and reverse countersinking
- Spotfacing and reverse spotfacing

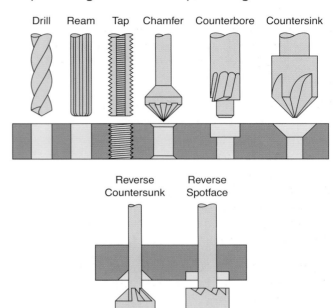

Fig 7.26 Examples of drilling operations

Drill jigs are subdivided into two general types: open jigs where only simple operations are to be carried out on only one side of the workpiece, and closed or box jigs where it is necessary to carry out machining operations on both sides of the workpiece.

Template jigs

These would be designed and made by a tool maker or fitter and are used for accuracy of production rather than speed. Template jigs are designed to fit over, on or into the workpiece and are not always clamped to the work. They are the least expensive and simplest type of jig. Some examples of simple template jigs (in engineering/orthographic drawing format) are shown in Figures 7.27 and 7.28.

In Figure 7.27a the jig is placed over the work and allows accurate drilling of a pair of holes that are at 180 degrees to each other; In Figure 7.27b the jig is placed on the top of the workpiece and allows for accurate drilling of four holes on a specified pitch circle diameter. In Figure 7.27c the jig fits into a recess in the workpiece, again allowing repetitive and accurate drilling to be carried out.

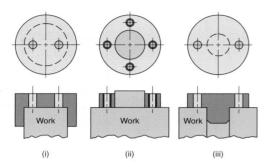

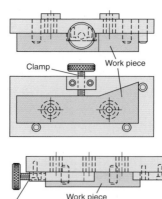

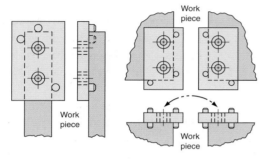

Fig 7.27

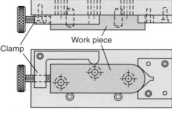

Fig 7.29 Plate-type jigs

Fig 7.28

In the case of the template jigs shown in Figure 7.28, locating pins are provided to hold the workpiece firmly in place while the machining operation is carried out.

Plate-type jigs

These are also simple template-type jigs. The only difference is that these have built-in clamps to hold the workpiece in place.

Presses and moulds

This material is covered earlier in the chapter.

Joining methods using fittings, adhesives, heat and common joints

This material is covered in Chapter 10 Manufacturing.

Section B	Engineering drawing techniques

Learning outcomes

By the end of this section you should have developed a knowledge and understanding of:

- Freehand sketching
- Isometric projection
- Perspective drawing
- Block diagrams
- Flow diagrams
- Schematic diagrams
- Circuit diagrams
- Third angle projection
- Assembly/exploded drawings and diagrams

Introduction

The fundamental techniques of any design process enable the manufacture of products or the associated engineering processes to be successfully developed. This is achieved by becoming familiar with a range of graphical methods that are available to design engineers, which enables them to select the method that is best suited for the application.

For example, simple freehand sketches, which are invaluable in communicating initial design ideas, would be of little use as an engineering workshop drawing. Similarly, such formal drawings would be inappropriate for presentation and marketing applications.

The convention for engineering drawings was, until recently, controlled by British Standard BS 308. This publication however, along with a range of BSI publications have been revised as part of a development of European and international standards. Although BS 308 (Students Version) will still be found in schools and colleges, it has been superseded by PP7307 and PP7308, the abridged versions of which are designed for school and college use.

These documents give guidance, for example, on the correct use of line types and the correct use of text.

- Only pencil or black ink should be used to produce formal engineering drawings
- Only two thicknesses of lines should be used – thick lines being twice the diameter of thin lines
- Centre lines should extend just past the outline of the drawing or the relevant feature and can, where required, be extended to form leader lines for dimensioning. They can cross one another but must end with a long dash
- Dashed lines and centre lines should meet and cross any other line on a dash rather than on a space

Table 7.3 The correct use of line types

Line type		Application
A	Continuous thick	Visible outlines and edges
B	Continuous thin	B1 Dimensions B2 Projection and leader lines B3 Hatching B4 Outline of revolved sections B5 Short centre lines B6 Imaginary intersections
C	Continuous thin (irregular)	Limit of partial or interrupted view
D	Continuous thin (straight with zigzags)	Sections and parts of drawings if the limit of the section is not on an axis
E	Dashed – thin	Hidden detail, outlines and edges
F	Chain – thin	Centre lines, pitch lines and pitch circles
G	Chain – thin (but thick at the ends)	Cutting planes for sectional views and changes of direction
H	Chain – thin (double dashed)	Outlines of adjacent parts and extreme position of moving parts

Text

Invariably on detailed engineering drawings text needs to be added for information and dimensioning. Any text used should be clear.

The drawing shown in Figure 7.30 is taken from PP7308 and is used to show the application of the various types of lines that are used in engineering drawing.

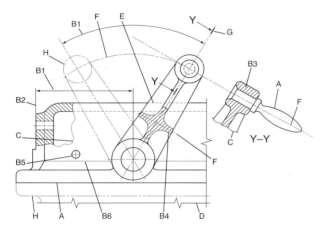

Fig 7.30 Engineering diagram showing typical use of line types

Freehand sketching

Sketches are invariably used as the first stage of a design process to develop and clarify initial ideas. They are used to communicate those initial ideas to other members of the design/engineering/manufacturing process.

Freehand sketching is like using a pencil and a notepad as an extension of your own brain and thought process. In 1959 for example, Alec Issigonis designed the original 'mini' car. His initial freehand sketches were said to have been made on a tablecloth in his local restaurant.

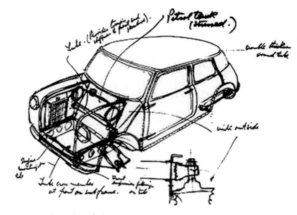

Fig 7.31 Sketch of the Mini

The ability to sketch fluently and clearly can be improved by learning a few simple techniques and by lots of practice. In order to learn to sketch fluently, start by drawing simple horizontal and vertical straight lines. Simple sketches can be built up from these flat geometric shapes by adding suitable curves. The secret to drawing straight lines is to concentrate on where you are going and not to look at the point of the pen/pencil that you are using. Begin by trying to draw long straight lines, both horizontally and vertically on a piece of paper. Place dots at each end of your lines as guides. Then progress to right angles and rectangles as shown in Figure 7.32.

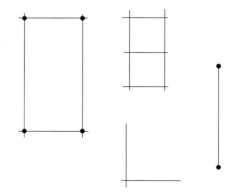

Fig 7.32

Curves and circles can be constructed with reasonable accuracy by drawing them inside a square or a grid. Again, practice is needed. Don't try to push the pen/pencil around the curve; instead turn the paper round so that you produce a nice flowing shape.

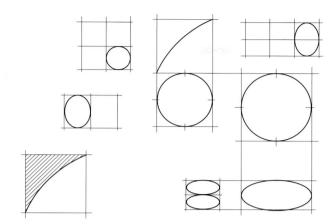

Fig 7.33 Examples of isometric and oblique curves and circles

It should be remembered that freehand sketching is probably the best way in which to communicate ideas. Initially you may just use them for yourself but, as the design develops, there will be a need to use them to communicate with other team members.

The diagram below shows a series of sketches that explore various ways of locating a hinge pin for a lid, cover or flap.

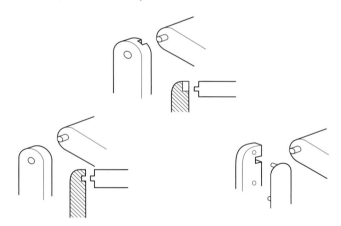

Fig 7.34

Isometric projection

Three dimensional or pictorial sketches often employ a technique where the object is drawn inside a cube-shaped structure or crate. To achieve this, a rectangular shape is drawn so that the object being drawn will just fit inside it. In this way, the correct overall proportions of the object being constructed are maintained, and the sides remain parallel to each other.

In engineering, pictorial drawings are either oblique or isometric. In oblique form, the object has the correct dimension and shape when viewed from the front, but the right-angled corners and side edges project backwards at an angle of 45 degrees as shown in Figure 7.35. This method does not represent the true dimensions of the object. It has to be shortened in order for it to look right.

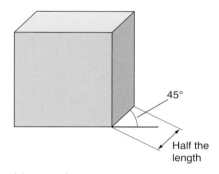

Fig 7.35 Oblique cube

Isometric drawings are the most popular of the two styles in engineering as they give a much better representation of the true proportions of the object being drawn. In this instance, the drawing is tipped forward slightly and the front and side edges are drawn at an angle of 30 degrees to the horizontal as shown in Figure 7.36.

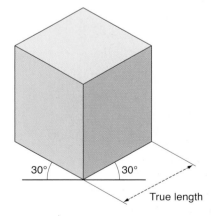

Fig 7.36 Isometric cube

Circles and curves are dealt with as before, but now the outer square is drawn in either oblique or isometric view and the circle or curve would then be drawn as an ellipse (or partial ellipse) as shown in Figure 7.37.

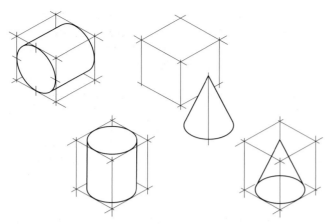

Fig 7.37

Perspective drawing

This is another method of representing objects pictorially and is widely used in architectural design applications.

The lines on this type of drawing are different from those in either isometric or oblique drawings. In this type of drawing, which is called two-point perspective, there are three types of lines:

- Those that converge at a point to the left of the object (the vanishing point left)
- Those that converge at a point to the right of the object (the vanishing point right)
- Vertical lines

VPL
(vanishing
point left)

VPR
(vanishing
point right)

Fig 7.38

When drawing in perspective you can vary your point of view. This is not true in either isometric or oblique drawings, both of which have a fixed point of view and appearance. Changing the location of the vanishing point causes a change in the appearance of the perspective drawing. For this reason it is rarely, if ever, used in engineering drawing applications.

Exploded drawings

Pictorial drawings of the types mentioned above (isometric and oblique) are often used in manufacturers' catalogues and data books, usually as a single item. However, when a number of separate items have been assembled there is obviously a problem with visibility, as some parts are hidden behind others.

This problem can easily be overcome by showing the respective parts of the assembly in an opened out or exploded view. Views of this type can be in either pictorial or orthographic projection as shown in Figures 7.39 and 7.40. The first is of a simple flashlight and is shown pictorially, and the second is of a simple fixing system comprising a bolt, two washers, a connecting lug and a securing nut. This assembly is shown in isometric projection.

Fig 7.39 Exploded drawing of a simple flashlight

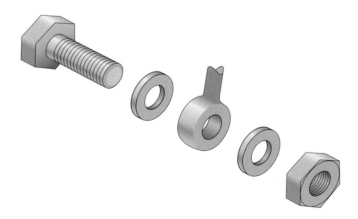

Fig 7.40 Exploded view in isometric projection of a nut and bolt assembly

Although many exploded views used in engineering are shown pictorially, they can also be shown orthographically. Obviously, the more complex the object being shown, the more complex the exploded views.

Assembly drawings

The diagram in Figure 7.41 is a typical example of an assembly drawing. In this case we see a bearing block that consists of a ball bearing race clamped in a split bearing housing, which is then securely bolted to a cast iron bed (it could possibly form part of the support for the drive shaft of a large generator or alternator).

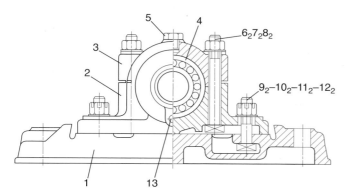

Fig 7.41 Assembly drawing showing a bearing block

Drawings such as these are often partially sectioned in order to show the inner details of the assembly. In this case, the cutting line for sectioning is through the centre line of the assembly. It would not be normal procedure to add dimensioning to this type of drawing other than maybe a few overall dimensions that will be of use for packaging, storing and transportation purposes.

The numbers shown around the periphery of the drawing are normally used as a reference for a parts list that may be included on the drawing. This would normally be in a separate table included on the same drawing sheet. Numbering is usually either from the bottom of the drawing going vertically upwards or from the top going downwards. This would enable additional parts to be added due to subsequent modifications or reissues.

Orthographic projection

This is the type of drawing that provides the engineer with the means of representing a detailed 3D object in a 2D format.

As mentioned above, pictorial drawings are not entirely accurate in that they do not show true sizes and right-angled corners are not correct. For manufacturing and fabrication purposes, these drawing methods can be confusing and are therefore unacceptable.

Consider the isometric drawing of a simple bracket shown in Figure 7.42. In order to present sufficient information to the engineer who will manufacture the bracket, three separate views are developed from the isometric drawing and placed in such a position that the relationship between them is clear.

Two such systems are in use in industry, namely 'first angle projection' and 'third angle projection; third angle projection is the more natural and the preferred mode.

These three views are referred to as 'elevations'. The three elevations are representations of the object when viewed directly in front and called the 'front elevation', viewed from the side and called the 'end elevation,' and viewed from above and called the 'plan'. The three views are placed as shown in Figure 7.42.

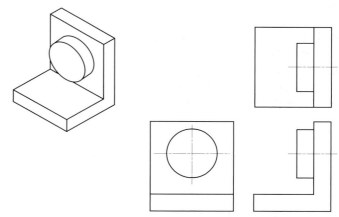

Fig 7.42

On all engineering drawings, the type of projection used (whether it be first angle or third angle) must be indicated on the drawing by using the correct British Standard symbol as shown below in Table 7.4.

You will find many examples of different drawing techniques and methods in the examples of coursework by students in Chapter 3 of this book, including many produced by CAD.

Table 7.4 British Standard symbols for first and third angle projections

Projection	Symbol
First angle	
Third angle	

Block diagrams

Block diagrams are widely used in engineering to give a graphical representation of concepts, processes and, often, organisational structures. Few rules apply to this type of diagram other than that it should be clear and that, if time is involved, then the diagram progresses either from left to right across the paper or from top to bottom.

Two typical block diagrams are shown in Figures 7.43 and 7.44. Figure 7.43 shows a 'concurrent' engineering process and, consequently, shows how the elements of the process would overlap along the time axis (in this case left to right) as they take place at the same time.

Figure 7.44 depicts a full product cycle and therefore does not show a time line. This particular diagram is used to indicate how and

where IT has been integrated into the production cycle.

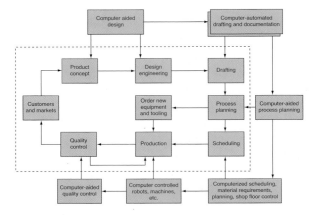

Fig 7.44

Flow diagrams

These are simply another form of block diagram that shows a sequence of events. They are frequently used when writing computer programs and when planning process operations where a 'yes/no' decision has to be made.

Flowcharts, like block diagrams with a time line, always flow either from top to bottom of the page (normally) or from left to right across the page. Arrows are often used to indicate the direction of flow of the diagram and to enhance clarity but they should not be overused. Any 'yes' outcome from a decision box follows the direction of the flow.

They are also used for diagnostic purposes and in fault-finding operations in the workplace where a strict set of instructions must be followed. Figure

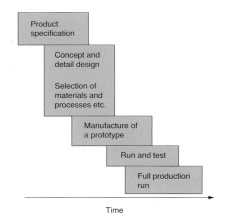

Fig 7.43

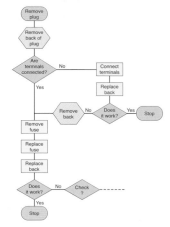

Fig 7.45

7.45 is a good example of simple flowchart that shows the process that is followed for checking a faulty cartridge fuse in a 13-amp plug top.

Schematic drawings and circuit diagrams

These diagrams are used to show the arrangement of components in electrical, electronic, hydraulic and pneumatic circuits and systems. They are used to indicate the relative points of interconnection of the components only. They do not show the positions of the components within the system as a whole.

The electronic circuit diagram in Figure 7.46 shows an example of this, where part a shows the actual position of the components on the face side of a printed circuit board, and part b shows the relative points of interconnection.

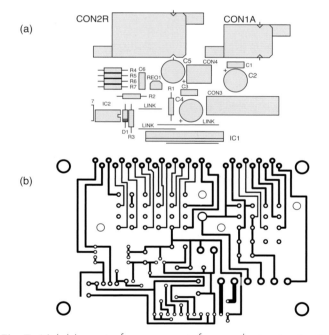

Fig 7.46 (a) Layout of components for an electronic circuit diagram; (b) The copper tracks on the reverse of the printed circuit board.

Pneumatic and hydraulic systems

The standard symbols used when drawing schematic diagrams for these systems are very similar so they are often grouped together. Valves, whether they control air or liquid, operate in a similar fashion, and so the same or very similar symbols are used for both.

❶ Valves X and Y are **three-port valves** (3PV), they are button operated with a spring return. The open arrow indicates that they exhaust to atmosphere. Pneumatic systems, unlike hydraulics, do not have to make a complete circuit, when the air has done its work it is allowed to escape. The valves are shown at rest.

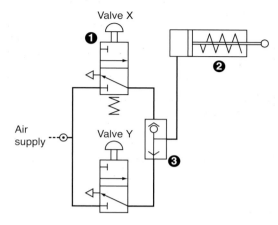

❷ The **single acting cylinder** (SAC) goes positive under operation and is then returned (to negative) by its own internal spring.

❸ The **shuttle valve's** function is illustrated by the symbol; a ball is pushed from end to end depending on the operation of the valves. The ball is pushed into a seat and acts as a cut-off valve so that the supply is directed towards the cylinder and not allowed to exhaust through the opposite valve.

Fig 7.47 A pneumatic system for operating a single acting cylinder from either one of two button-operated valves

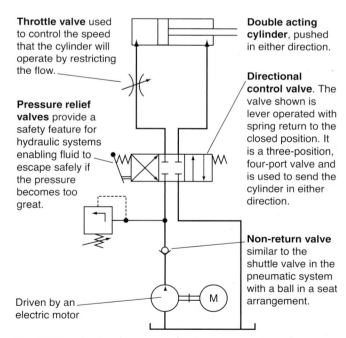

Throttle valve used to control the speed that the cylinder will operate by restricting the flow.

Pressure relief valves provide a safety feature for hydraulic systems enabling fluid to escape safely if the pressure becomes too great.

Driven by an electric motor

Double acting cylinder, pushed in either direction.

Directional control valve. The valve shown is lever operated with spring return to the closed position. It is a three-position, four-port valve and is used to send the cylinder in either direction.

Non-return valve similar to the shuttle valve in the pneumatic system with a ball in a seat arrangement.

Fig 7.48 A hydraulic system for lever operation of a double acting cylinder

A comprehensive range of standard symbols for electrical/electronic and pneumatic/hydraulic circuits are summarised in PP7308.

Learning outcomes

By the end of this section you should have developed a knowledge and understanding of:

- The selection and use of common ferrous and non-ferrous metals
- Mild steel, high-carbon steels, copper and alloys of brass
- Aluminium, tin, zinc, stainless steels and cast iron
- The properties of metal and metal products – strength, toughness, ductility and malleability, weight, durability, and thermal and electrical conductivity – in terms of suitability for specific consumer products
- The selection and use of common thermoplastics and thermosetting plastics – polystyrene, polyethylene, acrylic, polypropylene, PVC, ABS, PET, phenol resins, phenol formaldehyde, urea formaldehyde, melamine formaldehyde, epoxy resins
- The selection and use of common composite materials – Kevlar®, carbon fibre, GRP
- The properties of plastics – hardness, brittleness, tensile strength, plasticity, compressive strength, sheer strength, strength to weight ratio, chemical resistance, elasticity, stiffness and impact resistance
- Ceramics and ceramic composites
- Up-to-date developments of new materials and their potential applications (see Chapter 5 Core Knowledge)

Introduction

This section covers the selection, properties and use of common engineering materials. This subject is complex. To learn all of the facts you need to know without applying them to real products would be a very boring process!

You will need to identify a selection of products that may be familiar to you which contain a wide variety of materials suitable for analysis.

Fig 7.49 It's even got disc brakes!

The well-designed consumer product in Figure 7.49 contains many different materials. The materials must be strong, and the buggy sturdy so that it can be used in all terrains and in all weathers. The main purpose of the product is to protect and transport a small baby. The main frame needs to be light and compact when folded to store in a car boot. The body of the buggy doubles as a car seat and must protect the baby from any potential impact.

Material selection charts

Later in this section you will see in detail some of the materials the manufacturer selected for use in the child's buggy. You might however want to look at other products to identify the materials selected or want to select materials of your own for project work.

The selection of materials is a complex matter and quite often you will need to compare different issues to select one that is suitable for a given application. The older, more traditional method of doing this was to consult books on materials science and metallurgy, and look in detail at tables and graphs that showed complicated data and calculations. This subject cannot be addressed without some consideration of technical data.

A useful book if you need more information is *Materials for the Engineering Technician* by RA Higgins, (Butterworth): this has been a standard text book for many years.

Things have been made much easier however with the introduction of materials selection charts. OCR were on the steering group for this innovative project and have TEP permission to reproduce the charts in this book. The full range of charts can be found in the TEP book *Materials Selection and Processing – Product Analysis Tutorial and Case Studies*. The charts are available for you to use: you can see them in CD-ROM format or download them from the website (www.tep.org.uk) to use in your project work.

You will not be asked to draw the charts in an examination, but it is possible you could be asked to draw conclusions from a chart or select a suitable material or an alternative material for a given application. You might also need to cross reference different charts where a combination

Fig 7.50

of different properties is required. The main aim of this chapter is to enable you to analyse products and identify the materials used, and to suggest suitable materials for specific applications.

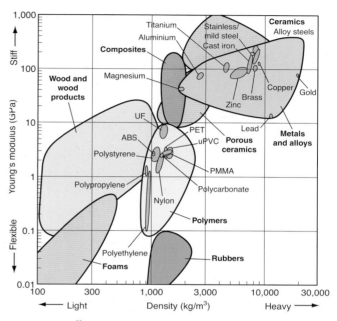

Fig 7.51 Stiffness vs Density

One of the main selling points for the Jane Slalom Pro (Figure 7.50) is its light weight. The chassis frame only weighs 8 kg. This is achieved by using aluminium extruded into an oval form. This gives a very light yet stiff construction. If you look at the material selection chart in Figure 7.51 you will see that aluminium is in the top left-hand section, which gives a high combination of lightness and stiffness.

Selection charts plot combinations of two properties. You need to notice that the scales are 'logarithmic' so a very small difference in position represents a very large change in properties. Similar groups of materials are shown in 'bubbles' with individual materials identified within.

The chart shows that an alternative group of material for the frame would be 'composites'. These look both lighter and stiffer than aluminium. Manufacturers need to consider performance and cost however. The next selection chart (Figure 7.53) shows the combination of strength and cost, and illustrates why composites such as carbon fibre reinforced plastic (CFRP) are only used for the most expensive of frames for racing bikes.

Fig 7.52 Oval frame concept chassis

Strength versus cost

What is the best choice for a pushchair or bike frame?

Steel is the cheapest structural material and would be strong enough for the frame. It is used for some less expensive pushchair and bicycle frames. It is much heavier than either aluminium or composites. Composites like CFRP would be a good choice for strength but would be very expensive for all but the most exclusive racing bike frames where cost is not a problem. Some bike frames are die cast from magnesium – another expensive option.

- Ceramics look strong but they are only strong in compression
- CFRP is as strong as many metals but very expensive
- Other metals could be used for frames but titanium would be expensive!
- Wood was used for the first bike frames before it became easy to process metals

Aluminium provides the best choice for optimum strength/cost.

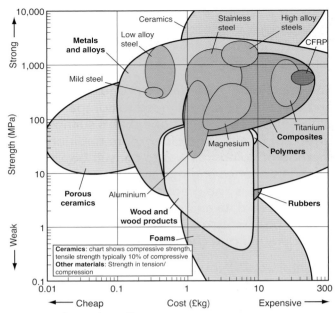

Fig 7.53 Strength vs Cost

Although manufacturers may refer to frames as being made from 100% aluminium, in practice most products will contain an aluminium alloy. Tubes, panels and 'hollow ware' are typically made from 99 per cent aluminium with 0.1 per cent manganese (Mn) and 0.5 per cent silicon (Si). This can give strength to weight ratio five times that of pure aluminium.

Duralumin, another alloy, contains 0.4 per cent copper (Cu): it is soft when heat treated then slowly age-hardens when cool. This property makes it useful for the manufacture of rivets. The tubes would be manufactured using the metal extrusion process and later anodised to protect against oxidisation.

Aluminium also comes in other forms: the brake lever could be made from die cast aluminium containing 10 per cent silicon. There is a reason for this, as a 10 per cent silicon alloy will change from a liquid to a solid very quickly and will not shrink.

Fig 7.54 Bicycle brake

Fig 7.55 Anodised aluminium key ring

Fig 7.56 Aluminium water bottle

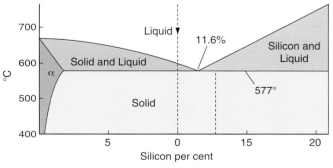

Fig 7.57 Aluminium–silicon thermal equilibrium diagram

The aluminium–silicon thermal equilibrium diagram is shown in Figure 7.57. Don't worry, you won't be asked to draw it! You do need to understand what it shows however: an aluminium alloy with 10 per cent silicon (shown by the arrow) will change from a liquid to a solid very quickly on cooling. This helps to speed up the die casting process.

When do we have to use steel?

Let's have a look at the disc brake and cable, and what about the footrest? What's the front bearing made from?

Looking at the disc brake (Figure 7.58), it appears shiny and hard. Positioned low down at the front of the pushchair it will be exposed to mud, dirt and salt. The disc will be subject to frictional forces and must have a hard surface so that it will not be worn down. Stainless steel would be appropriate for this application. It is much heavier than aluminium. The disc has many holes, both to improve performance and reduce weight. Stainless steel is a medium- to high-carbon steel alloyed with chromium (Cr); the chromium oxide film that forms on the surface behaves much like the oxide film that forms on aluminium to prevent corrosion. Chromium is also added to tool steel to prevent corrosion. Stainless steel is a ferrous metal: it contains iron.

The brake cable or 'Bowden cable' (Figure 7.60) is manufactured from thin strands of steel wire and hardened and tempered to give it the appropriate properties to act in both directions. The chart in Figure 7.63 will show you the carbon content.

The footrest (Figures 7.61 and 7.62) will obviously take a lot of knocks from little feet and is positioned in an exposed position at the front of the pushchair. It could be made from a variety of materials; pressed mild steel would be appropriate, with the strength coming from the profile of the pressing. As with the disc, weight is reduced with a number of holes in the top plate. Mild steel needs a protective finish.

The ball bearings in the 'monofork swivel wheel' would be made from high-carbon alloy steel, typically 1.0 per cent carbon, 0.45 per cent manganese and 1.4 per cent chromium, giving excellent wear resistant properties combined with excellent corrosion resistance.

The spanners (Figure 7.59) would be made from an alloy tool steel containing 0.45 per cent carbon, 0.9 per cent manganese and 1.0 per cent chromium, giving a good combination of strength and corrosion resistance. A higher carbon content would make them too brittle. Tool steels need to be 'heat treated'.

Fig 7.61

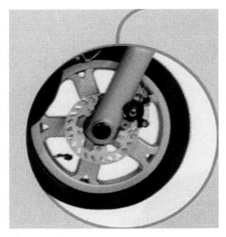

Fig 7.58

Fig 7.62

Fig 7.59

Key Points

'Ferrous' means iron; steel (an alloy of iron and carbon) is classified by its carbon content:

- Mild steel: up to 0.25%
- Medium carbon: 0.25–0.5%
- High carbon: 0.5–1.5%

Mild steel is tough and ductile but cannot be hardened (except by 'case hardening' – dipping red hot into a carbon-rich compound).

Key Points

Heat treatment: hardening and tempering
High-carbon steels can be hardened by quenching them in brine, water or oil depending on the hardness required and then 'tempered' by re-heating them to reduce brittleness. Ball bearings would be quenched in oil from 810°C and then tempered at 150°C. The socket set would be quenched in oil at 860°C then tempered at 550–700°C, making it less brittle in use.

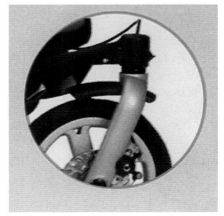

Fig 7.60

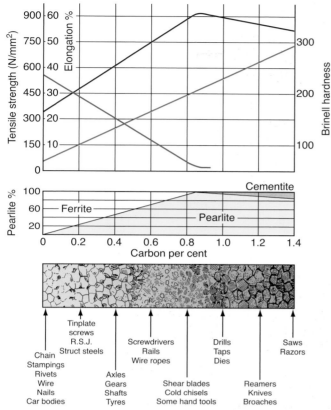

Fig 7.63 Properties of steels with different carbon content

Figure 7.63 shows the properties for steels with different carbon content. On the left side steels are known as 'mild steels', in the mid range 'medium carbon steels' and on the right side 'high carbon steels', also known as 'tool steel'. The more carbon, the harder the steel becomes. The tensile strength goes up but the ability to elongate goes down. Medium and high-carbon steels can be heat treated to improve hardness and mechanical properties.

The term ferrous comes from the name ferrite (iron) which is quite soft; the more carbon that is added causes the steel to become harder. The black and white 'pearlite' structure gives medium-carbon steels, and the mixture of pearlite and cementite produces steel that is really hard and suitable for razors and saws.

Other metals

You won't see them on any self-respecting pushchair; but these metals have their uses!

Cast iron

It was used by the Victorians for lamp posts and railings, but we still use it today for machine beds, engine castings and 'marking out equipment'. Why?

- Cast iron is cheap but very heavy
- It is very rigid and strong in compression
- It casts well and can be easily machined

Cast iron can contain the following elements:

- Carbon 3.0–4.0%
- Manganese 0.5–1.0%
- Phosphorus 1.0%
- Silicon 1.0–3.0%
- Sulphur 0.1%

Why do they slide? Cast irons are particularly useful because the carbon/graphite flakes trapped in the structure make them 'self lubricating' (see Figure 7.65). These flakes are sometimes spheroidal in shape.

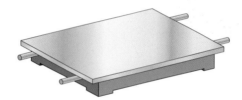

Fig 7.64 Cast iron surface plate

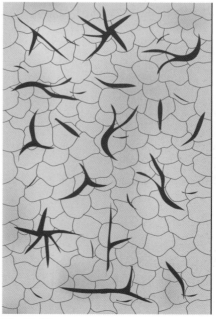

Fig 7.65

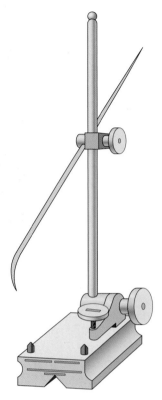

Fig 7.66 Cast iron surface gauge – made to slide!

Zinc

The 'altruistic' metal – it sacrifices itself for the good of others! The old watering can in Figure 7.67 has seen better days. Without its zinc coat it would have disappeared long ago! This process is only effective if it is not scratched.

Fig 7.67

Galvanising

The watering can was protected by the process of 'galvanising', where a mild steel product is dipped into hot zinc, which forms a protective coating. Zinc can also protect steel by 'sacrificing' itself. Attached to steel plates it will corrode itself

leaving the steel relatively intact. In the example in Figure 7.68, electrolytic corrosion would take place due to the proximity of steel and bronze in seawater, which acts as an electrolyte. This process is called 'sacrificial corrosion'.

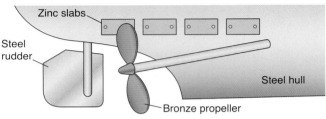

Fig 7.68

Key Point

Zinc and tin are non-ferrous metals.

Tin

This metal is rarely used just by itself. Looking at the 'tin' cans in Figure 7.69: are they really made from tin? What do you think solder is made from?

Fig 7.69

Tin cans are made from mild steel, but they wear a very thin coat of tin. This protects the inside from food contamination and the outside from corrosion.

Like galvanising, mild steel is dipped in a bath of molten tin; squeeze rollers then remove any excess. Most food tins are made from tin-coated mild steel.

The thermal equilibrium diagram for tin/lead in Figure 7.70 shows how this alloy cools down when different percentages of tin and lead are

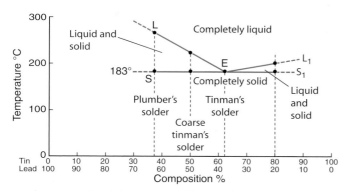

Fig 7.70 Tin–lead thermal equilibrium diagram

mixed. Plumbers solder contains 62 per cent tin and 38 per cent lead: this is because plumbers need to move 'mushy' solder around to 'wipe' a joint. If you are soldering small components with a soldering iron, the last thing you want is to be hanging around waiting for the solder to set; so a 60 per cent tin, 40 per cent lead mix would be better.

Copper and brass

These are two non-ferrous metals with good electrical conductivity.

Copper in its pure form is soft and can be beaten into shape. In the past many copper vessels were made on site by skilled coppersmiths (Figure 7.73). If you have a hot water tank in your house the chances are that it is made from copper. It has excellent corrosion resistance and thermal conductivity, and it is very malleable and ductile.

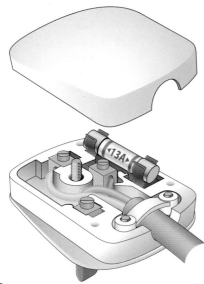

Fig 7.71

Its very high conductivity makes it suitable for electrical applications.

Brass is a harder alloy of copper and zinc.

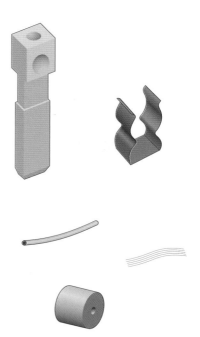

Fig 7.72 Close up of plug parts

Fig 7.73

Selection of material for the pins of a 13A plug (Figures 7.74 and 7.75)

This is a complicated business and a manufacturer would need to look at more than one materials selection chart to form a judgement. As well as electrical conductivity there are considerations of strength/toughness.

All of the best conductors lie in the bottom right-hand corner of the chart in Figure 7.74. Gold would make a good conductor but it would make it a very expensive plug!

Copper looks better than brass, so why isn't it used? The answer lies in Figure 7.75: brass has higher strength and better wear resistance.

Copper is fine for the fuse clips, the fuse ends and the wire: high conductivity combined with less wear.

Brass (60 per cent copper and 40 percent zinc) is used for the pins as it has relatively low cost, good conductivity and good wear resistance. Brass can also be extruded.

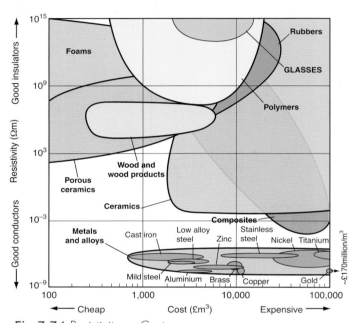

Fig 7.74 Resistivity vs Cost

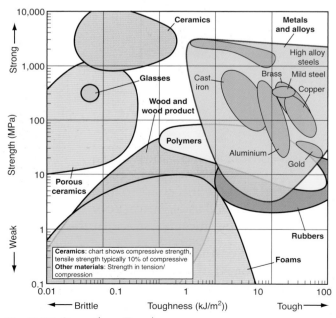

Fig 7.75 Strength vs Toughness

There are two other alloys you need to know about:

Cartridge brass & free-cutting brass

Cartridge brass

Another brass alloy is cartridge brass (70 per cent copper and 30 per cent zinc). It has to be 'deep drawn' so must have good elongation properties. The graph in Figure 7.77 shows why this proportion of copper and zinc is best.

Fig 7.76 Cartridge brass

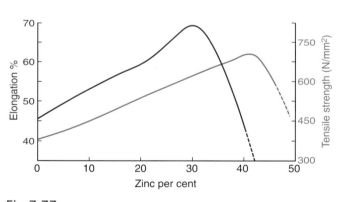

Fig 7.77

Free-cutting brass

Another very useful alloy of 58 per cent copper, 39 per cent zinc and 3 per cent lead; it helps chips of brass break away during high-speed machining processes.

Thermosetting plastics

Once these plastics have been shaped and hardened, they can not be reshaped or softened by heating. There are five main thermosetting materials in common use for products that you need to know about – they all have high safe working temperatures.

Table 7.5 Thermosetting plastics

Thermosetting plastic	Abbreviation	Safe working temperature	Characteristics	Uses
Urea formaldehyde	UF	80°C	Opaque, light in colour	Electrical fittings
Phenol formaldehyde	PF	120°C	Opaque, dark in colour	Domestic iron and saucepan handles
Melamine formaldehyde	MF	130°C	Opaque, multicoloured	Cups, plates, buttons, kitchen worktop surface
Polyester resin		95°C	Clear liquid resin	With GRP and CFRP; boats and vehicles
Epoxy resin – High melting point		200°C	Two resins in tubes	'Potting' of electronic circuits
Epoxy resin – Low melting point		80°C	One tube has activator	Adhesives

Urea formaldehyde

A light coloured thermosetting plastic. Where you find good conductors there will usually be good insulators nearby (Figure 7.78) to hold the components and protect users from electric shocks!

Urea formaldehyde (UF) is a thermosetting plastic that is shown in the top left hand section of the selection chart in Figure 7.79. It combines good insulation properties with relatively low cost. It can also be easily formed into complex shapes by compression moulding.

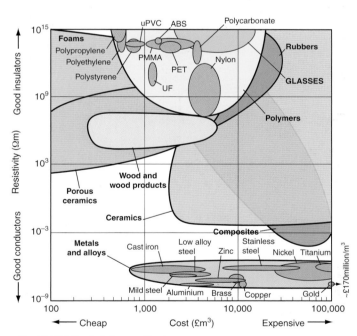

Fig 7.79 Resistivity vs Cost

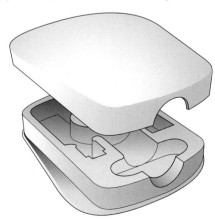

Fig 7.78

Phenol formaldehyde

A dark coloured thermosetting plastic (Figure 7.80). Figure 7.81 shows three points where a 'phenol' molecule can link to other molecules. Figure 7.82 shows where links are formed with 'formaldehyde' molecules. When heat is applied in compression moulding, 'cross linking' occurs (Figure 7.83) and the material sets hard.

Fig 7.80

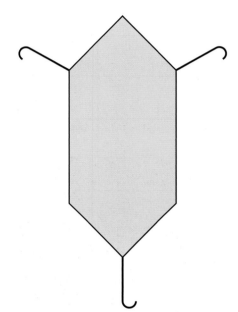

Fig 7.81

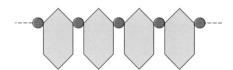

Fig 7.82

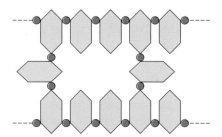

Fig 7.83

Phenol formaldehyde and urea formaldehyde are the main thermosetting plastics used in small consumer products. They have very similar properties. PF has a higher safe working temperature and is used in applications exposed to heat.

When urea reacts with formaldehyde in the first instance a 'syrup' is formed; this is mixed with filler and allowed to dry out. The powder formed is the raw material used in the compression moulding process. In the syrup stage the material can impregnate paper, cloth or cardboard and can also be used as a coating on metals on washing machines or fridges. In all cases the application of heat sets the material permanently.

Epoxy resins

Epoxy resins act in a similar way – a cross-linking agent is mixed from another tube. Epoxy resins are found in a popular two-part glue, they set hard and are particularly useful for joining metal where products are subjected to heat. Polyester resins are used in glass-reinforced plastic (GRP) and CFRP composites.

Composite materials

The choice for high performance – but can you afford them? Composite materials are a mixture of two or more materials. Particles, fibres or layers of one material are used to strengthen another. The resultant material has a better combination of properties than either of the separate materials.

> **Key Point**
>
> Use their full names when talking about composite materials. Don't call them fibre glass and carbon fibre!

Carbon fibre reinforced plastic (CFRP) and glass reinforced plastic (GRP)

The main bulk material in both of these materials is polyester resin. This material is a thermosetting plastic in liquid form. It sets when an activator is added. Polyester resin is quite strong in compression but weak in tension.

CFRP is much stronger than many metals; but takes longer to form. Figure 7.84 shows CFRP to be much stronger than mild steel and aluminium. It is very expensive and is only used where high performance is essential and cost is not an issue.

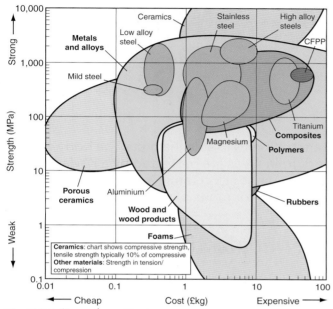

Fig 7.84 Strength vs Cost

Fig 7.85 Composite materials are used for high performance bicycles

Carbon fibres have an incredibly high tensile strength – 2900 N/mm^2 – but are of little structural use, resembling silk strands. Mixed with polyester they become strong and rigid.

Chopped strand glass mat (a ceramic material) can be mixed with polyester resin (Figure 7.86) to form a strong and durable composite used for car and boat bodies (GRP).

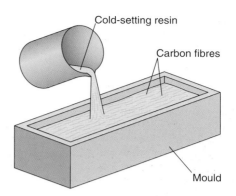

Fig 7.86

Ceramics and ceramic composites: cermets

The structure shown in Figure 7.87 is a composite of tungsten carbide and cobalt cermet. The particles of tungsten are white and the cobalt black. Tungsten carbide is used for high-speed cutting tools. Tool inserts are supported firmly in a special tool bit holder.

Fig 7.87

Ceramics and cermets have very high compressive strength but much lower tensile strength (about 10 per cent of compressive value).

Alumina, chemical name aluminium oxide, is a ceramic material used in fuses, sparking plugs, cutting tools, and crucible and foundry instrument casings (Figures 7.88 and 7.89). Its melting point is 2050°C!

Fig 7.88

Fig 7.89

Fig 7.90

penetration than steel helmets. The body, neck and groin areas are protected by a Kevlar® vest with 25 mm thick ceramic inserts. Look in Chapter 13 for more information on Kevlar®.

Kevlar® and ceramic insert materials

Peacekeeping forces need to protect themselves from mines, bomb fragments and bullets. The modern material Kevlar® is used in helmet manufacture to protect the head from injury.

Kevlar® is up to 40 per cent more resistant to

Thermoplastics

This group of plastics deform whenever heated. They lose their rigidity and can be remoulded many times. This makes them suitable for vacuum forming, injection/blow moulding and

Table 7.6 Thermoplastics

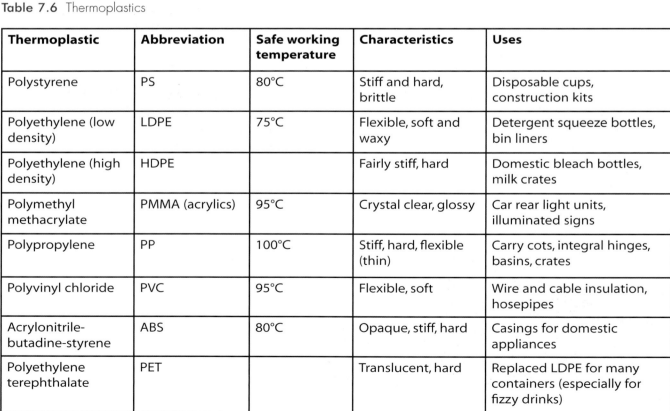

Thermoplastic	Abbreviation	Safe working temperature	Characteristics	Uses
Polystyrene	PS	80°C	Stiff and hard, brittle	Disposable cups, construction kits
Polyethylene (low density)	LDPE	75°C	Flexible, soft and waxy	Detergent squeeze bottles, bin liners
Polyethylene (high density)	HDPE		Fairly stiff, hard	Domestic bleach bottles, milk crates
Polymethyl methacrylate	PMMA (acrylics)	95°C	Crystal clear, glossy	Car rear light units, illuminated signs
Polypropylene	PP	100°C	Stiff, hard, flexible (thin)	Carry cots, integral hinges, basins, crates
Polyvinyl chloride	PVC	95°C	Flexible, soft	Wire and cable insulation, hosepipes
Acrylonitrile-butadiene-styrene	ABS	80°C	Opaque, stiff, hard	Casings for domestic appliances
Polyethylene terephthalate	PET		Translucent, hard	Replaced LDPE for many containers (especially for fizzy drinks)

recycling. You need to know about the following thermoplastics; they generally have a lower safe working temperature than thermosets.

Why use a thermoplastic instead of a thermosetting plastic?

ABS is used when products are subject to impact and weight isn't a critical factor.

If you look at the chart in Figure 7.91 you will see that ABS combines good qualities of strength and toughness. UF and other thermosetting plastics are not as tough as ABS and other thermoplastics. The chart shows UF on the left hand 'brittle' side of the chart. In applications where products might be subject to impact, thermosets are rarely used. ABS would be the natural choice for a pre-fitted, injection-moulded plug (Figure 7.92) for a vacuum cleaner – where the plug is usually dragged along behind!

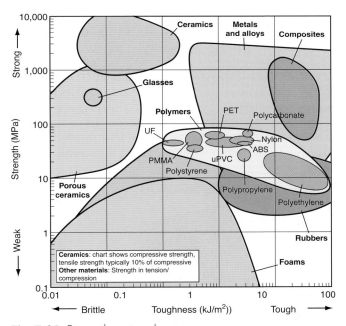

Fig 7.91 Strength vs toughness

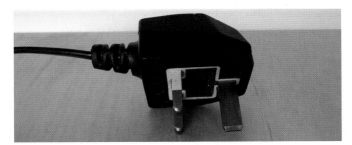

Fig 7.92

Fig 7.93

ABS has been used for the hard shell of the backpack shown in Figure 7.93, designed to carry a laptop in hazardous country where impact due to rugged terrain or falling might be a problem. A product like this would be useful for scientists or journalists during expeditions. ABS is ideal for this purpose: it can be injection moulded, and additional strength is built in through its monocoque construction. Weight is not a problem as only a laptop and associated accessories are being carried.

Polypropylene

A small baby is even more precious than a laptop! So why is this a suitable thermoplastic to protect a baby in a carrycot or car seat? Is any additional protection needed?

Fig 7.94

Additional protection is provided using expanded polystyrene on the headrest, back and sides. This is also very light and absorbs energy on impact.

Fig 7.97

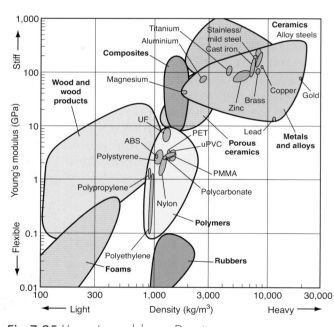

Fig 7.95 Young's modulus vs Density

Car seats and carrycots have a primary purpose of protecting a baby or small child against head-on impacts and side impacts when travelling in cars. Their secondary purpose is to keep the child comfortable during everyday activities.

Another very important point is that they have to be convenient and light to carry. Nobody is going to buy a product which is too heavy to lift.

The two main requirements are therefore toughness combined with being lightweight.

Fig 7.98

The selection chart in Figure 7.91 shows polypropylene to be as tough as ABS. The selection chart in Figure 7.95 shows polypropylene to be one of the lightest polymers. Remember the scales are logarithmic – a small distance on the scale represents a very large difference in value.

Polypropylene also has the ability to absorb energy on impact. This combination gives the

Fig 7.96

best strength to weight ration and makes polypropylene the most suitable polymer for this application.

Polypropylene is also used where 'integral hinges' are required. An integral hinge is a feature injection moulded into components making a mechanical joint unnecessary. If you look at the top of a 'one-piece' washing up detergent bottles you might see the recycling mark with the letters PP.

Fig 7.99

The body of the cap and the pop-on top are moulded in one piece. The mould is made with a very small gap between the two parts. The polypropylene squeezes through the small space and the polymer arranges itself in lines across the gap. An incredibly flexible hinge is formed.

POLYMERS

Polyethylene, polyvinyl chloride, polystyrene . . . why do they all start with 'poly'? What on earth is a 'mer'?

'Poly' simply means many, and 'mer' is the name of the simple single units from which thermoplastics are made. Polymer is therefore a name for many of these units joined together.

They are usually joined together in long chains. These chains can become entangled during a chemical reaction and a solid is formed that is both strong and rigid. When the solid is heated, the forces of attraction between the molecules decreases, making the material less rigid. This makes it easy to mould and is the source of the name 'thermo' plastic.

Long-chain molecules behave in an entirely different way to the cross-linked structure found in thermosetting plastics. The process of forming these long-chain molecules is called 'polymerisation'.

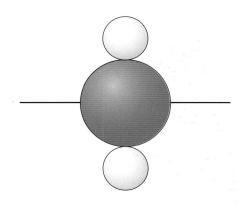

Fig 7.100

Fig 7.101

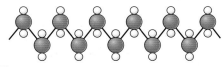

Fig 7.102

The single 'mer' of 'polythene' is shown in Figure 7.100. The gas ethylene is 'polymerised' to form polyethylene (Figures 7.101 and 7.102). Individual carbon atoms are shown in red, with two hydrogen atoms attached.

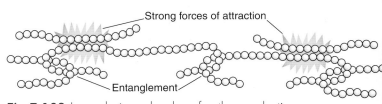

Fig 7.103 Long-chain molecules of a thermoplastic are attracted to each other and become entangled

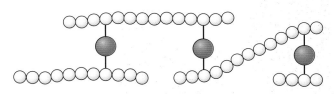

Atoms forming strong chemical bonds

Fig 7.104 Do not confuse this with the permanent cross-linking that occurs in thermosets

Polyethylene (PE)

This is a very widely used plastic and is available in two forms:

- Low-density polyethylene (LDPE)
- High-density polyethylene (HDPE)

LDPE is fairly flexible and has a soft, waxy feel. It was almost exclusively used for detergent bottles before the development of PET. It is still used for fertiliser bags, bin liners and the insulation for co-axial television aerial leads (this is the only application it monopolises from PVC).

HDPE is fairly stiff and hard and is used for many domestic bleach bottles. PE has excellent chemical resistance.

Fig 7.106 PVC

Fig 7.107

Fig 7.105 PE

Polyvinyl chloride (PVC)

PVC is available in two forms:

- Rigid/unplasticised
- Plasticised

In its plasticised form it is widely used for hosepipes and wire/cable insulation (excluding aerial cables). In its unplasticised form it is used for rainwater guttering and down pipes – replacing cast iron. Before the introduction of PET it was used for clear soft drinks bottles, transparent shampoo bottles and liners for chocolate and biscuit boxes. PVC has good chemical resistance.

Polystyrene (PS)

This is a very common plastic and is available in two forms:

- Polystyrene (PS)
- Expanded polystyrene (EPS)

It is stiff, hard and, in its PS form, is used for many food containers such as yoghurt cartons. The most common applications are in throwaway 'plastic' cups, and construction and modelling kits.

In its EPS form it is used as expanded light foam to protect or package delicate items and to aid thermal insulation in cups. Polystyrene is a low-cost plastic with fairly good chemical resistance.

Fig 7.108 PS

Polymethyl methacrylate (acrylic, PMMA)

Acrylics, sometimes known by their trade name Perspex, are a very important group of plastics. PMMA, the abbreviation of the full, correct name is made by the polymerisation of methyl methacrylate.

Acrylic was developed in the late 1930s for use in aircraft windscreens. It is much tougher than glass and has the advantage that it can be easily moulded. It was used for scooter windscreens all through the swinging sixties and was only replaced for machine guards when tougher polycarbonate became available.

One of the main properties of acrylics is the ability to transmit more than 90 per cent of daylight; this makes acrylic sheet valuable for roofing panels (it can eventually degrade in sunlight after many years however).

Due to excellent optical properties it is used extensively today for motor car rear light units. The ability to be multicoloured makes it particularly useful for this purpose. Another very visual use is for illuminated advertisement signs.

Acrylics have a fairly good chemical resistance and can be used for baths, sinks, dentures and toilet articles.

Fig 7.110 PET

The 'fizzy drinks' market alone is huge, and you will rarely see any other plastic used for the very large detergent market where clear bottles advertise the contents. Why has PET replaced these other plastics? The answer again lies in the material selection chart shown in Figure 7.112.

Fig 7.109 PMMA

Polyethylene terephthalate (PET)

The natural choice for all clear and fizzy drinks bottles! PET has been left until last as it is a relatively new development and its unique qualities have led it to replace many of the functions performed by other polymers for many years. PET has taken over from polyethylene and PVC for the following applications:

- 'Fizzy' drinks bottles
- Clear washing up detergent bottles
- Milk containers
- Chocolate box liners.

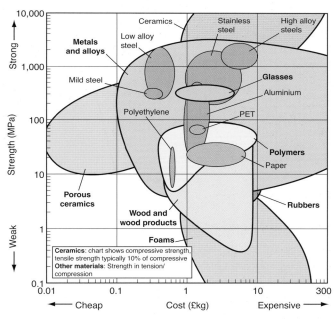

Fig 7.111

PET is placed at the top of the polymer bubble showing it to be one of the strongest plastics. It is more expensive than PE but its much larger strength means that much less of it is needed to withstand the gas pressure in fizzy drinks and huge cost reductions can be made in the mass market. PET bottles are thin and could be even thinner if the gas didn't seep through very thin sections.

PE bottles are used for still drinks and opaque detergent bottles although HDPE is now used because of the increased strength needed for the density of packing in modern supermarkets. PET was unknown in the 1980s but it is now one of the most-used plastics.

Mechanical properties of plastics

The material selection charts will give you most of the information you need to know about the comparative mechanical properties of the plastics you need to study.

In addition to the selection charts you need to know how plastics perform under tension and understand about elasticity and plasticity. Figure 7.112 shows comparative information on different groups of plastics on a stress–strain graph.

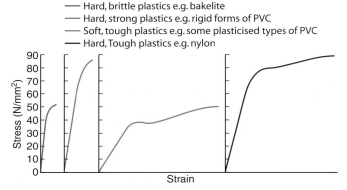

Fig 7.112

These are nearly, but not quite, the same as force–extension diagrams used when metals are tested. In general terms, the more the plastic elongates before breaking the softer the plastic. Very hard plastics have steep graphs then suddenly break without elongating. Most plastics, as their name suggests, behave in a plastic fashion – they elongate for a long time before breaking. This property is called plasticity. This property also makes thermoplastics suitable for a variety of moulding processes.

Elasticity is the ability to return to an original state when a load is removed. This condition only usually occurs during 'the straight line' part of the graph. After this point plastic deformation occurs, which is non-reversible.

Plastics are mainly known for their ability to withstand corrosion and impact. Most thermosetting plastics are brittle in tension but stronger in compression. Care is needed to reinforce designs where this might be a problem (the body of the plug is a good example of this). Most thermoplastics cut very easily and are therefore susceptible to shear (slicing) forces.

Key Terms

Hardness – resistance to deformation
Brittleness – little deformation before fracture
Plasticity – permanent deformation under loading
Tensile strength – resistance to deformation under tension
Compressive strength – resistance to deformation under compression
Sheer strength – resistance to deformation under a shear stress
Elasticity – temporary deformation and return to original shape
Stiffness – resistance to deformation or extension
Impact resistance – ability to withstand impact without breaking or shattering

Mechanical properties of ferrous and non–ferrous metals

Figures 7.113 and 7.114 show force–extension graphs for carbon steel and a non-ferrous alloy. You can see by the shape of the graphs the properties associated with that material.

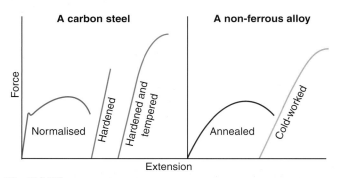

Fig 7.113

Normalised steel – which has been softened by heating and cooling in air – shows the characteristics of the graph in Figure 7.114. Up to point A, which is called the elastic limit, the steel will spring back to its original shape if the force is removed. After point Y – the 'yield point' – plastic deformation takes place and the material stretches until it breaks at B.

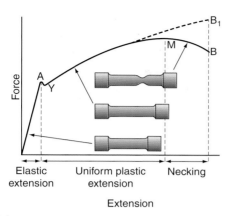

Fig 7.114

Hardened tool steel does not elongate much before it breaks. Hardened and tempered, it becomes much less brittle. A non-ferrous alloy annealed (softened by heat and left to cool slowly) behaves in a plastic fashion and elongates considerably before it breaks. Cold working – by hammering or rolling – will cause an alloy to be much harder.

You can compare the ferrous and non-ferrous graphs on the same axis. The plastics graphs are shown at a different scale.

Developments in new materials and their potential applications

This material is covered in Chapter 5 Core knowledge.

Section D	Engineering components

Learning outcomes

By the end of this section you should have developed a knowledge and understanding of:

- Mechanical components: nuts, bolts, screws, springs, rivets, pins, keys, drive mechanisms, knock-down fittings

Introduction

Engineered products, especially those made in the automotive sector (motor vehicles), the aeronautical sector (aircraft) and products made for the domestic market, are assembled using a variety of methods. Some are permanent (welding for example) and some are non-permanent.

Mechanical components

If access is needed for servicing or repair purposes, or if components need to be replaced, then a range of simple mechanical fixing components including some or all of the following, would be used:

- Nuts and bolts and machine screws
- Wood screws and self-tapping screws
- Studs and grub screws
- Rivets
- Springs and spring washers
- Keys
- Knock-down fittings

For most nuts and bolts the International Organization for Standardisation (ISO) metric screw thread system is now used in most countries for general engineering purposes and has virtually replaced the old British imperial screw thread sizes, although they can still be found on older manufactured products and replacement parts.

These thread patterns include:

- British Standard Whitworth Thread (BSW)
- British Standard Fine Thread (BSF)
- Unified Course Thread (UNC)
- Unified Fine Thread (UNF)

Together with the ISO metric system, there is also a range of small, fine thread sizes that are used extensively on electrical equipment and instrumentation. These are referred to as the 'British Association' or 'BA' thread.

Nuts and bolts

For many engineered products, this is the most common method of securing two components together.

A bolt, usually with a hexagonal head, comprises of a fixed length of hardened round bar – the shank – and threaded for only part of its total length (as opposed to a set screw, which would be threaded for the entire length of the shank).

Thread sizes are often specified on engineering drawings using a standard notation. For example, a simple metric bolt together with a hexagonal nut may be described as follows: 'Steel, Hex Hd Bolt – M8 × 1.25 × 50' or 'Steel, Hex, Nut – M8'.

- Hex Hd – indicates a hexagon-shaped head
- M – indicates an ISO metric thread pattern
- 8 – indicates the diameter of the bolt in mm
- 1.25 – indicates a thread pitch of 1.25 mm
- 50 – indicates that the bolt has a length of 50 mm (from the underside of the head)

The diameter of a bolt chosen for a specific purpose depends on the load that the bolt would have to carry. Generally, bolts are designed so that, in tension, there would be an equal chance that, on failure of the bolt:

- The head would shear off, or
- The thread would be stripped

When fixing two component parts together using a bolt and a nut, it is usual for the unthreaded part of the shank of the bolt to extend through the face of the joint so that the threaded part would not be subject to a shearing load (see Figure 7.115).

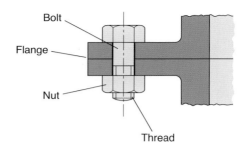

Fig 7.115 Simple flange secured with a nut and bolt

Although it is not clearly shown in Figure 7.115, it is usual practice to place a flat washer between the face of the component and the nut. Additionally, washers are sometimes also placed under the bolt head in order to spread some of the load and to prevent damage to the surfaces of the components being joined.

Machine screws

These are basically the same shape as a bolt but have the thread extended from the tip of the screw to the underside of the hexagonal head.

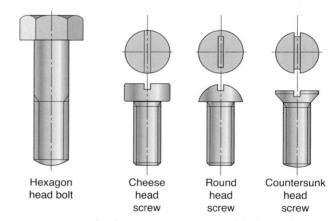

Fig 7.116 Standard BS representations of bolts and machine screws

Nuts

There is sometimes a tendency for nuts to loosen when the components that they secure are subjected to vibration. In order to overcome this there are a wide range of special nuts and locking devices. These are loosely divided into two categories: friction locking devices and positive locking devices.

Friction locking devices

These include lock nuts, various types of spring washers and special friction nuts with nylon or fibre inserts.

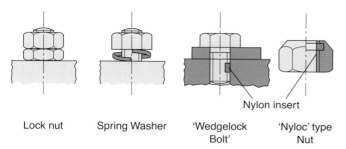

Lock nut   Spring Washer   'Wedgelock Bolt'   'Nyloc' type Nut

Nylon insert

Fig 7.117 Friction locking devices

Positive locking devices

These are used where the loosening or loss of a nut or bolt could cause serious damage. They are used in conjunction with a split pin or a locking plate.

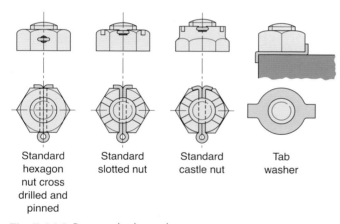

Standard hexagon nut cross drilled and pinned   Standard slotted nut   Standard castle nut   Tab washer

Fig 7.118 Positive locking devices

It is essential that, when any of the above devices are removed during a servicing or maintenance programme, split pins and tab washers are replaced with new ones on reassembly.

Split pins (cotter pins)

These are simple metal fasteners comprised of two tines that are bent during installation (see Figure 7.119). These pins are typically made of a relatively soft steel wire with a semi-circular cross section, making them easy to install and remove. They are available in a variety of sizes and types.

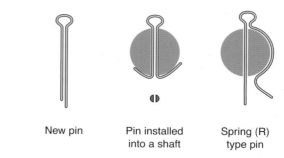

New pin   Pin installed into a shaft   Spring (R) type pin

Fig 7.119 Split/cotter pins

Spring-type cotter pins, sometimes known as R-pins (due to their shape), are also available. These are not designed to be permanently bent. With this type of design, only one section of the pin passes through the shaft to be secured, the other section being curved to wrap around the outside of the shaft. This type of pin is usually made of a round wire of a harder metal than is appropriate for traditional cotter pins. Split/cotter pins should not be used where they are likely to be subjected to strong shear forces.

Also, as mentioned previously, it is advisable to always replace them after a maintenance operation rather than reuse them, in case metal fatigue causes them to fail while in use.

Set screws

These are often used to fasten parts together where it would be difficult to use a bolt. In these cases the larger part of the component pair is drilled and tapped with a suitable sized screw thread. The smaller part is then 'spot faced' (machined so that the contact area of the screw head is perfectly flat) so that the screw head sits properly. The smaller part may also be

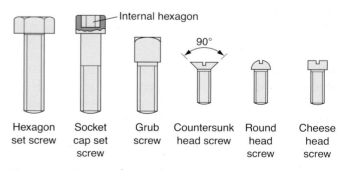

Internal hexagon

90°

Hexagon set screw   Socket cap set screw   Grub screw   Countersunk head screw   Round head screw   Cheese head screw

Fig 7.120 Range of typical set screws

counterbored (so that the screw head is below the top surface of the component) or countersunk in applications where a flat surface is needed. Examples of set screws are shown below in Figure 7.120.

Studs

In applications that require the frequent removal of the fastening devices, the threaded portion of the bolt may become damaged over time. In this case a 'stud' would be used in preference to set screws or machine screws along with the nut.

A stud is a length of round bar that is threaded at each end. One end of the stud would be screwed tightly into the body of the main component (for example the cylinder head on your car or an inspection chamber on a large pump) as shown in Figure 7.121.

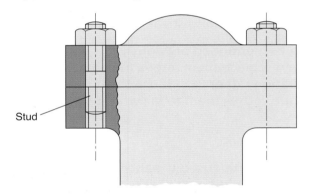

Fig 7.121 Studs used to secure an inspection cover

The advantage of using these devices is that any wear takes place only on the outer portion of the stud thread and not in what may be a very expensive component body. If the wear becomes excessive the stud is simply replaced.

Self-tapping screws

These are more often used with sheet metal and plastic components. They may be of a 'thread cutting' or 'thread forming' type depending on the nature of the material being joined.

When using self-tapping screw it is necessary to drill a small pilot hole first. The diameter of the pilot hole would be equal to the 'root diameter' (the diameter at the bottom of the thread) of the screw thread being used.

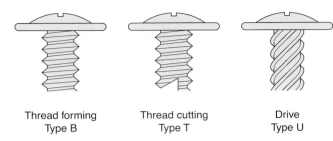

| Thread forming | Thread cutting | Drive |
| Type B | Type T | Type U |

Fig 7.122 Self-tapping screws

- Type B is used on relatively soft materials and forms a thread in the material by simply displacing the softer material
- Type T cuts away the material in which it is being inserted, and so is more suitable for harder materials
- Type U is designed to be hammered into the component and would be used in applications where removal is not intended; It has more than one spiral grove forming the thread meaning that fewer turns are required to drive it home

Wood screws

This material is covered in Chapter 10 Manufacturing.

Springs

A spring is a flexible device that exhibits elastic properties and is used to store mechanical energy. Springs are usually made from hardened steel. Small springs can be wound from a source of pre-hardened stock while larger ones are manufactured from specially annealed steel that is hardened after they are fabricated.

A range of non-ferrous metals are also used in the manufacture of springs. These include phosphor bronze for parts that require some degree of corrosion resistance and beryllium copper, where they are required to carry an electrical current due to its low electrical resistance (brush springs in electrical motors and generators for example).

Spring types and uses

The most common types of spring are:

- Compression springs – designed to become shorter when they are subjected to a load. The coils of the spring would not usually be touching when in the unloaded position and, in general, they do not require anchor points to assist their operation. Examples include front suspension springs in a motor vehicle

- Tension springs – designed to become longer when subjected to a load. The turns (coils) of these springs are generally touching in the unloaded position and they have some form of eye or hook or other device at each end as a means of attachment (for example simple closure springs on garden gates)
- Torsion springs – these apply in a turning/rotational direction and are widely used in watches, clocks and electrical instruments. Larger versions are used in recoil mechanisms for electrical cables in vacuum cleaners and in the starter pull cord mechanism on lawn mowers and other petrol-powered garden equipment
- Leaf springs – a series of strips of flat springy sheet; the are mostly used in commercial vehicle rear suspensions systems

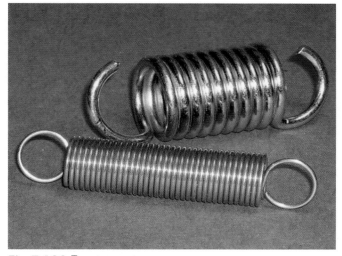

Fig 7.123 Tension springs

Rivets

Rivets are used where a permanent joint is required. The type and size of the rivet chosen for a particular task depends on the thickness of the plates, sheets or components being joined. Riveted joints are not designed be used where there are large tensile forces in the joint. Their strength lies in their ability to withstand the

'shearing forces' present in the joint that act across the 'shank' of the rivet.

Rivets are manufactured from a wide variety of materials (steel, aluminium, brass, copper and so on) and with a range of different shaped heads (see Figure 7.124). The only requirement is that the material used is reasonably malleable, allowing the head to be formed when making the joint, and that it should be manufactured from the same material as those being joined to prevent electrolytic/galvanic action at the surface of the joint should moisture be present.

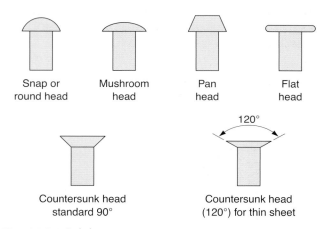

Fig 7.124 Solid rivets

Snap head and pan head rivets are used where maximum joint strength is required; flat head and mushroom head rivets are used where the head of the rivet should not protrude too far above the surface of the joint; countersunk rivets are used where a flush surface is required on the joint.

Where access can only be gained from one side of the joint it is possible to join thin plate or sheet metals using 'pop rivets' (used on commercial vehicle body panels, aircraft panels and so on). Where a sealed joint is required, a solid type of pop rivet is available. Further details about pop rivets can be found in Chapter 10 Manufacturing.

Keys

These are used to prevent relative movement between two parts and are usually made from medium- to high-carbon steel because of the stresses placed on them in use. Their main use is to prevent slippage when a coupling, hub assembly or pulley is attached to a shaft, and to transmit torque from a component to the shaft.

There are several types of keys in general use, some of which are described below:

Simple square and parallel keys

The hub is made so that it is very slightly larger than the shaft and key, allowing it to slide over the shaft during assembly. Often a small set screw is used to take up the slack between the hub and the key. The resulting friction provides resistance to axial motion. Thread adhesive is also sometimes used to ensure that vibration doesn't cause the set screw to loosen.

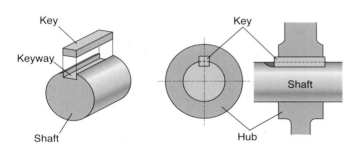

Fig 7.125 General arrangement of a key securing a hub or pulley to a drive shaft

Tapered keys

These are designed to be inserted from one end of the shaft after the hub is in position. The taper on the key imparts a compressive contact pressure between the hub and the shaft. Friction then helps to transmit the torque and to provide resistance to axial motion of the hub relative to the shaft. Tapered keys do not require set screws by virtue of their design. Access to both ends of tapered keys is required so that the key can be inserted and driven out when the key is being removed.

Fig 7.126 Examples of simple taper keys

Gib-head keys

Installation is similar to standard tapered keys. The extended head provides a holding method for removing the key by pulling instead of driving it out.

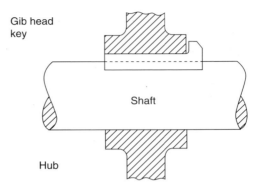

Fig 7.127 Use of a Gib head key

Woodruf keys

In this case a circular groove in the shaft holds the key in place while the hub is slid over the shaft. The Woodruff key is easily adjustable due to its semi-circular shape. It is most often used in applications that have tapered shafts and hubs and as a location device only. The main drive between mating parts would be provided due to the taper of the shaft and the hub. It has less shear strength than a rectangular or square key.

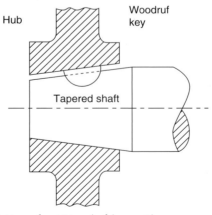

Fig 7.128 Use of a Woodruf key with a tapered shaft

Drive mechanisms

In engineering applications it is often necessary to transmit power (torque) from a main drive system (motor) to a variety of ancillary equipment, for example pumps, generators and alternators. The most common form of drive mechanism in use in the engineering and automobile industries is the belt drive system.

Belt drives

This is a relatively simple system that comprises a looped strip of a flexible material which is used to link two (or more) parallel shafts. In a simple two-pulley system, the belt can either drive the pulleys in the same direction, or the belt may be crossed so that the direction of the driven shaft is opposite to that of the drive shaft.

Flat belt drive

The belts are flat and rectangular in section. Flat belt drives are now less common than other belts because the belts slip off very easily, especially when worn.

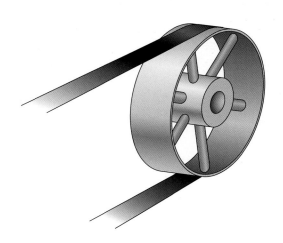

Fig 7.129 Example of a flat belt and drive pulley

Round belts

As the name implies, these belts are circular in cross section and are designed to run in a pulley with a semi-circular (or near semi-circular) groove. They are generally used for low-torque operations and can be purchased in various lengths. They are used in sewing machines, domestic appliances such as vacuum cleaners, and hi-fi equipment.

Vee belts

Vee belts are a solution to the slippage and alignment problem. The V shape of the belt tracks in a mating groove in the associated drive pulley with the result that the belt cannot slip off. The belt also tends to wedge into the groove of the pulley as the load increases — the greater the load, the greater the wedging action – thus improving torque transmission.

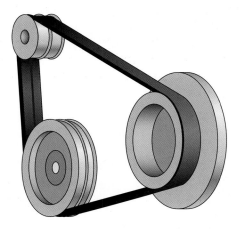

Fig 7.130 A vee belt and pulley system

They are usually supplied in various fixed lengths depending on their use. For high-power requirements, two or more vee belts can be utilised side-by-side in an arrangement called a multi-V, running on matching multi-grooved pulleys.

The strength of these belts is increased by reinforcing them with fibres such as polyester. Along with washing machines and other domestic appliances, pillar drills and other workshop machinery use vee belts.

Fig 7.131 A toothed or notched belt

Toothed or notched belts

Toothed or notched belts are a type of positive-transfer belt that can track relative movement. They have teeth that fit into a matching toothed pulley. When correctly tensioned they have no slippage and are often used to transfer direct motion for indexing or timing purposes (hence their name). They are commonly used on camshafts of automobiles; stepper motors often utilise these belts.

Drive mechanism: Gears and gear systems

What you need to know

You should be aware of the below gears and gear systems and their use; both in machine applications and in everyday consumer products. You should understand that the difference in the speed of gears is called the velocity ratio (VR).

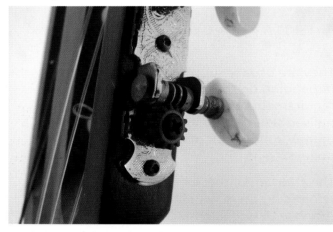

Fig 7.132 Worm gear system

$$VR = \frac{\text{number of teeth on driven gear}}{\text{Number of teeth on driver gear}}$$

The velocity of ration of a gear system is often referred to as the gear ratio.

Knock-down (KD) fittings

This material is covered in Chapter 10 Manufacturing.

1. A hexagonal-headed steel bolt is specified in the manufacturer's catalogue as M10 × 1.75 ×75. What do these numbers represent?
2. On what type of engineering application would you use a screw thread with the designation British Association (BA)?
3. Why is it desirable to use a flat washer in conjunction with a nut and bolt?
4. State, and describe, two different forms of friction locking devices that could be used to prevent a screwed fastening from becoming loosened due to vibration.
5. State under what conditions a positive locking device would be used in preference to a friction type?
6. Name two basic types of self-tapping screw and state what types of material they would be used with.
7. What type of rivet head would be used where the finished joint is required to have a flush finish?
8. For what applications would you use a pop rivet?'
9. Why is it advantageous to use a stud and a nut when securing an inspection cover to an engineering component?
10. Name the type of spring that would be used in conjunction with the McPherson strut suspension on a modern motor vehicle.

Learning outcomes

By the end of this section you should have developed a knowledge and understanding of:

- Tolerances
- Fit
- Performance
- Finish

Product dimensions and tolerances

It is not always possible to repeatedly make all components to precisely the same size, nor is it cost effective to try. There will always be some very small variation, even with a CNC machine, due to wear in the tooling and variations in the materials being used.

When the individual parts of an assembly or product are brought together, it is only important to know that they will fit as required without the selection of specific component parts having to be made.

Component parts are therefore manufactured within dimension guidelines known as tolerances. Tolerancing of dimensions is concerned with what is acceptable in respect of the component parts being able to fit together to form an assembly, or to function as the designer intended.

Datum measurements

The dimensions that are needed to make a particular engineered product are specified in the accompanying engineering drawing. These dimensions are usually given in terms of the distance from a fixed reference that is referred to as the datum face or datum edge, or from a datum line or datum point. This is in order to avoid over dimensioning or multiple or cumulative errors occurring during production.

The overall dimensions of the component or product may not always be given on the

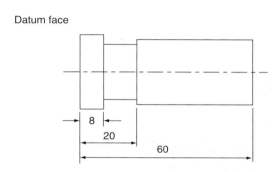

Datum face

Fig 7.133 Dimensioning from a datum face

drawing, but can be obtained by simply adding together the values of the intermediate dimensions. Dimensions on modern engineering drawings are normally given in millimetres (mm) unless otherwise stated in the information box. Diameters of shafts are always prefixed with the symbol Ø.

A toleranced dimension therefore defines the allowable limits of the size of a component part of an assembly, that is, the dimensional accuracy to which it must be produced. The tolerance of a component often dictates the engineering processes used to make it.

Depending on the overall size of the workpiece, for a well-maintained machine in a production engineering workshop, machining limits of around +0.03 mm should be within the capability of a lathe and around +0.01 mm for a milling machine. If the stated limits of a component were, say +0.008 mm, then precision grinding would be the most appropriate material removal process to finish the component.

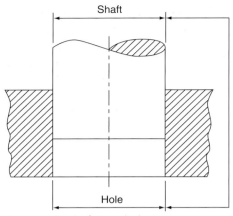

Fig 7.134 A simple shaft in a hole

The following would have to be considered during the manufacture of the two components shown in Figure 7.134:

- Assume that the diameter of both hole and shaft were to be, say, 50.000 mm and the design function of the two component parts requires that they be a sliding fit, that is, there would be a very slight difference in the sizes of each of the components. Therefore the hole should always be larger than the shaft
- The design engineer would then determine what constitutes an acceptable clearance for the component to function correctly; this could typically be a minimum clearance of say 0.025 mm and a maximum of say 0.095 mm, so a tolerance range of 0.070 mm
- At this point the engineer has to determine the engineering tolerances that would ensure that the shaft and hole would fit under any combination of assembly conditions without being outside of acceptable limits (0.070 mm)
- Slightly more than half of the tolerance would be assigned to the boring of the hole (it is often easier to be more accurate when turning a shaft than when boring a hole)
- This means that the hole size can be anywhere between Ø 50.000 mm and Ø 50.040 mm and still be acceptable
- The largest size for the shaft can now be determined by considering the smallest hole size together with the minimum clearance: Ø 50.000 minus 0.025 mm (Ø 49.975 mm); the smallest acceptable shaft size would therefore be the largest size of hole minus the largest clearance: 50.040 mm minus 0.095 mm (49.945 mm)

In order to assist the engineer in making informed decisions regarding tolerancing to allow for specific functions, British Standards publish tables of limits and fits in BS4500.

The associated basic engineering drawing would look something like that shown in Figure 7.135.

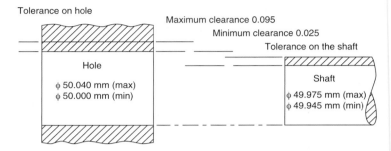

Fig 7.135 Final shaft sizing

Finish – the use of comparison standards

Finish in engineering terms refers to the quality of surface finish resulting from a machining process not the surface finish which is applied to the engineering component. The quality of surface finish resulting from a machining process is established using a 'surface roughness comparison standards'. These give a comparative μ 'micron' figure for the standard of finish. 12.5 μ is a very rough finish, with grooves clearly visible, whereas 0.4 μ is a mirror finish. Comparisons are literally made by comparing the samples with the engineered component by appearance and feel. μ Micron charts are labeled in μ mm and μ inches.

Accuracy of calibration – The use of gauge plates

When looking at a finished component it is also important to establish whether it has been made to an acceptable standard of accuracy. Simple gauges are used to establish whether a component is within a required tolerance. Final testing is often carried out on an inspection bench where micrometers and vernier calipers are used which have been calibrated accurately using gauge plates. The '**fit**' is tried under simulated working conditions. Internal measurements can be accurately made with 'internal micrometers' or by using the interior jaws of a vernier caliper.

Performance – A specialised engineered component is usually produced to a very high standard of accuracy and finish to a high specification. Following detailed scrutiny on an inspection table a sample of components will be subjected to performance tests. An example of this from the engineering company who provided the test equipment is that they produce valve stems and valves seats to a very high specification then pressure test them to see if they perform under working conditions.

It follows that any quality control system would involve random sampling of the measurements of the component parts in order to check conformity of production and anticipate tool change intervals. This would also give an indication of whether the finished assembly was likely to be fit for the purpose for which it was designed.

Section F — Engineering sectors

Learning outcomes

By the end of this section you should have developed a knowledge and understanding of:

- Why engineering sectors exist
- The advantages of belonging to a sector
- The relationships between sectors
- The modern technologies and materials used in sectors

Introduction

The engineering industry can be said to be an environment that transforms or converts raw materials into finished assemblies or products. It is a very complex organisation in which a wide range of products are being made – far too many to say that they belong to a global 'engineering facility'. To make it easier to classify products it has become common practice to place different product ranges into different sectors.

One way to classify the general engineering industry and, indeed, the manufacturing industry, is in terms of the products and materials that they handle and produce. For example:

- **Heavy engineering** – companies such as these produce products that are not necessarily sold on in large quantities on a regular basis. Rather, each output is a major work of engineering and, as such, could be high cost. It includes steel making, the transport industry such as shipping and railway transportation (engines and carriages), power generation and the like
- **Medium engineering** – products sold by these companies are somewhat less costly but are produced and sold on a more regular basis. Examples include machine tools and cars
- **Light engineering** – products made in this sector are widely used by both industrial and domestic customers and are generally produced in large batches. Such products include television sets, washing machines and small components used by other engineering companies
- **Processing** – this includes companies that produce products using some form of continuous process. There is a continuous mixing of raw ingredients to form a continuous supply of finished product. Good examples include the food and drink industries, the manufacture of chemicals and textiles, and so on

Another way that companies can be categorised in terms of their main products is to allocate them to a particular sector. The advantage of belonging to a particular sector means that each can then specialise in their own product ranges while still interacting or being reliant for materials and component parts on other sectors.

The customer also has a good idea of what they are getting. To this end, the engineering industry is divided into the following common sectors:

- **Mechanical engineering** – manufactures machine tools, construction and earth moving equipment, some office machinery, pumps, valves and maybe agricultural machinery
- **Electrical/electronic** – production of radio and electronic components, radio equipment, electrical appliances for the domestic market such as washing machines, dryers, television sets, video and audio equipment
- **Automotive** – manufacture and maintenance of motor vehicles and commercial vehicles
- **Aerospace** – companies involved in the construction of or the supply of component parts for the manufacture of aeroplanes, space satellites, satellite launch vehicles and the like
- **Chemical engineering** – companies concerned with the refining of oil, petroleum and the production of plastics
- **Metal production** – companies that produce components for other sectors, for example, small items such as engineer's small hand tools, nuts, bolts, screws, rivets, cutlery and so on

It is not unusual for some or all of these sectors to interact with each other in the production of their product. It means that each sector not only specialises in its own product range as mentioned earlier, but also develops technologies to suit its own ends.

Section G — Engineering systems and control

Learning outcomes

By the end of this section you should have developed a knowledge and understanding of:

- Computer integrated engineering (CIE)
- Programmable logic controllers (PLCs)
- Embedded systems in industrial appliances
- Robotics and automation

Computer integrated engineering (CIE)

We are familiar with the term computer aided design (CAD). However, this is only a very small part of a fully integrated engineering or manufacturing system.

CIE is, in many respects, very similar to computer aided engineering (CAE) in that it is all about the automation of all the associated stages that go into providing an engineered product or service.

When applied effectively, CIE unites all of the functions within an engineering company into one system, with information (data) being transferred from one computer-aided process to the next. It is an environment in which computers are a common link that ties together all of the various stages of producing an engineered product from the initial design using, say, a CAD system, to final product testing.

While all of these abbreviations can be a little confusing, it is worth noting that they are all related to some form of computer process. Indeed, in modern engineering, the boundaries between CAD and computer aided manufacture (CAM) are becoming increasingly blurred.

CAE is a relatively new and significant branch of engineering that utilises the more specialist applications of computers in engineering. It

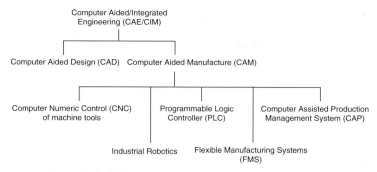

Fig 7.136 CIE system

includes CIE, embedded systems, engineering system modelling and simulation, systems integration, robotics and metrology.

Using this system computer analysis may be conducted to investigate and predict, for example, mechanical, thermal or fatigue stress in the production/process materials, fluid flow and heat transfer, together with any vibration or noise characteristics that may limit the performance of the final product.

Additionally, all of the machine tools within a particular engineering company may be linked to a central management control centre, which would then monitor each of the machining operations.

Systems of this type are used extensively in the steel and iron making industries, and in the automotive industry, which relies quite heavily on the use of computer-controlled robot devices for assembly, handling and transportation of the final product.

Programmable logic controllers (PLCs)

A good description of a PLC would be a digitally controlled operating system designed to be used in an industrial situation. They are used extensively in both the mechanical and chemical industries to control a range of automated processes.

They use a programmable memory for the storage of user-orientated instructions that are then used to initiate specific functions within the engineering/manufacturing process, such as sequencing, timing, counting and controlling of a wide range of machinery and systems. Inputs and outputs can either be a digital or analogue instructions depending on the response

required. They can be linked to computers, other PLCs or can be of a stand-alone construction.

Although not strictly an engineering example, we frequently come under the control of a PLC each time we approach a set of traffic lights. In this case they are placed in a metal cabinet at the side of the road and are used to control the sequence of light changes and the timing of traffic flow in each direction. Because they are 'programmable' they can readily be altered from time to time to cater for different demands in traffic patterns.

This is only a very simple example of the use of PLCs compared to the often quite complex operations that have to be controlled in large engineering facilities, especially in production line operations in the automotive industry, catering for models with differing specifications, for example:

- A choice of different body colours and interior trims
- Engine size and configuration
- Optional accessories

A simplified block diagram of PLC is shown below in Figure 7.137.

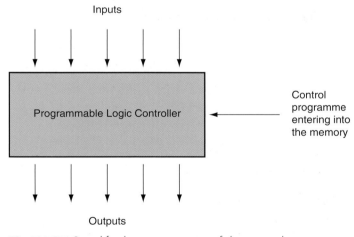

Fig 7.137 Simplified representation of the control action of a PLC

Figure 7.137 shows how a simple control action could be achieved. Input signals, which could be derived from limit switches, the closing of mechanical contacts, proximity devices and so on, and output signals to, say, motors controlling robotic arms, relay coils, indicator lamps and so on, would be connected to the relevant terminals on the PLC. A 'user' program comprising a sequence of instructions would then be fed into the PLCs memory.

The PLC continually monitors the state of the input signals, producing outputs in accordance with the algorithm defined in the user program.

Because the stored program can be easily modified, new control features can readily be added or old ones removed or changed without the requirement of physically rewiring the input and output devices. This results in an extremely flexible system that can be used to control a number of tasks that vary in nature and complexity.

The major differences between a conventional microcomputer-controlled system and a PLC are that:

- Programming is predominately concerned with simple control logic and function block diagrams
- Interfacing circuits are invariably integrated into the controller
- They are rugged, being suitably packaged to withstand vibration, wide temperature ranges, humidity and (electronic) noise

Typically PLCs range from:

- Simple, stand-alone, single-unit systems – these have a small, fixed number of input/output (I/O) points and are really only suitable for relatively small automation tasks; they are often hand-held units and are directly connected to the PLC
- Modular PLC systems – these comprise a series of self-contained modules that are plugged into a propriety racking system. The minimum configuration for this type of PLC system is a CPU unit, a power supply unit and a number of I/O modules catering for both digital and analogue input/output signals. This modular approach to the construction of PLC hardware means that the numbers and types of I/O units can be expanded as and when required

Special function modules that could be used for operations such as computer linking (networking) could also be incorporated. Additionally, a host computer would normally be incorporated through which the program codes for the PLCs would be developed.

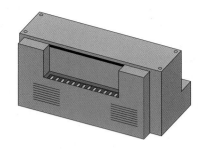

Fig 7.138 Simple, stand-alone unit

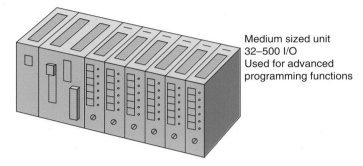

Medium sized unit
32–500 I/O
Used for advanced
programming functions

Fig 7.139 Modular-based PLC system

Embedded systems

These are special-purpose computer systems designed to perform one or more dedicated functions, often with real-time computing constraints. As the name implies, they are usually built into a product as part of a complete device and include any associated hardware and mechanical parts.

Since an embedded system is dedicated to a specific task, design engineers can optimise it by reducing the size and cost of the product, or increasing the reliability and performance. Some embedded systems are mass-produced, making them relatively cheap.

The complexity of the devices range from low, with maybe just a single microcontroller chip, to very high with multiple units, peripherals and networks mounted inside a large chassis or enclosure.

Embedded systems are being used increasingly in a wide range of applications, for example:

- Telecommunications systems employ numerous embedded systems from telephone switches for the network to mobile phones
- Computer networking uses dedicated routers and network bridges to route data

- Consumer electronic devices include personal digital assistants (PDAs), MP3 players, video game consoles, digital cameras, DVD players, GPS receivers and printers
- Many household appliances, such as microwave ovens, washing machines and dishwashers, include embedded systems in their design and construction to provide flexibility, efficiency and features
- Modern home automation systems use wired and wireless networking systems that can be used to control lights, temperature, security, audio/visual and so on, all of which use embedded devices for sensing and controlling

Robotics

Industrial robots are officially defined by the ISO as: 'an automatically controlled, reprogrammable, multipurpose manipulator which can be programmed to operate in three (or more) axes.'

Typical applications of these robots include repetitive operations such as welding and paint spraying of machine and car parts, assembly, pick and place in electronic circuit production, packaging and palletising, together with product inspection and testing. The advantage of using industrial robotic systems is that all of the above functions can be accomplished with high speed, precision and endurance.

There are various ways in which these robots can be programmed:

- They can be controlled by a simple program that has been developed manually or on a computer in a similar way to those used on CNC machines
- They can be 'taught' by a skilled operator who physically moves the robot arm though the movements required to complete a task; the movements are recorded in the computer that controls the robot. These techniques are widely used for robots carrying out repetitive tasks such as those mentioned above
- The robot can be 'driven' through its movements step by step using a hand-held

control keypad coupled to the robot's computer by a trailing cable. Again, the movements of the robot arm are saved in the memory of the computer controlling the robot

Simplified line diagrams of such devices are shown in Figure 7.140.

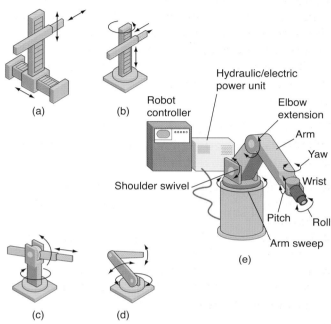

Fig 7.140 Simplified representation of industrial robot arms and manipulators. (a) Cartesian co-ordinate; (b) cylindrical co-ordinate; (c) spherical (polar) co-ordinate; (d) revolute (angular) co-ordinate; (e) typical robotic arm and associated control system.

Fig 7.141 Staubli RX-Paint robotic arm teach pendant

Learning outcomes

By the end of this section you should have developed a knowledge and understanding of:

- Dynamics – linear, angular and rotary acceleration/deceleration, centripetal motion, momentum
- Statics – basic principles of forces and moments
- Stresses – compression, torsion, tensile and shear

Statics

As the title suggests, this concerns the study of bodies that are at rest but under the influence of forces. Static systems are the simplest of engineering systems as they are at rest. Some examples of simple static systems include:

- Simple balances, for example a set of old-fashioned kitchen scales
- A simple beam that supports a load

If all of the forces acting on the body were such that the body remained at rest with no tendency to move in any direction (static), then the body is said to be in static equilibrium.

This state of equilibrium can loosely be sub-divided into three categories: stable, unstable and neutral.

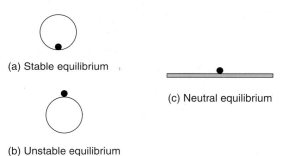

(a) Stable equilibrium

(b) Unstable equilibrium

(c) Neutral equilibrium

Fig 7.142 Static equilibrium

Stable equilibrium

This is when a body, which is initially at rest, is disturbed either by pushing or pulling it; its centre of gravity is raised and it will move. When the disturbing force is removed the body will return to its original position. This can be likened

to a simple ball bearing placed in the bottom of a hollow spherical cup (Figure 7.142a).

Unstable equilibrium

In this case, if a disturbing force is applied the centre of gravity is lowered and the body will move away from its original position. This is the opposite of stable equilibrium and can be likened to a simple ball bearing sitting on top of a sphere (Figure 7.142b).

Neutral equilibrium

This applies when a body is disturbed and the centre of gravity remains at the same height. When the disturbing force is removed, the object does not move. An example of this is a simple ball bearing situated on a flat horizontal surface (Figure 7.142c).

Basic forces and reactions

One of the most important factors that must be considered in mechanics is that, for every force acting on one body, there must be an equal and opposite force (or reaction) acting on that body (Newton's third law of motion). This seems like a very simple statement, but is sometimes difficult to understand in practice.

Suppose that a piece of metal of mass m kg were to be freely suspended on a length of string as shown in Figure 7.143.

In this case, the force due to gravity – weight of the metal – exerts a downward pull (or force, W) on the string, and the string in turn exerts an upward force, F, on the piece of metal.

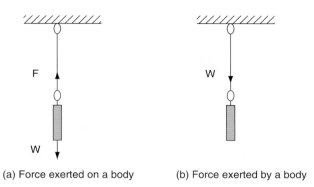

(a) Force exerted on a body   (b) Force exerted by a body

Fig 7.143

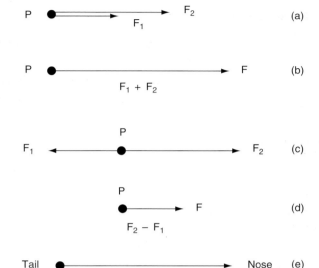

Fig 7.144

equal to the arithmetic sum of the individual forces. This single force is known as the resultant force or simply the resultant (Figure 7.144a + b)

Since the body is stationary, it is said to be in a state of equilibrium. Therefore, force F is exactly equal in magnitude and opposite in direction to force W. A body cannot, then, remain in equilibrium under the action of a single force. For instance, if we were to cut the string in Figure 7.143a, then the only force acting on the piece of metal would be its weight, and the piece of metal would fall to the surface with its velocity increasing as it fell.

The forces shown in Figure 7.143a are those exerted on the piece of metal only. The force of gravity W is acting downward; force F is therefore acting upward and is the reaction to W. The downward force W is exerted by the piece of metal on the string, as shown in Figure 7.143b. (The reaction to W on the string is also F).

When a number of forces are all acting in the same plane, they are said to be coplanar. If they also act at the same time and at the same point, they are said to be concurrent.

A force is therefore a 'vector quantity', which means that it has both magnitude and a given direction. Vector forces can be represented graphically on a force diagram by drawing a straight line (to scale) with an arrow head at one end to show direction of action.

Resultant of coplanar forces

For two forces acting at a point, there are a number of possibilities:

1 If the forces are acting in the same direction and also have the same line of action, they can be represented by a single force that would be

2 If the forces act in opposite directions but still have the same line of action (figure 7.144c), then the resultant would be equal to the arithmetic difference between the two individual forces (figure 7.144c + d).

3 When the forces do not have the same line of action then the value of the resultant force must be found by using a procedure called the vector addition of forces. There are two ways in which this can be achieved:

■ The triangle of forces method
■ The parallelogram of forces method

Triangle of forces method

This is a simple and effective method and works as follows:

1 Draw a vector representing one of the two forces to an appropriate scale and in its direction of action.
2 From the 'nose' of that vector, using the same scale draw a vector representing the second force in the direction of its line of action.

3 The resultant is then represented in magnitude and direction by a line drawn from the nose of the second vector to the tail of the first vector, that is, the line that closes the triangle.

WORKED EXAMPLE – TRIANGLE OF FORCES METHOD

Using the triangle of forces method, determine both the magnitude and direction of the resultant of a 20 N force acting horizontally to the right and a force of 30 N acting at an angle of 60 degrees to the 20 N force.

Using the information given above, proceed as follows:

1 Draw line ab horizontally equal to 20 N (to a suitable scale).

2 From point b, draw line bc equal in length to 30 N (same scale and inclined to the horizontal at 60 degrees).

3 Draw line ac to close the triangle and measure it to your scale: this represents the value of the resultant force.

4 Using a protractor, measure the value of the angle bac. This is the direction of action of the resultant force.

This gives a resultant force of 43.6 N acting at an angle of 36.5 degrees to the horizontal. Note that these values could also be calculated by making a sketch of the arrangement and then using the sine and cosine rules).

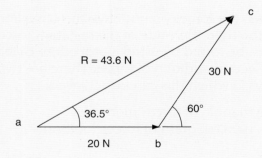

Fig 7.145

The parallel of forces method

A simple method for using the parallelogram of forces method is as follows.

1 Draw a vector to represent one of the forces, again to a suitable scale.

2 From the tail of that vector draw a second vector (to the same scale) to represent the second force.

3 Complete the parallelogram using the two vectors drawn in steps 1 and 2 above.

4 The resultant force is given by the length of the diagonal of the parallelogram drawn from the tails of the vectors in 1 and 2 above.

WORKED EXAMPLE – PARALLEL OF FORCES METHOD

Using the parallelogram of forces method, determine the resultant of a force of 250 N acting at an angle of 60 degrees to the horizontal and a force of 400 N acting at an angle of −45 degrees to the horizontal.

Again, using the information given above, proceed as follows:

1 Draw a horizontal line ax to any convenient length.

2 Draw line ab to represent the 250 N force (to a suitable scale) at an angle of 60 degrees.

3 Draw line ac from the tail of the first vector (to the same scale) to represent the 400 N force at an angle of −45 degrees.

4 Using suitable drawing equipment (a ruler and a set square are best), draw line cd parallel to and the same length as line ab.

5 Draw line bd parallel to line ac as above.

6 Draw a diagonal line to connect point a to point d.

7 Measure the diagonal ad to your scale: this represents the magnitude of the resultant force.

8 The angle that the resultant makes with the horizontal represents its line of action.

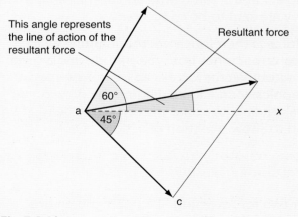

Fig 7.146

Where there are more than two forces acting at a point a similar approach to that outlined above in the parallelogram of forces method would be used. This entails the construction of a polygon of forces.

In this case, each progressive force would be drawn from the nose of the proceeding force, again to a suitable scale and in the appropriate direction. The line that closes the polygon represents the resultant force for the system.

Moments and centres of gravity

Moment of a force about an axis

Suppose that a uniform strip or rod of wood, such as a metre rule, is pivoted at its mid point C as shown in Figure 7.147. If no external force is applied to the rod, there is no tendency for the rod to rotate in either direction about point C.

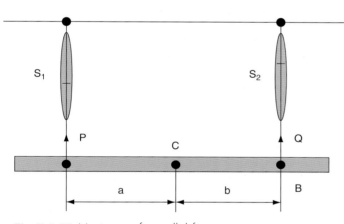

Fig 7.147 Moments of parallel forces

If we now attached the hook of spring balance S_1 to a cord passing through the hole at point A, then any upward force applied by this balance would cause the rod to rotate in a clockwise direction about point C.

On the other hand, if we apply an upward force at point B through spring balance S_2, the rod would obviously rotate in an anticlockwise direction.

If two upward forces were simultaneously applied by means of both S_1 and S_2 the rod would stay in a horizontal position. If P and Q are these forces exerted by S_1 and S_2 respectively, and if a and b are the distances of points A and B from point C, it can be demonstrated that:

$$P \times a = Q \times b$$

This relationship can be confirmed by making a note of the balance readings for a range of upward forces exerted by the balances at differing values of distances a and b.

This tendency of a force to produce rotation of a body about a fixed axis is termed the moment of a force about that axis or, more simply, just a moment.

The magnitude of this moment depends on two factors:

- The magnitude of the force acting at right angles to a given point
- The perpendicular distance from the point of application of the force to the pivot

For example, in Figure 7.147, if P is the applied force (in Newtons) exerted by the spring balance S_1 such that the rod tends to turn clockwise about an axis through point C perpendicular to the paper, and if a is the distance (in metres) of the line of action of force P from the axis through point C, then the clockwise moment of force P about the axis is given by:

$$P \text{ (Newtons)} \times a \text{ (metres)} = Pa \text{ Newton metres}$$

Similarly, the anticlockwise moment of force Q about the same axis is given by:

$$Q \text{ (Newtons)} \times b \text{ (metres)} = Qb \text{ Newton metres}$$

When the moments are such that the rod remains in the horizontal position, then:

clockwise moment   =   Anticlockwise moment
(due to P)   (due to Q)

So:

$$Pa = Qb$$

This relationship is called the principle of moments.

Levers

A lever is simply a rod or bar that is capable of turning about a fixed axis called the fulcrum. This can be a spindle or a simple knife edge. This lever can be straight or curved or can be cranked, and the forces acting upon it can be parallel or otherwise.

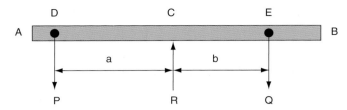

Fig 7.148 A simple, straight lever

Consider a straight uniform lever AB resting on a fulcrum at point C as shown in Figure 7.148. By arranging the fulcrum to be in the middle of the lever, midway between A and B, we can avoid any complication resulting from the weight of the lever, since the gravitational pull on the right end of the lever (tending to turn it in a clockwise direction about point C) is perfectly balanced by the left end of the lever (tending to turn it anticlockwise). Consequently, when the lever is not loaded in any way, there is no tendency for it to turn about the fulcrum at point C.

A downward force P, applied at a distance a metres from the fulcrum, would cause an anticlockwise moment Pa about C. Similarly, a downward force Q, applied at distance b metres from the fulcrum causes a clockwise moment Qb about point C. The lever therefore remains horizontal if the clockwise and the anticlockwise moments are equal.

So, If Pa = Qb, then the supporting or upward force R exerted by the fulcrum on the lever must be equal to the sum of the downward forces P and Q. So:

$$R = P + Q$$

It should be noted that, in this case, moments have been taken about the axis through which the supporting force R acts upon the lever. This force R has no moment about the axis C and therefore has no affect on the balance of the lever. Had moments been taken about any other axis, then, of course, the moment of R about that axis would have had to be considered.

WORKED EXAMPLE – LEVERS

A uniform lever is pivoted at its mid point C. A body exerting a force of 150 N is suspended from a point E, 120 mm to the right of point C. Calculate the value of the force that has to act at a point D, 300 mm to the left of C, in order that the lever is balanced.

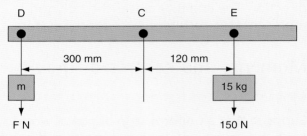

Fig 7.149

The clockwise moment about point C = 150 × 0.12 = **18 Nm**

If F is the force in Newtons that has to act at point D, then:

The anticlockwise moment about point C = F × 0.3 = **0.3 F Nm**

For balance to occur, clockwise moments must equal anticlockwise moments, so:

18 = 0.3 F

Therefore:

$$F = \frac{18}{0.3} = \mathbf{60\ N}$$

Dynamics

Unlike the static systems described above, dynamic systems are systems in which some or all parts of the system are in motion. Dynamic systems can be roughly divided into two categories:

- Linear motion – motion in a straight line
- Rotary motion – motion in a circular manner about a fixed point

Simple linear motion

When considering any form of motion in a straight line, we need to be aware of five basic quantities or variables. These are:

- The initial velocity of the body, u
- The final velocity of the body, v
- The acceleration of the body, a
- The time that the body is in motion, t
- The distance through which the body has moved, s

These five variables are related to each other in a series of simple equations:

$$v = u + at$$

$$s = \frac{v - u}{2} \times t$$

$$s = ut + \frac{1}{2}at^2$$

$$v^2 = u^2 + 2as$$

By using an appropriate equation from those listed above, we are able to determine any one of the five variables provided that we know at least three of the others.

In practice, since the system is in motion, we often need to know the value of the final velocity v or the distance through which the system has moved s. This is the reason that the above equations are written in that particular format.

Of course, any of the above equations can be rearranged to find any of the other variables to suit the conditions.

The basic definitions of the terms used in calculation involving both linear and angular motion, together with their basic and derived units are listed below.

Linear velocity (v or u)

This is defined as the rate of change of linear displacement (distance, s, in a straight line) with respect to time (t).

$$\text{Linear velocity} = \frac{\text{change of displacement (m)}}{\text{change in time (sec)}}$$

i.e.,
$$v = \frac{s}{t} \qquad \text{(Equation 1)}$$

The unit of linear velocity is the metre per second (m/s or ms^{-1}).

Linear acceleration (a)

This is defined as the rate of change of linear velocity (v) with respect to time (t). So, for an object whose acceleration is increasing in a linear manner:

$$\text{Linear acceleration} = \frac{\text{change of linear velocity (m/s)}}{\text{change in time (sec)}}$$

i.e.,
$$a = \frac{v - u}{t} \qquad \text{(Equation 2)}$$

Units are metres per second per second (metres per second squared, or m/s^2 or ms^{-2}).

Linear momentum

This is defined as the product of the mass of a body and its velocity.

$$\text{momentum} = \text{mass} \times \text{velocity} \ (m \times u)$$

Mass m is in kg and velocity u is in m/s. The unit of momentum is therefore the kilogram metre per second (kg ms^{-1}).

Since velocity is a vector quantity (it has both magnitude and direction) momentum must also be considered to be a vector quantity.

Newton's first law of motion states that a body will continue in a state of rest or in a state of uniform motion in a straight line unless acted on by some external source.

From this we can say that the momentum of a body must also remain the same unless it is subject to some form of external source.

The principal of the conservation of momentum for a closed system (that is, one that is not being subjected to any external force) states that the total linear momentum of a body, or system, is constant.

This means that if a body or system is moving and it is involved in a collision with another body or system, then the total momentum of the body or system after the collision is the same as it was before collision.

So if m_1, m_2, u_1 and u_2 represent the mass and initial velocities of two bodies or systems before a collision, and m_1, m_2, v_1 and v_2 represent mass and final velocities of those systems after a collision, then:

$$m_1u_1 + m_2u_2 = m_1v_1 + m_2v_2$$

The radian

Unlike linear models where the displacement of a body is in a straight line and can be readily measured in metres, the same cannot be said for angular displacement where any displacement is circular.

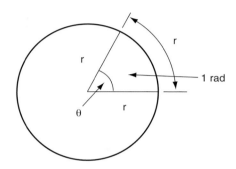

Fig 7.150

The unit of angular displacement is known as the radian. It is a constant and is the measurement of the angle subtended at the centre of a circle by an arc whose length is equal to the radius of that circle. See Figure 7.150.

The relationship between this angle θ in radians, the arc length s and the radius of the circle r is given by:

$$s = r\theta \qquad \text{(Equation 3)}$$

Since the length of the arc of a complete circle is given as $2\pi r$, and the angle subtended by this arc is 360 degrees, then for a complete circle

$$2\pi r = r\theta, \text{ so } \theta = 2\pi \text{ radians}$$

Hence, $\qquad 2\pi \text{ radians} = 360° \qquad$ (Equation 4)

Angular velocity (ω)

In many practical applications, it is assumed that the speed of rotation of a wheel or a shaft is given in terms of revolutions per minute (or revolutions per second). However, these units do not form part of the SI (System Internationale d'Unites) system of measurements.

Angular velocity is simply defined as the rate of change of angular displacement with respect to time t for a body that is rotating about a fixed axis at a constant speed.

$$\text{Angular velocity} = \frac{\text{angle moved through}}{\text{time taken}}$$

i.e., $\qquad \omega = \dfrac{\theta}{t} \qquad$ (Equation 5)

The unit of angular velocity is therefore the radian per second.

So, a body that is rotating at a constant speed of n revolutions per second would subtend an arc of $2\pi n$ radians in one second. Therefore its angular velocity is:

$$\omega = 2\pi n \text{ rads/sec} \qquad \text{(Equation 6)}$$

Combining equations 3 and 5, we get:

$$s = r\omega t \quad \text{or} \quad \frac{s}{t} = \omega r \qquad \text{(Equation 7)}$$

However, from Eq 1, we see that $v = \dfrac{s}{t}$ (linear velocity)

Hence: $\qquad v = \omega r \qquad$ (Equation 8)

This gives us a simple relationship between linear velocity and angular velocity.

Angular acceleration (α)

Angular acceleration is defined as the rate of change of angular velocity with respect to time. For a body whose angular velocity is increasing in a uniform manner:

$$\text{angular acceleration} = \frac{\text{change of angular velocity}}{\text{time taken}}$$

i.e., $\qquad = \dfrac{\omega_2 - \omega_1}{t} \qquad$ (Equation 9)

The unit of angular acceleration is then the radian per second squared or rad/s² or rads⁻².

Rewriting the above equation and making ω_2 the subject gives:

$$\omega_2 = \omega_1 + \alpha t \qquad \text{(Equation 10)}$$

From equation 8, $v = \omega r$. For any motion in a circular path having constant radius r metres:

$$v_1 = \omega_1 r \text{ and } v_2 = \omega_2 r$$

So equation 2 can be rewritten as:

$$a = \frac{\omega_2 r - \omega_1 r}{t} = \frac{r(\omega_2 - \omega_1)}{t}$$

From equation 9, $a = r\alpha$

From equation 5, $\theta = \omega t$; assuming that the angular velocity is changing in a uniform manner from ω_1 to ω_2, then:

$$\theta = \text{mean angular velocity} \times \text{time}$$

Giving:

$$\theta = \left(\frac{\omega_1 + \omega_2}{2}\right)t \quad \text{(Equation 11)}$$

Deceleration

The term deceleration means that a body is slowing down. It is a negative acceleration. If you calculate a to be negative then it is a deceleration.

Centripetal motion

When a body moves in a circular path at a constant speed, its direction of motion must be continually changing. Consider Figure 7.151:

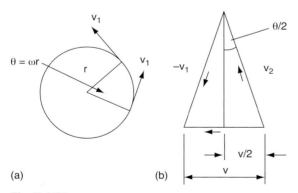

(a)   (b)

Fig 7.151

Since velocity is a vector quantity and is expressed as both a magnitude and direction, then it must be continually changing; and, since acceleration is a measure of the rate of change of velocity with time, then the object must be accelerating.

In Figure 7.152a, assume a body is moving with a constant angular velocity ω and a tangential velocity v. Also, let the change in tangential velocity for a small displacement say, of angle θ be represented by V (that is, $v_2 - v_1 = V$).

The relevant vector diagram is shown in Figure 7.151b, and, since v_1 and v_2 are of the same magnitude, the vector diagram can be likened to an isosceles triangle. Remember, the magnitude of $v_1 = v_2 = v$

Bisecting the angle between v_2 and v_1 gives:

$$\sin\frac{\theta}{2} = \frac{V/2}{v_2} = \frac{V}{2v_2}$$

i.e., $V = 2v_2 \sin\frac{\theta}{2}$ (Equation 1)

But, since $\theta = \omega t$, then $t = \frac{\theta}{\omega}$ (Equation 2)

Dividing equation 1 by equation 2 gives:

$$\frac{V}{t} = \frac{2v_2 \sin(\theta/2)}{\theta/\omega} = \frac{2v_2\omega \sin(\theta/2)}{\theta}$$

For very small angles, $\frac{\sin(\theta/2)}{\theta}$ is equal to $\frac{1}{2}$, hence

$$\frac{V}{t} = v_2\omega = v\omega$$

Hence: $\frac{V}{t} = \frac{\text{change of velocity}}{\text{change in time}} = \text{acceleration, a}$

So, $a = v\omega$

But, $\omega = \frac{v}{r}$ thus acceleration $a = \frac{v^2}{r}$

This is the acceleration of the body and is directed towards the centre of the circle (along path V) and is called the centripetal acceleration.

If the rotating body has mass m in kg, then by Newton's second law of motion, the centripetal force involved in the motion is:

$$m\frac{v^2}{r}$$

And its direction of action is also directed towards the centre of the circle.

Stress and strain

When any material is subjected to a force, the material is said to be stressed or in a state of stress.

Tensile stress

If, for example, a metal rod is subject to a pulling force (tension) then the force per unit area of the cross section of the rod is referred to as tensile stress. Examples of materials or bodies that would be subject to tensile stress include:

- The rope or cable of a crane that is carrying a load
- Simple rubber bands when stretched
- A bolt (when a nut is tightened onto a bolt, the bolt is in tension)

Compressive stress

Conversely, if the rod is subject to a pushing force (compression) then the force per unit area is referred to as compressive stress. Examples of materials or bodies that would be subject to 'compressive stress' would include:

- A concrete or steel pillar supporting a bridge
- The sole of your shoe when you are walking
- The jib of a crane that is loaded

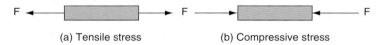

(a) Tensile stress   (b) Compressive stress

Fig 7.152 Tensile and compressive stress

In general we can state that:

$$\text{Stress} = \frac{\text{force}}{\text{cross-sectional area}}$$

Tensile and compressive forces are sometimes referred to as normal stresses since they act at right angles to the cross-sectional area that is used for calculating the individual values of the necessary stress.

The SI unit of stress is the Newton per square metre (N/m^2) which is known as the Pascal (Pa) named after the French philosopher Blaise Pascal (1623–62). He carried out a number of important experiments connected with hydrostatics and pneumatics.

It is often more convenient to express stress in kilonewtons per square metre (kN/m^2) or kilopascals (kPa), or in meganewtons per square metre (MN/m^2) or megapascals (MPa).

$1\ kN/m^2$ or $1\ kPa = 10^3\ N/m^2$ or $10^3\ Pa$

$1\ MN/m^2$ or $1\ MPa = 10^6\ N/m^2$ or $10^6\ Pa$

EAXMPLE 1

A tie-bar has a cross-sectional area of 125 mm^2 and is subjected to a pull of 10 kN. Calculate the value of the stress in meganewtons per square metre or megapascals.

Force = 10 kN = 10,000 N

Cross-sectional area = 125 mm^2 = $125 \times 10^{-6}\ m^2$

Therefore: Stress $= \dfrac{10000}{125 \times 10^{-6}} = 80 \times 10^6\ N/m^2$

$= 80\ MN/m^2$ or 80 MPa

EXAMPLE 2

A steel wire, 1.5 mm in diameter, supports a mass of 60 kg. Calculate the value of the stress in the wire.

Tension in the wire = mass $\times$ 9.81 = 60 $\times$ 9.81
$= 588.6\ N$

Cross-sectional area of the wire =

$\dfrac{\pi d^2}{4} = \dfrac{\pi \times 1.5^2}{4} = 1.767\ mm^2$

$= 1.767 \times 10^{-6}\ m^2$

Stress $= \dfrac{588.6}{1.767 \times 10^{-6}} = 333.5\ MN/m^2$ or 333.5 MPa

Shear forces and stress

A shear force is one that tends to slide one face of a material over an adjacent face. Examples of shear stress include:

- A rivet joining two plates together – it would

be subject to shear stress if a tensile force were to be applied to the joint

- A guillotine that is cutting sheet metal
- Garden shears or scissors when they are cutting
- A simple horizontal beam
- Transmission joints on motor vehicles are subject to shear forces and therefore shear stress

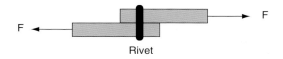

Fig 7.153 Shear stress

Torsion

In solid mechanics torsion is a twisting action of an object due to an applied torque (turning force). In circular sections the resultant shearing stress is acting perpendicular to the radius of the section.

It should be noted that the largest shear stress on a shaft is at the point where the radius of the surface of the shaft is maximum. High stresses at the surface may be compounded by what are referred to as stress concentrations, such as rough spots.

Thus, shafts for use in high-torsion applications (the drive shafts of a ship for example) are polished to a very fine surface finish to reduce the maximum stress in the shaft and therefore to increase its service life.

If the shaft is loaded only in torsion, then one of the principal stresses will be in tension and the other in compression. These stresses are at a 45 degree helical angle to each other around the shaft.

If the shaft is made of a relatively brittle material then the shaft will fail. This is initiated by a fine crack appearing at the surface, which will then propagate through to the core of the shaft, fracturing it in a 45 degree angle helical shape. This can sometimes be demonstrated by

twisting a piece of blackboard chalk between one's fingers.

A torsion spring is a spring that works by torsion or twisting, that is, it is a flexible elastic object that stores mechanical energy when it is twisted. The amount of force (torque) that it exerts is directly proportional to the amount it is twisted.

A torsion spring is often made from a wire, ribbon, or bar of metal or rubber. A good example of the use of a torsion spring is shown in Figure 7.154: a simple mousetrap.

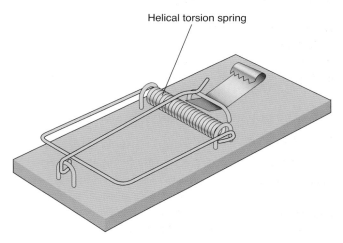

Helical torsion spring

Fig 7.154

Strain

When a rod or a wire is pulled, it will stretch. The total amount of stretching, or the elongation, expressed as a fraction of the unstretched length, is called direct strain or just strain, where the change in length is caused by tension or compression.

$$\text{Strain} = \frac{\text{extension}}{\text{original length}}$$

Thus, if the pull on a brass rod having an unstretched length of 1 m produces an extension of, say, 2.5 mm, then:

$$\text{strain} = \frac{2.5 \times 10^{-3}}{1} = 0.002$$

It is most important that the extension and the original length be expressed in the same units.

Hence, strain is merely a ratio of like terms and, as such, will have no units and is expressed simply as a number.

CHAPTER 8

Food

Learning outcomes

By the end of this section you should have developed a knowledge and understanding of:

- **Protein** – The functions and working properties of protein. The sensory functions and properties of meat, fish, eggs, milk and cheese. The effects of processing on these foods.
- **Lipids/fats & oils** – The chemistry, functions and working properties of lipids/fats. The nutritional, physical and sensory functions of fats and oils. The effects of processing on these foods.
- **The functions and working properties of carbohydrates** – Monosaccharides, disaccharides and polysaccharides. The effects of processing on these foods.
- **Vitamins and minerals** – Vitamins: A, B, C and D. The minerals iron, calcium, phosphorus and sodium. Why they are needed, where they are found and how they can be lost through processing.
- **The fortification of food products**
- **Daily Recommended Values (DRVs)** – How this information is used by food manufacturers.
- **New developments** – Novel foods, functional foods, pre and probiotics, modified foods, lipid/fat replacers and meat analogues.

Introduction

Food provides the energy we need to survive. All bodily functions depend upon the energy and trace elements found in the food we eat. Having a healthy balance of foods is essential for growth and to maintain good health.

Nutrients are divided in to two types:

- Macronutrients – fats, carbohydrates and proteins. These are needed by the body in relatively large quantities and form the bulk of our diet.

- Micronutrients – vitamins and minerals. These are found in food and are vital to health but are required in very small quantities.

The government provides guidelines to help ensure we have the right nutrients and eat a healthy diet.

The government guidelines for what we should eat:
- A good variety of foods
- Meals based on starchy foods: bread, pasta, rice, potatoes
- Five to six helpings of fruit and vegetables each day
- Protein foods such as fish, meat and pulses in moderation
- Low amounts of saturated fats such as lard, butter or ghee
- Small daily quantities of salt and sugar

Fig 8.1 The eatwell plate – a guide to the types and proportions of foods that make up a healthy diet

Protein

Protein is one of the macronutrients essential for growth and repair of body tissue and is crucial to the healthy functioning of the body. Protein is made up of complex chains of molecules called amino acids; there are 20 different types of amino acid, each having a specific function in the body. Enzymes, which are vital for metabolism, are proteins, as are some hormones, which regulate some important bodily functions. When there is too little carbohydrate or fat in our diet

to provide sufficient energy, energy is provided by taking protein from muscles and other organs. If there is too much protein in the diet, it is broken down into its amino acids and transported by the blood to the liver where it may be stored as fat.

Key Point

The human body needs all 20 amino acids for the maintenance of health; 11 of these can be made by the body itself but the others have to be obtained through the food we eat and are called **essential amino acids**.

HBV and LBV proteins

Foods that contain all of the essential amino acids are said to have a high biological value (HBV). Most of these come from animal sources (meat, fish, poultry and dairy products) plus the vegetable source, soya.

Vegetable sources of protein include peas, beans, pulses, nuts and seeds, but because they do not contain all of the essential amino acids they are called low biological value (LBV).

Vegetarian, vegan and other restricted diets rely on combining LBV proteins, for example beans on toast, dhal and rice, to form proteins of higher value.

Denaturing

During food preparation and processing the structure of proteins can be changed. This process is called denaturing.

Applying heat changes protein structure. Think of the difference between a cooked and a raw egg. In a raw egg the white is transparent and liquid; after cooking it is opaque and solid. The

Table 8.1 HBV and LBV proteins

HBV proteins	LBV proteins
Meat Fish Eggs Cheese Milk Milk products Soya bean products	Peas Beans Pulses, e.g. lentils Nuts Seeds Rice Wheat

skin on top of boiled milk is also an example of denaturing.

Adding an acid also changes protein structure, for example adding lemon juice to milk will curdle it.

Fish protein is also denatured by lemon juice: it is effectively 'cooked'. This process is used to produce *gravadlax,* a Scandinavian dish similar to smoked salmon, or *cerviche* another dish 'cooked' without heat.

Marinating meat in vinegar or alcohol is also a form of denaturing; the structure of the meat proteins breaks down partially, making the flesh tender.

Mechanical denaturing, for example beating egg white or beating meat with a mallet, permanently changes the structure of the protein.

1 Why does the body need protein?
2 Explain the difference between HBV and LBV proteins.
3 Why can combining proteins be important? What individuals might be most at risk?

Design a HBV ready meal using non-animal sources. Use computer software to carry out a nutritional analysis of the finished dish.

Meat

Lean meat consists of the muscle tissue of animals. Muscles are long fibres bunched together and wrapped in connective tissue. Connective tissue is made from the basic proteins collagen, elastin and reticulin. Each of these proteins behaves differently when cooked. Collagen is weakened and forms gelatine, which makes the meat tender. The other two become firmer, holding the meat together.

Meat from young animals is tender because their muscles have done less work. It can be cooked quickly and will remain tender. Tougher meat from older animals or from the areas of the animal that do the most work, such as the neck and legs, needs gentle slow cooking to make it tender. Tougher cuts are often minced to break down the fibres. This helps to utilise all of the animal and enables quick cooking.

Key Points

The colour and flavour of meat depends on:
- The type of animal
- Its diet and lifestyle
- How much fat is present
- The cooking method

Table 8.2 Types of meat

Types of meat	Examples
Red meat	Beef, lamb, mutton
White meat	Pork, veal
Poultry	Chicken, duck, turkey
Game	Pheasant, partridge, rabbit, venison
Offal	Kidney, tongue, liver
Cured meat	Bacon, ham, salami

Fish

Table 8.3 Types of fish and seafood

Types of fish and seafood	Examples
Fresh water	Fresh water salmon, trout, perch, eels
Sea water	Cod, haddock, plaice, sardines
Crustacea	Prawns, lobster, scampi, crab
Shellfish	Cockles, mussels, squid, octopus

Fish can also be classed as oily and non-oily (see Vitamins and minerals).

The muscle of fish is made up of short bundles of fibres separated by connective tissue. This tissue is very fragile and converts to gelatine quickly. This gives fish its characteristic tenderness and flaky texture and is why it falls apart when overcooked. Fish deteriorates very quickly once it is caught and is usually put on ice immediately (see Preservation).

Eggs

Eggs are a rich source of protein and vitamins. Hens' eggs are the most commonly eaten but quail and duck eggs are readily available in the shops. Eggs are classified as free range, organic, battery or barn depending on how the birds live and what food they eat.

There is no nutritional difference between eggs with white and brown shells.

The shell of an egg is porous: if it is stored too long bacteria may enter and cause the egg to go bad. Eggs should be kept away from other foods because of the risk of their shells being contaminated with salmonella and cross-contamination occurring (see Preservation).

When cooked, eggs turn from a viscous liquid to a solid as the protein coagulates.

Fig 8.2 Eggs in food

Milk and milk products

Milk is available with different levels of fat:

- Whole milk contains up to 3.9g per 100ml
- Semi-skimmed contains up to 1.7g per 100ml
- Skimmed contains no more than 0.2g per 100ml
- '1% milk' contains no more than 1g fat per 100ml; this milk is a new development and is a blend of skimmed and semi-skimmed milk

It is also preserved as:

- Sweetened condensed milk
- Evaporated milk (less concentrated than condensed; no sugar added)
- UHT (ultra heat treated)
- Dried milk powder

Milk is a good source of protein, minerals and, in whole milk, vitamins A and D. Milk and its products should be stored chilled to keep them fresh.

Table 8.4 Properties of eggs in food production

Thickening	Custards and sauces can be thickened by adding eggs and heating
Emulsification	Egg yolk contains lecithin, which can be used to emulsify immiscible liquids, e.g. oil and vinegar to make mayonnaise
Binding	Eggs can be used to bind ingredients together, e.g. veggie burgers; the protein in the egg coagulates on heating and holds the burgers together
Coating	Eggs help coatings to hold on to products, e.g. breadcrumbs on fish fingers
Glazing	Baked goods can be glazed with whole egg prior to cooking to give a golden finish; a glaze of egg white with sugar gives a sparkly finish
Foaming	Whisking an egg white produces a foam which, when heated, will remain stable, e.g. meringues
Enriching	Because eggs are rich in protein and vitamins they enrich the foods to which they are added
Industry	Eggs are pasteurised and used in bulk

Table 8.5 Products made from milk

Cream	Cream is a fat-in-water emulsion made by separating the fat and solids from milk; it can be classed as double, single or whipping according to the percentage of fat present and each has slightly different characteristics
Butter	Butter is made from full-cream milk; it is a water-in-oil emulsion that is solid when cool
Yoghurt	Yoghurt is milk that has been coagulated by lactic acid produced from 'good bacteria'; bio-yoghurts contain live bacteria (see Probiotics and prebiotics)
Crème fraiche, sour cream	Slightly fermented creams with an acidic flavour; low-fat varieties are available
Natural buttermilk	Resulting from the removal of milk solids during the production of butter; it is used in baking
Cultured buttermilk	Fermented as sour cream and thickened through evaporation
Cheese	Cheese is produced from milk that is curdled following the addition of an enzyme, rennin; the solid curds are drained from the liquid whey and used as the basis for different varieties of cheese

Fig 8.3 Milk and its products

Investigate a range of milk products available in the supermarket and compare the nutritional content. Report your findings and suggest specific target groups for each product.

Lipids/fats and oils

Fats and oils are derived from animal, fish or vegetable sources. All fats and oils have different flavours and melting points so are suitable for different purposes:

■ **Shortening** – fats help to give pastry and cookies their crumbly texture

■ **Moisture retention** – fats help to keep cakes moist and increase the shelf life of baked goods

Key Points

'Lipids' is a general term for both fats and oils. Oils are fats that are liquid at room temperature. Fat is one of the macronutrients essential to health. All fats and oils have similar chemical structures and functions, and are all high in calories.

Fig 8.4 Lipids

The chemistry of fats

Fats are large molecules made up of only the elements: carbon, hydrogen and oxygen.

Glycerol is a molecule that consists of three carbon atoms, each connected to a hydroxyl (OH) group and represented as $C_3H_8O_3$.

Fatty acids are long chains of carbon and hydrogen (CH_2–CH_2–…) with a carboxyl (–COOH) at the end.

When any fatty acid combines with glycerol, fat is formed; the nature of the fat depends on the length of the fatty acid.

If all the carbon atoms are joined by a single bond (C–C–C etc.) they are known as saturated fats; if two or more carbon atoms are joined by a double bond (C–C–C=C–C etc.) they are known as unsaturated fats.

Saturated fats

Generally speaking saturated fats are solid at room temperature and have a high melting point. Too much saturated fat in the diet has been linked to high blood cholesterol.

Cholesterol

Cholesterol has the consistency of soft wax and is produced in the liver and transported round the body in the blood. It has been found that when too much cholesterol is in the blood it is deposited on the walls of the arteries, narrowing them and making them less efficient. Narrowed arteries are one of the major causes of coronary heart disease.

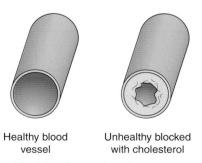

Healthy blood   Unhealthy blocked
vessel   with cholesterol

Fig 8.5 A healthy blood vessel and one narrowed by cholesterol deposits

Unsaturated fats

There are two types of unsaturated fats: polyunsaturated and monounsaturated. Unsaturated fats are usually soft or liquid at room temperature and have a lower melting point. These fats, monounsaturated in particular, are considered healthier because they can help to lower blood cholesterol. Some of these fats contain **essential fatty acids** and are an important part of a healthy diet.

> ### Key Points
>
> - Fats are used by the body for energy and also form part of the structure of cells
> - Fat stored under the skin helps to insulate the body against the cold
> - Fat is a source of the fat-soluble vitamins A, D, E and K (see Vitamins and minerals)
> - Fat in our diet helps to promote a feeling of satiety (feeling full or satisfied after eating)

Animal fats

Examples are suet made from the shredded fats of cattle and lard made from rendered (melted) pig fat. Ghee is made by heating and clarifying butter. Other examples include cream, cheese, butter, yoghurt, etc. which are all made from milk.

Vegetable fats

Vegetable fats can come from fruit, nuts and seeds. There has been a significant increase in the oils and fats available from vegetable sources as consumers and manufacturers look to follow the latest dietary advice.

Fig 8.6 Cod liver oil products

Fish as a source of fat

Some oily fish are a very good source of dietary fat, for example salmon, trout, mackerel and sardines.

Table 8.6 Sources of vegetable fats

Fruit	Nuts	Seeds
Olives, avocados	Peanuts (groundnut oil), walnuts, cashews	Sunflower, oil seed rape, hemp, soya, sesame

Cod is a white fish that stores very little fat in its body, but its liver contains oil and this is used as a dietary supplement.

Hydrogenation and trans-fats

Hydrogenation is a process used by food manufacturers to turn oils into spreads. Hydrogen is pumped through the oil and changes the structure of the fat molecules creating a solid. Although the oil is unsaturated the result of the process is that the new product is a saturated fat. During the hydrogenation process trans-fats are formed. It has been recognised that trans-fats produced in partially hydrogenated fats are more damaging to health than the fully saturated hard fats they were designed to replace.

It is now accepted by the government that trans-fats pose a real risk to health and some supermarket chains and manufacturers are banning their use in their products.

1 Why does the body need fat? Give three animal and three vegetable sources of fat.
2 Explain the differences in the nature and function of saturated and unsaturated fats.
3 What is cholesterol? Explain the effect it can have on the body.
4 What are trans-fats? Explain how they are formed and why they are a risk to health.

> **Draw up a chart to show the difference between the energy value of low fat and comparable regular products available in the supermarket.**

Carbohydrates

Carbohydrates are important macronutrients.

> ## Key Points
>
> Functions of carbohydrates:
> - They provide the body with energy for physical activity
> - They provide the body with energy to maintain bodily functions
> - They provide non-starch polysaccharide (NSP) dietary fibre to help digestion
> - They sweeten and flavour foods

They are divided into sugars and starches also known as simple and complex carbohydrates.

Table 8.7 Carbohydrates

Simple		Complex
monosaccharides	disaccharides	Polysaccharides /NSP
Glucose/fructose	Sucrose/ lactose/ maltose	Starch/fibre
		Vegetable/ fruit starches

Fig 8.7 Sources of carbohydrates

Carbohydrates are molecules made from carbon, hydrogen and oxygen.

Monosaccharides

Monosaccharides are also known as simple sugars. The simpler the carbohydrate the more quickly it can be absorbed in the body and the faster energy can be provided.

Glucose is one of these simple sugars and although it is found in some fruits and vegetables it is often used by athletes in tablet or powder form to provide a fast energy boost.

It is also used by industry in confectionary, jam making and in the brewing industry.

Fructose is similar in structure to glucose and is found naturally in the juices of some fruits and plants, but mainly in honey. It can be bought in crystalline form. As it is the sweetest of all sugars it is used to replace sucrose, enabling less sugar to be used to provide the same level of sweetness.

Disaccharides

Lactose is the disaccharide found in milk, which some people think gives milk its slightly sweet taste.

Maltose, another of the disaccharides, results from the fermentation of cereal grains. Malt is used in some food production and as a dietary supplement.

Sucrose is the most common disaccharide, a white crystalline substance used in homes and industry. It provides the body with energy but has no other benefits in the diet. It contains no other nutrients.

The most common problems relating to sucrose are obesity and tooth decay. If more sugar is taken than is needed for energy it is converted by the body to fat and stored under the skin and in the liver as glycogen. A diet containing too much sugar can lead to obesity and diabetes. Bacteria in the mouth convert sugar to acid. This can attack the enamel on teeth contributing to dental caries (tooth decay).

Polysaccharides

Polysaccharides or complex carbohydrates provide the body with energy and fibre, and are thus important nutrients. Modern dietary advice recommends that the dietary intake of fibre-rich complex carbohydrates be increased to provide at least 50 per cent of our daily energy needs.

These carbohydrates are found in grain products like bread, rice, cereals and pasta, and in fruit and vegetables.

Because of their complex structure these carbohydrates take longer to for the body to digest and so provide a feeling of satiety for longer, helping to avoid over-eating and obesity.

NSP (fibre) is a non-digestible carbohydrate found in plant foods. As it cannot be digested it passes straight through the digestive system absorbing moisture and providing bulk. NSP helps to 'push' other food through the system and helps to 'clean' the walls of the intestine. The efficient removal of waste products from the body is vital to health.

Too little NSP (fibre) in the diet can cause constipation and, in extreme cases, diverticular disease, where the lining of the intestine becomes distorted and inflamed.

Foods high in dietary fibre include whole grains, nuts, peas, beans and pulses. Vegetables and fruit also provide fibre, especially if eaten with their skins.

Fig 8.8 High fibre foods

The Glycemic Index

The Glycemic Index (GI) is a recently developed method of categorising carbohydrates by the speed with which this breakdown occurs. This new system questions some of the previous thinking about carbohydrates. It measures the rate at which the blood sugar level rises after a food is consumed.

A food with a high GI, such as processed carbohydrates, causes the blood sugar level to rise rapidly. Such rises have been linked to diabetes and heart disease.

Foods with a low GI are much slower to digest, cause less fluctuation in blood sugar levels and provide a longer period of satiety and help to control diabetes.

It is suggested that a diet rich in low GI foods can help general health and well-being, and control obesity.

1 Why does the body need carbohydrates?
2 Describe the differences between simple and complex carbohydrates.
3 Although NSP does not provide the body with nutrients, why is it important?

> **Research GI foods. Design a low GI savoury product suitable for a family meal, showing detailed ingredients and sketches. Carry out a nutritional analysis and compare your findings with a similar commercial product.**

Vitamins and minerals

Vitamins

They are called micronutrients because they are needed only in very small quantities, measured in milligrams (mg) or micrograms (μg). Although they all have chemical names they are usually referred to by letters.

Vitamins are divided into two main groups, fat-soluble and water-soluble:

- **Fat-soluble vitamins** – vitamins A, D and K are stored in the liver and used by the body when needed.
- **Water-soluble vitamins** – the B group vitamins and vitamin C are excreted unchanged in the urine. It is important, therefore, that foods containing these vitamins are consumed regularly.

Keeping vitamins in food

To maintain the highest level of vitamins in food it is important to take care with preparation and cooking.

A high proportion of vitamins in fruit and vegetables are found near their skins, so peeling strips away vital nutrients.

Prolonged exposure to heat and water also causes vitamin loss, so fast methods of cooking like stir frying or microwaving helps retention of vitamins.

Cooking in water is best done with a small amount of boiling water for a short time; steaming is preferable.

Functions of vitamins

Table 8.8 Fat-soluble vitamins A, D and K

Chemical name	Why is it important?	Where to find it		How it can be lost through processing
A Retinol Beta carotene	Keeps eyes healthy Improves night vision Helps maintain skin	Retinol in liver and oily fish Dairy products Beta carotene in red, green and orange vegetables and fruit		Destroyed by very high cooking temperatures or prolonged cooking
D	Builds and maintains strong bones and teeth	Dairy products Animal fats Cereals sunlight		Not normally affected
K	Blood clotting and wound healing	Dairy products Leafy vegetables		Not normally affected

Vitamin B is a group of vitamins all having similar functions. The most important B vitamins are listed in Table 8.9.

Table 8.9 Water-soluble vitamin B

Chemical name	Why is it important?	Where to find it		How it can be lost through processing
B_1 Thiamine	Nervous system Muscles	Whole grains Peanuts Offal Dairy products Some vegetables		Milling of grains Cooking of meat and vegetables
B_2 Riboflavin	Metabolism and growth Wound healing	Offal Almonds Dairy products Green vegetables		Destroyed by prolonged exposure to sunlight

Table 8.9 continued

Chemical name	Why is it important?	Where to find it		How it can be lost through processing
B$_3$ Niacin	Metabolism, growth and energy release	Cereals grains Peanuts Dried fruit Dairy products Pulses		
Folic acid	Red blood cells Foetal development	Offal Cereal grains Pulses Dark green vegetables		Exposure to sunlight and air Cooking in water

Table 8.10 Water-soluble vitamin C

Chemical name	Why is it important?	Where to find it		How it can be lost through processing
C Ascorbic acid	Helps wound healing and calcium absorption Prevents scurvy	Citrus and soft fruits Green vegetables Potatoes Peppers		Cooking and exposure to air and heat Blanching before freezing Canning Use of fertilisers

Minerals

Mineral elements are inorganic compounds found in many foods. They, like vitamins, are micronutrients vital to bodily function but are only required in very small quantities.

Some are needed in such small amounts they are called trace elements and their precise function is uncertain.

Minerals work together, each contributing to the growth, development and efficient functioning of the body.

1 Why is it important to have some fat in our diet?
2 What are the functions of vitamins?
3 What is meant by 'trace elements'?
4 Explain the functions of calcium in the diet and why requirements for this mineral change with age.

Design the meals for a young child for a day. Make a chart of the foods and detail the nutrients present.

Table 8.11 Minerals

Mineral	Function	Sources		Vulnerable groups
Iron	Production of red blood cells to carry oxygen in the blood	Red meat/offal Eggs Bread Green vegetables Drinking water		Pregnant, nursing and menstruating women
Calcium	Blood clotting Nerve and muscle function Heart regulation Combines with phosphorus to harden bones and teeth	Dairy products Fortified white bread Oily fish Green vegetables Citrus fruits		Growing children The elderly Pregnant and nursing women
Phosphorus	Metabolism Bones and teeth (in combination with calcium) Muscle function	Dairy products Nuts Meat Fish Foods rich in calcium		Deficiency rare
Sodium	In combination with potassium it is essential for maintaining the electrolytic balance of the blood Nerve transmission	Cheese Bacon Smoked meats Fish Processed foods		Athletes Deficiency rare Excess may cause high blood pressure

Fortification of foods

There are two main reasons why food is fortified:

- Government intervention to prevent illness and improve the general health of the nation
- Manufacturers wanting to provide consistent products and increase sales by appealing to a larger target audience

Fortification of certain foods was introduced by the government between the First and Second World Wars to help with some of the problems caused by poor diet and continues to be used to help the health of the nation.

Fluoride is added to drinking water to help combat tooth decay, vitamin A in margarine,

Fig 8.9 Fortified foods

vitamin D in milk, and calcium in flour. One of the recent recommendations by the Food Standards Agency (FSA) is to fortify flour with folic acid (see Vitamins and minerals) to help prevent neural tube defects in developing embryos.

Examples of fortification by manufacturers

Orange juice – The quality of the oranges varies through the year. At their peak the fruits' vitamin C content is high. When they are not so ripe the vitamin C content is lower. When they are not so ripe the manufacturers add vitamin C to the juice to ensure the nutrient content of the end product remains the same.

Bread – Wholemeal is recognised to be 'healthier' and provides more fibre, but not all consumers enjoy it. To appeal to the 'healthy' market the manufacturers developed a high-fibre white bread with the qualities of wholemeal while keeping the consumer appeal of white bread.

Dairy products – Some dairy products now have cholesterol-lowering ingredients added so that they to appeal to consumers anxious to follow current healthy eating advice.

1 Explain the term 'fortification' and give reasons why some foods are fortified
   (a) By the government.
   (b) By manufacturers.

- **Research folic acid and write a detailed argument for its addition to flour.**
- **Research a supermarket website for fortified products. Make a chart of the food and the nutrients with which it is fortified. Suggest reasons and the likely target market.**

Daily Recommended Values

Different groups of people need differing amounts of nutrients to stay healthy and provide them with the energy they need. Age, sex, lifestyle and body size all affect the quantity of nutrients needed.

COMA (Committee on Medical Aspects of food and nutrition policy), a government committee in the Department of Health, put together lists of the quantities of different nutrients and energy from food needed by the different groups of people. This list is:

DRVs, also known as dietary reference values, include:

- **Reference Nutrient Intake (RNI)** – shows estimated quantities needed for protein, vitamins and minerals
- **Estimated Average Requirements (EAR)** – lists estimates of requirements of energy from food

Energy is measured in kilocalories (kcal.) All the food and drinks we consume contribute to the total daily amount. If you eat and drink foods higher in energy than the body needs, the energy is stored as fat and you gain weight. If you use more energy than calories consumed you will lose weight.

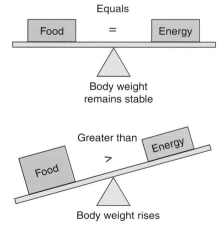

Fig 8.10 Foods consumed = energy used = weight steady

Foods consumed > energy used = weight increases

Ages and stages

Fig 8.11 Stages of life

During the different stages of life people require different foods and quantities of nutrients to keep healthy, and different nutritional theories exist on how best to do this (see Nutrients section). From a manufacturer's point of view this offers the opportunity to target groups with products designed for their particular needs.

Babies initially drink only milk but, as they grow, more energy is required so they are weaned on to solid food.

Toddlers are growing fast so they require a lot of energy from their food. They need a balanced diet but a high proportion of complex carbohydrates to provide this energy.

Adolescents still require a lot of energy from their food, particularly boys during their growth spurt.

Girls have a greater need for the mineral iron to replace that which is lost during menstruation.

Puberty is also the time when peak bone mass is reached; without sufficient calcium and phosphorus in the diet at this crucial time there will be a weakening of the bones leading to rickets and osteoporosis in later life.

Adults needs vary the most depending on their lifestyle and occupation. For many adults consuming too much energy from food leading to weight gain is the problem and they look to buy products to help with weight loss.

The elderly still require a balanced diet but they often suffer from a loss of appetite. They require appetising products in smaller quantities.

1 It is common for people to take packed lunches. Design a packed lunch for two days for:
   (a) A teenage student.
   (b) An office worker.
   What factors do you have to consider? Give reasons for your choices.
2 You have been asked to design a range of ready meals for the elderly. What factors do you have to consider? Suggest some suitable dishes giving reasons for your choices.

■ **Design some promotional material to convince young adolescents of the importance of drinking milk.**
■ **Visit the supermarket and investigate products aimed at adults on slimming diets. Using the information from the labels, compare the calories and nutritional values with similar non 'diet' products.**

New developments

As a result of consumers' need for convenience and a greater understanding of the role diet plays in health, scientists and food manufacturers continue to develop new products.

Novel foods

Enzymes

Enzymes are proteins that speed up reactions in physiological processes. They are used in industry in the production of cheese. One of the recent enzyme developments is chymosin, an artificial 'rennet'. Rennet is a naturally occurring

Fig 8.12 Products containing artificial sweeteners

enzyme used in cheese production; it is extracted from the stomach linings of cows. Increased consumer demand required an alternative source to be found and chymosin was developed using genetic modification. This development has resulted in an increase in the availability of vegetarian cheeses.

Artificial sweeteners

Artificial sweeteners fall into the category of novel products. There are two types: intense sweeteners such as aspartame, which is much sweeter than sugar and is used particularly in 'diet' drinks; or bulk sweeteners, such as sorbitol, which has a similar intensity of sweetness to sugar and is used in sweet manufacture.

Modified starches

Thickened (gelatinised) sauces, for example custard or cheese sauce, change if left to stand for some time or if they are reheated. They can become less viscous (thick) or more rubbery and start to exude liquid (syneresis). To provide consumers with dependable ready-made products a new type of starch was needed. Modified starches provide this and they can be adapted to meet the demands of the product. They are used in cook-chill meals, pot snacks, instant custard and soups, hot drinks and salad dressings. Modified starches:

- Prevent syneresis (weeping) in thickened sauces
- Act as a fat replacer in low-fat products
- Thicken sauces without forming lumps
- Stabilise oil in water emulsions

Functional foods

Key Points

Functional foods are those containing ingredients which give the product health-promoting qualities above their natural nutritional value (see also Fortification of foods). Many new products now exist in this category and the area is expanding all the time.

Cholesterol-lowering products

These contain plant sterol or sterol esters which are natural chemicals found in grains such as wheat and maize. These chemicals are able to absorb cholesterol in the digestive system. Benecol is probably the best-known brand name and produces yoghurts and spreads. These products have been found to be helpful in lowering blood cholesterol when used as a substitute for similar products as part of a healthy balanced diet.

Probiotics and prebiotics

Probiotics are foods that contain live, so-called 'good' bacteria. It is thought that eating more of these good bacteria helps the digestive system. There are numerous products on the market. They are consumed in the belief that they promote energy, help to prevent bowel problems and strengthen the immune system. Their critics however suggest that because they are 'live' cultures they may not always stay as effective after they have gone through the digestive process.

Prebiotics do not have that problem. They are not live cultures but they feed the healthy bacteria in our systems causing them to multiply. They are made from a form of starch found naturally in some fruits and vegetables. Because they are not 'live' they can easily be added to manufactured foods. They are already added to some breakfast cereals, bread, baby food and soft drinks.

They boost the healthy bacteria in our systems but also improve the absorption of calcium, making them particularly good for the young, pregnant women and the elderly.

Modified eggs

Eggs contain cholesterol. A diet high in saturated fatty acids is known to increase blood cholesterol, which is bad for the heart. Changing a hen's diet can change the nutritional profile of the egg however. Hens fed on a diet containing 10 to 20 per cent flax seed lay eggs that are higher in omega three fatty acids, which are known to help lower blood cholesterol. It is also possible to produce vitamin E enriched eggs.

Lipids/fats

Low-fat spreads made from polyunsaturated fats (for example sunflower oil) or monounsaturated fats (for example olive oil) are often used instead of butter and margarine to help reduce the consumption of saturated fats and overall energy intake. The spreads appear to be low fat because a percentage of their bulk is water. They are in fact water-in-oil emulsions fortified with vitamins. They are generally unsuitable to replace butter or margarine in cooking because they do not function well at high temperatures.

Low fat 'butter' is produced using similar methods, with water forming part of its bulk (see also Cholesterol-lowering products).

Fat replacers

The growth in the 'low fat' market has been huge in recent years. It follows concerns about increased levels of obesity and heart disease. Fat replacers are produced by modifying starches and proteins found in grains. They can be an effective way of reducing calories in commercially produced goods.

Meat analogues

HBV proteins are generally obtained from animal sources. They are often expensive and can be high in fat. Modern dietary advice suggests that we should eat less saturated fat.

Meat analogues made from vegetable sources such as soya beans and mycoprotein have been developed:

- Textured vegetable protein (TVP) – an example is Realeat's Vegemince® based on processed soya beans
- Quorn is made from a fermented mycoprotein, distantly related to the mushroom.

These products are sold as products resembling 'ham' or 'chicken' or loose in packets like 'mince' or 'meaty' chunks that can be used in recipes instead of meat.

They are used to by manufacturers to produce ready meals like lasagne and pies, or made into products like sausages and burgers. They are also used to 'bulk' or extend meat products, which lowers manufacturing costs. Alternative meat-like protein is an area of continuing development.

Carbohydrates

Fructose products are now readily available in the shops. They are produced to resemble granulated sugar and can be used in the same way. As fructose is the sweetest of the sugars, less is needed to achieve the same level of sweetness. Fructose is used by consumers trying to reduce their calorie intake.

1 Explain the difference between novel and functional foods.
2 How are modified starches used in modern food production?
3 What is the difference between probiotic and prebiotic foods?

Investigate products available in the local supermarket that have probiotic or prebiotic qualities. Make a note of the health benefits claimed.

Learning outcomes

By the end of this section you should have developed a knowledge and understanding of the importance of the sensory qualities of food in product design and analysis.

Introduction

The sensory analysis of food is one of the most important aspects of food product design and manufacture.

The organoleptic qualities of food

Organoleptic refers to the sensory properties of food: taste, smell, sight, sound, mouth-feel and touch.

> **Key Points**
>
> When analysing food products the senses – sight, hearing, feel and taste – work together.

Taste

There are four basic tastes: sweet, sour, bitter and salt. In addition, umami – savoury – has been suggested as a fifth taste.

When we eat food taste and smell work together to help us appreciate it fully.

We taste with our tongue; different areas are sensitive to different tastes. The food mixes with saliva in the mouth, which enables the flavour to develop.

Areas of taste

Fig 8.13 The tongue – areas of taste

Certain factors affect taste:

- Taste buds – the number of taste buds we have diminishes with age so flavours appear much stronger to young children than to adults, an important factor in product design.
- Temperature – flavours appear strongest around blood temperature, 37°C. Very cold foods, like ice cream, need to be flavoured more intensely as, once frozen, the flavour appears to diminish.

Smell

The smell of food enables the taste to be fully appreciated. If our sense of smell is reduced, as with a cold, we cannot taste food as well.

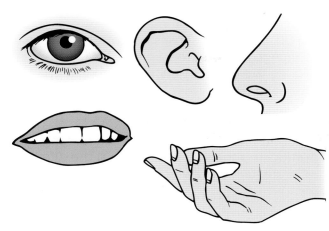

Fig 8.14 The senses

Sight

Pictures of appetising food is often enough to stimulate our taste buds. The colour of food is important. Different foods are associated with different colours and the depth of colour with the intensity of flavours. Artificial colours are frequently used in industry to enhance the appearance of food products (see Additives used in industry).

Sound

The sound made during the preparation of food, though not necessarily the most important aspect, also affects our enjoyment. The noise of frying for example can whet the appetite, as can the sound of food crunching in the mouth or a fizzy drink being opened.

> ### Key Term
> Mouth-feel

Mouth-feel and touch

Mouth-feel is the term used to describe the way food feels in the mouth. The texture and consistency, such as moist, dry, rough, smooth, slimy, rubbery, lumpy, crumbly, as well as cool and hot, are all qualities of mouth-feel and have a direct effect on our enjoyment of eating. Touch can also affect our judgement of foods: a firm apple, a crusty loaf.

Sensory analysis and industry

> ### Key Points
> Sensory analysis in design and development of food products is used to:
> - Compare products with competitors' products
> - Try out new products on the public
> - Monitor prototypes during the design process
> - Check consistency of quality during mass production
> - Assess the quality and shelf life of products

Sensory panels

Sensory panels, made up of highly trained individuals, are also used by food companies to assess the organoleptic qualities of products in design and manufacture.

Using their different senses they evaluate foods, producing results using descriptors (words that describe the product's characteristics).

- **Sight descriptors** – words about appearance colour and texture: shiny, dull, golden, lumpy
- **Smell descriptors** – odour or aroma: burnt, acid, fruity
- **Touch descriptors** – temperature, texture or mouth-feel: dry, rough, smooth, slimy
- **Flavour descriptors or food type** – lemon, orange, chocolate, sweet, sour, spicy
- **Sound descriptors** – noise in cooking or eating: popping, sizzling, fizzing, crunching

These panels are trained in sensory analysis and their assessments are made under strictly controlled conditions. Similar methods can be used by students engaged in product development in schools.

In schools, tests should be made as fair as possible. Tasters should be given water to sip between sampling.

Samples being assessed must be:

- In identical containers
- At the same temperature
- The same sized portion
- Numbered or named to identify them clearly and avoid confusion

Clear instructions about the feedback required should also be given.

Tests

Rating tests – foods are scored from one to five for the required judgement, for example one is poor and five is excellent.

Ranking tests – asks testers to rate foods in order of preference

Triangle tests – asks for the odd one out in a sample of three

Star Profiles – used to assess several sensory descriptors and create a profile of the product

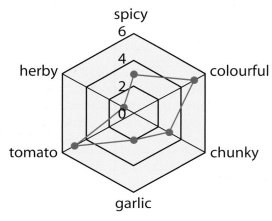

Fig 8.15 A star profile used in sensory analysis

1 What is 'organaleptic'?
2 Explain how our senses contribute towards the enjoyment of food.
3 Name the different tastes, giving examples for each one.
4 What is sensory analysis and why is it used in the food industry?
5 Explain the differences between ranking, rating and triangle tests. Explain the different circumstances in which they might be used.

■ Using descriptors for sight, sound, taste and smell, analyse two sweet and two savoury products. Suggest a possible target group, giving reasons for your choice.
■ Carry out a comparison between similar products such as yoghurt, drinks, burgers and chips. Draw a star profile for each of the products analysed.
■ Compare low and full-fat or diet and regular products such as butter-type spreads, cheeses, yoghurts, fizzy drinks and sweets. Draw star profiles for the products analysed.

Section C — Working characteristics of materials and components

Learning outcomes

By the end of this section you should have a knowledge and understanding of the range of materials, components and additives used in commercially produced foods to achieve thickening, aeration, emulsification, coagulation, gelation, shortening, flavouring, colouring, stabilising and extending shelf life.

Introduction

In this context **materials** refers to any basic ingredient, for example flour. **Components** refers to more than one ingredient combined to form a product, for example sauce mixes.

Thickening

Food products, particularly sauces and soups, are thickened to change their consistency and appearance. There are four main ways in which products can be thickened:

- The addition of starch
- Blending
- Evaporation
- The addition of a protein (egg)

Starch

The starches most commonly used for thickening, by the process of gelatinisation, are wheat and corn flour, but potato flour, rice flour and arrowroot are also used in certain circumstances.

> **Key Term**
>
> Gelatinisation

Starch consists of minute granules. When heated in a liquid the granules swell; when the temperature reaches 80°C the granules rupture releasing the starch and thickening the liquid. A gel has been formed; this is called gelatinisation. The thickness of the gel depends on:

- The proportion of starch to liquid
- The type of starch
- The effect of other ingredients, for example acids

Fig 8.16 The use of starch

Arrowroot starch granules are particularly large and create a clear gel. It is used in confectionary and packet mix glazes, for example Quick Gel.

To avoid some of the problems relating to gelatinised sauces, modified starches have been developed (see New developments).

Blending

Vegetables cooked in a sauce or soup and blended together result in a thickened product.

Evaporation

Continued cooking of a blended liquid at high temperatures causes excess liquid to be driven off as steam, resulting in a thicker product.

Addition of a protein

Eggs can be used to thicken sauces, an example being custard. With gentle heating the egg proteins coagulate, thickening the sauce.

A firmer set and lower coagulation temperature can be achieved by:

- Increasing the quantity of egg to liquid
- Adding salt
- Adding an acid

Aeration

Aeration lightens the texture and increases the volume of baked goods.

> **Key Points**
>
> Aeration of food is produced in three main ways:
> - Mechanical methods
> - Raising agents
> - Steam

Mechanical methods for introducing air

- Sieving, for example flour
- Whisking, for example egg white to make meringues
- Rubbing in, for example fat into flour during cake and pastry production
- Creaming, for example butter and sugar in cake production

- Beating, for example batter for Yorkshire puddings
- Rolling and folding, for example puff pastry

Raising agents

Raising agents work on the principle of gases expanding during heating.

Fig 8.17 Products using raising agents

Steam

Steam is produced during the baking process in foods containing water. As the mixture is heated, the liquid evaporates and steam is formed, which raises the product.

The volume of any liquid present is multiplied 1600 times as steam as a result, so even small quantities of liquid are important in the raising process.

Steam is the major factor in the production of batters and choux and flaky pastries.

Emulsification

Immiscible liquids like oil and water will not mix together; emulsifiers are substances that help to stabilise such a combination by lowering the surface tension of the droplets of liquid, enabling them to combine. In industry additives are identified by 'E' numbers; all emulsifiers have numbers between 322 and 499, for example lecithin (E322). Emulsifiers are added to a whole variety of products including chocolate, sauces, salad dressings and ice cream.

Fig 8.18 The use of emulsifiers

Coagulation

Coagulation refers to the changes that take place in liquid proteins following the application of heat. The proteins become 'denatured' and they become solid. An example is the difference between a raw and cooked egg (see Protein).

Gelation

The production of some foods requires gelation or 'setting'. The substances most commonly used are:

Table 8.13 Raising agents

Chemical raising agents		Biological raising agent
Bicarbonate of soda	**Baking powder**	**Fresh and dried yeast**
Used for strongly flavoured cake-like products owing to the residual taste of soda after baking. It works by releasing carbon dioxide.	A mixture of chemicals, including bicarbonate of soda, with differing qualities, ensuring a dependable outcome in baked goods with no residual flavour. It works by releasing carbon dioxide.	A natural raising agent used for making sweet and savoury bread products. It is a single-celled organism which, when mixed with warm water and sugar, reproduces and releases carbon dioxide bubbles that expand when heated, causing the dough to rise. Yeast is destroyed at high temperatures.

- Gelatine, a protein extracted from animal carcasses
- Agar, a substance extracted from marine plants
- Gums derived from natural plant sources
- Starch, which when boiled and cooled, becomes a gel

Gels absorb the liquid in a product creating a gel; when heated it becomes a liquid called a sol. When the sol is cooled it sets and again becomes a gel.

Gelation is used in the production of jellies, some cheesecakes and mousses, yoghurts, low-sugar jams and low-fat ice creams.

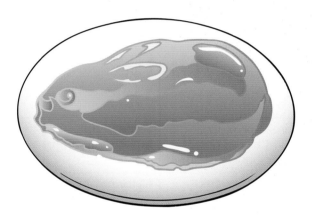

Fig 8.19 Products that use gelation

Pectin is a complex carbohydrate found in some fruit. This performs a similar function in traditional jam making.

Shortening

Shortening is a term used to describe the effect of fats on the protein, particularly gluten, found in flour. In cake and pastry production the fat coats the flour granules preventing the gluten forming and effectively 'shortens' the protein, making the product more crumbly and 'tender'. The higher the proportion of fat to flour the 'shorter' the product.

When the word shortening appears in American recipes, however, it refers to cooking fat.

Fig 8.20 Products with shortening

Additives used in industry

Table 8.14 Additives used in industry

Types	Reasons for use	Products	Possible disadvantages
Flavourings do not have E numbers and are not always listed specifically	Create flavours; often many used together to mimic natural flavours	Wide range of products, e.g. crisps, soups, pot snacks, juices yoghurts, desserts	Can signal a reduction in the natural product, e.g. the fruit in juice
Flavour enhancers	Enhances the flavour of food products without adding flavours of their own	Sweet and savoury products	Mono sodium glutamate (MSG); E621 can cause allergic reactions in sensitive individuals
Artificial and natural colourings E100–180	Improve the appearance of food; replace colour lost in processing	Sweets, desserts, juices, meat products, gravy mixes, squashes	Can cause hyperactivity ADHD or allergies in sensitive children
Stabilisers and emulsifiers E322–495	Prevent the separation of emulsions; gives products a consistent texture	Low-fat products, salad dressings, sauces, dessert mixes	Can affect the flavour of foods
Preservatives E200–285	To extend shelf life of foods	Numerous products including baked goods, sauces, yoghurts, meats	Some people are sensitive and have a reaction
Anti-oxidants E300–321	To prevent rancidity of fats and colour loss	Numerous products, particularly those containing fats and oils	As for preservatives

Enzymic and non–enzymic browning

Enzymic browning is particularly noticeable in fruit with white flesh and potatoes. The cut surface, when exposed to the air, turns brown or black. Enzymes, which cause this effect, are present in all living cells; they break down animal and plant tissue but the process speeds up once the cells are damaged.

Methods to prevent or slow down enzyme activity include:

■ 'Blanching' vegetables or fruit for a short time in boiling water
■ Adding an acid such as lemon juice
■ Adding vitamin C to fruit
■ Storing prepared fruit under water

Fig 8.21 Enzymic browning

Non-enzymic browning happens through the Maillard reaction and caramelisation. The Maillard reaction refers to the browning of meat in cooking. It is caused by a series of reactions between the sugars and amino acids in the meat. The whole process is not fully understood but it results in the formation of polymers that affect the flavour of the meat and turn it brown.

The flavour, appearance and aroma of meats are enhanced. This reaction will not occur below 140°C; as this is too high a temperature to cook meat for a long period, it is therefore seared or browned over a high heat before being cooked more slowly. In the production of ready meals a blow torch is often used.

Caramelisation occurs when sugar is heated. The sugar melts and gradually dehydrates. Coloured polymers are formed, initially golden, turning brown and eventually black and developing a bitter taste.

Caramel is used in sweet and dessert products, and as a natural colouring in savoury products and gravy browning.

1 Give four methods of mechanical aeration, giving examples of products made using each method.
2 What is gelatinisation? Explain the process by which sauces are thickened and the factors that can affect the results?
3 How does gelatinisation differ from gelation? What different products are used?
4 What are immiscible liquids? Suggest how the problems they create can be solved.
5 Explain how the fat in pastry helps to make it light and tender.
6 Give the main reasons why additives are used in food and drink products. Suggest any problems they might cause.
7 Explain the difference between enzymic and non-enzymic browning and give examples.

■ Investigate fresh and dried yeast. Suggest advantages and disadvantages for each type.
■ Explain in detail how yeast functions.
■ Give examples of products made using yeast.

Section D — Principles of preservation and prolonging the shelf life of food products

Learning outcomes

By the end of this section you should have developed a knowledge and understanding of:

- **Preservation** – what is food preservation? Types of micro-organisms and factors affecting their growth. Some aspects of food hygiene and safety.
- **Methods of preservation** – the underlying principles of preservation. Some methods of preservation. The effects preservation can have on the properties of food and materials. Pasteurisation, sterilisation, UHT, canning, drying, freeze drying, freezing, chemical preservation.
- **Food packaging** – the main materials used in food packaging, such as glass, paper, board, metal, foil and plastics. The methods used in packaging, such as shrink wrap, aseptic, modified atmosphere (MAP) and vacuum.

Preservation
Micro–organisms

Food preservation is the process of treating, storing and handling foods in such a way as to slow down deterioration and help prevent food-borne illness.

It involves inhibiting the growth of micro-organisms including bacteria and fungi that can cause food poisoning and spoilage.

Bacteria

Bacteria are micro-organisms too small to be seen by the naked eye. There are three main types:

- Helpful bacteria
- Spoilage bacteria
- Pathogenic bacteria

Helpful bacteria are used in the production of cheese, yoghurt, wine and beer, and are added to some functional foods (see Functional foods).

Spoilage bacteria, as their name suggests, cause food to spoil and deteriorate.

Pathogenic bacteria are the most dangerous; they can contaminate food without affecting its appearance or aroma. They cause food poisoning and occasionally death, especially in the very young and old.

Other factors affecting bacterial growth:

- pH – the level of acidity is a factor in bacterial

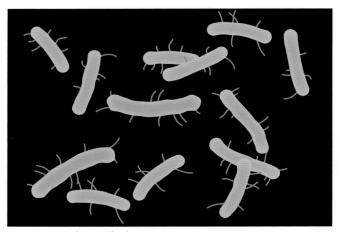

Fig 8.22 Salmonella bacteria

Key Points

Bacteria are living organisms and generally need certain conditions to thrive and multiply:
- Food
- Moisture
- Warmth
- Time

growth; bacteria cannot thrive in acid or alkaline conditions.

- Oxygen – some bacteria, known as aerobes

Table 8.15 Pathogenic bacteria that can cause food poisoning

Bacteria	Source	Found in	Onset/symptoms
Salmonella	Human and animal intestines, sewage and pests	Raw poultry, eggs, meat and milk	12–36 hours, abdominal pain, diarrhoea, vomiting, fever
Clostridium perfringens	Animal and human excrement, soil dust, insects	Cooked meat	12–18 hours, abdominal pain, diarrhoea
Bacillus cereus	Cereals, soil dust	cooked rice, cereal grains	generally1–5 hours but can be up to 16 hours, abdominal pain, some diarrhoea, vomiting
Staphylococcus aureus	Human body (skin, nose, cuts, spots)	Cold meats and dairy products, anything touched by hand	1–6 hours, abdominal pains and cramps, vomiting, low temperature
Clostridium botulinum (botulism)	Soil, water	Meat, fish, vegetables, canned foods	12–36 hours, difficulty in breathing and swallowing, paralysis, death

require oxygen. Others, called anaerobes, do not.

Botulism is a very serious condition caused by an anaerobic bacteria; it is usually fatal. It is very rarely found in humans and usually results from poor canning techniques.

Key Points

Bacteria thrive on so called high-risk foods; these are often moist and high in protein. They can be raw or cooked. High-risk foods include:
- Meats and poultry and their products
- Dairy products and dishes made with them, such as mousses, mayonnaise and sauces
- Fish and seafood
- Soft cheeses and pâtés
- Cooked rice

Moisture

Bacteria need moisture but cannot survive where there is a high concentration of salt, sugar or acid. Dried foods are therefore low-risk foods. Once moisture is added to them, however, they behave like fresh food and will be susceptible to bacterial contamination.

In food where there is a high concentration of salt, acid or sugar, the moisture is effectively absorbed and not available to the bacteria so they will not thrive (see Methods of preservation). Such foods include pickles and sauces, jams and preserves, and salted meats and fish.

Temperature

Bacteria need warmth to grow. At low temperatures they lie dormant but will start to

multiply once conditions improve; only at prolonged high temperatures are they killed.

During the production of cooked foods thermometers and probes are used ensure the centre of the product has reached 70°C, the temperature required to kill most bacteria. Particular care should be taken when cooking foods that have been frozen to ensure they are thoroughly defrosted.

In industry the monitoring systems are often automated and temperatures are displayed digitally on the equipment concerned.

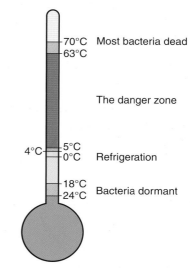

Fig 8.23 Important temperatures

Key Points

The danger zone
Temperatures between 5°C and 63°C are called the *danger zone* because within this range bacteria are most likely to multiply. Ambient temperature, around 37°C, is the optimum for multiplication.
It is important to note that there are some exceptions: bacteria such as Listeria will continue to multiply below 5°C.

Table 8.16 The effect of temperature on bacterial growth

	Temperature	Effect on bacteria
Freezing	–18°C to –24°C	Bacteria dormant
Chilling	5°C to 1°C	Minimal bacterial growth
Danger zone	5°C to 63°C	Bacteria can multiply
'Hot holding' (cafeteria and carvery)	minimum 63°C	Minimal bacterial growth
Cooking	not below 70°C	Most bacteria dead (core temperature must be held at 70°C for two minutes)

Some bacteria form spores, a protective mechanism to avoid being destroyed at high temperatures. The bacteria will emerge however and continue to multiply when conditions improve. Storage outside the temperature danger zone is crucial.

Time

Bacteria multiply by binary fission. In ideal conditions, one bacterium becomes two, two become four, and so on, every five to ten minutes. It does not take long for there to be sufficient bacteria to contaminate food and become a threat to health. Food must remain within the danger zone for only a short period of time.

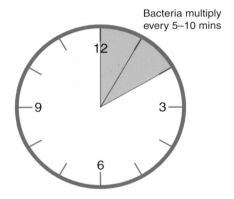

Bacteria multiply every 5–10 mins

Fig 8.24 Bacteria multiply every five to ten minutes

Bacteria can be aerobic or anaerobic. Anaerobes do not require oxygen to multiply and can continue to grow in airless conditions such as cans and vacuum-packed foods.

Key Points

Correct food handling procedures:
- Ensure good personal hygiene
- Wear protective clothing
- Wash hands frequently and thoroughly
- Keep all food preparation and storage areas clean
- Store and prepare raw and cooked foods separately
- Keep foods covered until needed
- Store fresh food below 5°C
- Keep freezers below -18°C
- When cooking, the core temperature of food must reach a minimum of 70°C and be held at that temperature for at least two minutes
- Keep the time food spends in the danger zone to a minimum

Food hygiene and safety procedures

It is important to follow correct food handling procedures to avoid contamination and prevent the growth and spread of bacteria.

Vehicles of contamination

Bacteria cannot move on their own, they rely on 'vehicles' to transport them. These include:

- Hands
- Work surfaces and containers
- Equipment and tools
- Clothing and cloths
- Insects and animals

All these are possible sources of contamination and, if hygienic practice and cleaning routines are not followed, contamination can occur.

Cross-contamination

Cross-contamination is when pathogenic bacteria are transported from a contaminated source such as raw meat to high-risk, ready-to-eat food. Bad practice, resulting in such contamination, is the greatest cause of food-borne illness.

Viruses

Some viruses and other microbes can cause food-borne illness. Unlike bacteria they do not multiply on foods but in living cells. Food and water just provide the vehicles for their distribution. When eaten they start to multiply in the body cells.

Key Points

Viruses are transported on food and in water but do not require them to survive.
- The main sources of viruses are polluted water and sewage
- Viruses can be spread easily through poor personal hygiene

Table 8.17 Viruses

Virus	Source	Found in	Onset/symptoms
Listeria	Soil, water, sewage, humans	Soft cheeses, pâté, salads, unpasteurised dairy products	Flu-like symptoms
Campylobacter jejuni	Animals, sewage, impure water	Raw meat, poultry and milk, untreated water	Abdominal pain, diarrhoea, nausea, fever
Escherichia coli (E-coli)	Human and animal intestines, sewage, impure water, raw meat	Beef (especially mince), raw milk and impure water	Abdominal pain, diarrhoea, fever, kidney damage, in extreme cases kidney failure

E-coli has been the cause of recent deaths; the infection was linked to poor standards of hygiene in meat processing.

Fig 8.26 Mould

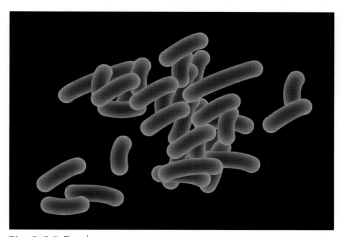

Fig 8.25 E-coli

Moulds

Moulds are a type of fungus that can grow on both moist and dry food. Humidity will quicken their growth. They are most often found on baked goods and citrus fruits. They cause spoilage but generally do not cause food poisoning. Mouldy food is usually destroyed. Some 'blue' cheeses have mould introduced into them, such as stilton and cambazola.

Yeasts

Yeasts are micro-organisms that need oxygen. They will grow in foods high in sugar or salt. They can cause spoilage, making jams or fruit juices ferment. They have positive uses in the production of bread, wine and beer.

(a) Describe some of the micro-organisms that can be found in foods.
(b) List ways in which micro-organisms can be:
    (i) Harmful
    (ii) Beneficial
(a) Name three bacteria that cause food poisoning.
(b) Describe briefly the symptoms they cause.
(c) What safeguards can be put in place to minimise the risk of bacterial contamination?
1 What is the difference between food-borne illness and food poisoning?
2 What is the danger zone and why is it important?
3 Explain the term 'high-risk food' and why it can pose a threat.

■ Investigate HACCP and risk assessment.
Explain their importance in industry.
■ Draw up a HACCP chart for the production
of a fresh cream trifle and a shepherd's pie.

Methods of preservation

Pasteurisation/sterilisation/UHT

These processes are most often associated with milk but can be used on other products to kill bacteria and help prevent food-borne illness and deterioration. Eggs yolks bought in bulk and used in industry for mayonnaise production are pasteurised to avoid the dangers of salmonella.

Canning

Canning and bottling work using the same principles: cooked foods are sealed in containers that are heated to 121°C to kill any remaining bacteria. They are held at this temperature for a period of time, depending upon the product, before being cooled in chlorinated water and labelled. Cans used for acidic products like tomatoes and some fruits are lined with plastic

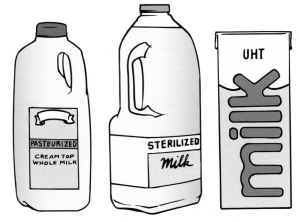

Fig 8.27 Pasteurised, sterilised and UHT milk

to prevent corrosion. If undamaged, the containers prevent further contamination and the products will remain fresh, if unopened, for years.

Fig 8.28 Canned and bottled food

Table 8.18 Pasteurisation, sterilisation and UHT

	Process	**Effect on pathogens**	**Effect on product**
Pasteurisation	Food is heated to 72°C for 15 seconds and cooled rapidly	Pathogenic bacteria are destroyed	Products taste fresh, needs refrigeration, short shelf life
Sterilisation	Milk is sealed in a glass bottle and heated to 115–130°C for 10–30 minutes	All micro-organisms are destroyed	Flavour and colour is affected, needs no refrigeration, shelf life of two to three weeks if unopened
UHT	Food is heated to 132°C and held for one second	All micro-organisms are destroyed	Flavour affected, shelf life of up to six months if packaging is undamaged

Drying

Drying is one of the oldest methods of preservation. Removing moisture from a food product creates conditions in which bacteria cannot multiply, thereby increasing their shelf life. Foods are often naturally dried in the air and sun.

Drying reduces weight and bulk, which can lower the cost of transportation for manufacturers.

Fig 8.29 Dried food

Hot air bed drying

Foods are spread on perforated trays and are dried in a current of hot air. Temperature and time is carefully controlled. Foods often dried in this way are fruits, herbs, meat, fish and grains.

Accelerated freeze drying

Accelerated freeze drying is a form of drying that creates less deterioration in the taste, texture and nutritional value of the food. The food is first frozen and then subjected to a vacuum that evaporates the water in the food. This method is most commonly used to produce instant coffee and complete meals but is also successful for fish and some fruit and vegetables.

Spray drying

Food in a liquid form is sprayed into a blast of hot air; the droplets dry forming a powder. This method is used in the manufacture of some dried milks and soups but can form lumps when reconstituted.

Fluidised bed drying

This form of drying produces granules rather than powder. Dried food particles are formed into clumps which are then dried in hot air. The clumps reconstitute easily and are less prone to lumping. This method is used in the manufacture of products such as instant mashed potato and instant coffee.

Chilling

- Refrigeration is a short-term method of preservation
- Bacterial growth and enzyme action is slowed but not stopped
- Chilled foods have a short shelf life
- Storage temperature is between 5°C and 0°C
- Following research into Listeria (a pathogenic bacteria) commercial fridges do not rise above 4°C

Chilled foods have shown the greatest increase in popularity with consumers and encompass virtually all areas of food production from ready meals, fresh pasta and ready prepared fruit and vegetables, to desserts, snack foods, dairy products and juices.

> **Key Points**
>
> - 'Use by dates' for chilled products only apply if foods have remained at 5°C or below
> - Foods that have reached room temperature should be consumed as soon as possible.

> **Key Term**
>
> Cook chill

Cook chill refers to food dishes that are prepared in bulk, portioned and rapidly chilled for consumption later. Careful portioning during manufacture helps reduce costs.

Rapid chilling helps preserve the flavour and appearance of the food and ensures the longest possible shelf life.

- Dishes are prepared and blast chilled to 4°C within 90 minutes
- They must continue to be stored at this temperature until they are used

Table 8.19 The three main types of freezing

	Method	Used for
Multi-plate freezing	Foods are pressed between freezing plates	Flat foods and food products such as burgers and lasagne
Blast freezing	Foods are subjected to a blast of freezing air	Vegetables and solid fruits
Cryogenic freezing	Foods are immersed in or sprayed with liquid nitrogen	Delicate products and fruits, such as raspberries and meringue desserts

- With normal packaging such foods have a commercial shelf life of three to four days
- The system is used to produce domestic ready meals and in the mass catering industry

Freezing

Freezing is probably the most commonly used form of preservation commercially and domestically. It is suitable for a wide range of products both fresh and cooked.

Domestic freezers run at –18°C, short-term fast freezers are set at –26°C, while commercial freezers run at –29°C. The three main types of freezing used in industry are shown in Table 8.19.

Before freezing some foods are blanched to halt enzyme activity. This helps retain the colour and quality of the food.

It is important that foods are frozen as quickly as possible to avoid the formation of large ice crystals; these damage food cells and can affect the texture, flavour and nutritional value of the food when defrosted.

Foods with a naturally high water content, for example lettuce and cucumber, are not suitable for freezing.

Star ratings on food products indicate how long food can be safely stored in a domestic (****) freezer:

* = one week
** = one month
*** = until the best before date

Chemical preservation

Chemical preservation (see Additives used in industry) is used to:

- Slow down bacterial growth
- Maintain the appearance of food
- Prevent the rancidity of fats

Rancidity

When fats and oils spoil they are said to be rancid. They develop an unpleasant smell and taste. Rancidity is caused by:

- **Enzymes** – heat destroys these enzymes but they remain active in frozen food; this shortens the storage life of foods containing a high proportion of fat, such as bacon.
- **Bacteria** – although unable to grow in fat, bacteria can be present in foods containing moisture or protein, such as butter and ham.
- **Oxidation** – this occurs more quickly in high temperatures or strong sunlight and particularly affects unsaturated fats and oils. Antioxidants are added to foods by manufacturers to help prevent this deterioration (see Additives used in industry).

The most commonly used chemicals are:

- Benzoates
- Nitrites
- Sulphites

Food labels inform consumers which chemicals have been used in production.

> **Key Points**
>
> Check the temperature of fridges and freezers regularly.

1 After reading this section what do you consider to be the principles behind preservation? Why have these techniques been developed?

2 Explain the differences between pasteurisation, sterilisation and UHT treatment of milk.

3 How does home freezing differ from that used in industry?

4 Describe the rules that govern the 'cook chill' process. How is it of benefit to the consumer?

Research irradiation as a method of preservation.
- **Explain the principles behind the method**
- **List the advantages for the consumer**
- **Give reasons why there might be consumer concerns**
- **State the foods most likely to be treated**

Food packaging

There are two categories of packaging: primary and secondary.

- **Primary packaging** holds the food itself and is designed to protect, promote and inform about the product.
- **Secondary packaging** is used for protection and security in transit and distribution.

Packaging materials are chosen to suit the requirements of the product. Some of the factors that determine choice are: will it need to be visible? How delicate is it? Might it be eaten from the container? Does it need to be resealed?

Key Points

Packaging is used for six main purposes:
- Physical protection
- Environmental protection
- To prevent tampering
- To provide information
- Advertising/marketing
- As a container for cooking

Protection

Products need protection from damage during transportation and storage, and products such as liquids or powders need containment to avoid them being spilled.

In terms of environmental protection, packaging can help to protect foods from physical and bacterial contamination caused by moisture, dirt or insects and animals.

If packaging is undamaged it also proves that the product has not been tampered with. Some products such as baby foods have tamperproof lids, the centre of the lid 'popping up' when opened.

Information

Labelling on packaging provides information to help the consumer make informed choices. Much of this information is governed by the Food Safety Act and is required by law.

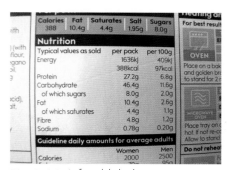

Fig 8.30 A food label

Key Points

Food labelling regulations require:
- Name of the food (not the brand name)
- List of ingredients (in weight order)
- Precise percentage of food specifically mentioned, such as tomatoes in soup
- Claims: energy value, protein value and so on; these are subject to tight controls
- Nutritional content
- Allergen labelling
- Genetically modified foods and ingredients (newly introduced)
- Lot marking to enable a manufacturer to recall lots if contaminated
- 'Best Before' or 'Use By' date
- Storage instructions
- Instructions for use
- Name and address of manufacturer or packer and/or seller within the EU
- Place of origin (in some circumstances)

Weight need not be included by law but most foods include an indication of their weight or volume.

Table 8.20 The advantages and disadvantages of different materials used in packaging

Material	Advantages	Disadvantages
Glass	■ Contents can be seen ■ Cheap to produce ■ Can withstand heat ■ Easy to reseal ■ Rigid ■ Fluid resistant ■ Can be coloured ■ Easy to clean and sterilise ■ Can withstand pressure ■ Can be recycled repeatedly ■ Can be fitted with tamper-proof lids ■ Will not taint foods	■ Easily broken ■ Heavy to transport
Paper and paperboard	■ Cheap to produce ■ Biodegradable ■ Sustainable ■ Lightweight ■ Rigid ■ Can be recycled ■ Easy to print on ■ Easy to form ■ Can be produced in different weights ■ Has insulating qualities ■ Can be treated to withstand grease and moisture ■ Can be coated in PET to withstand high and low temperatures	■ Affected by moisture if untreated ■ Can be crushed
Steel	■ Strong ■ Can be recycled ■ Can extend shelf life (canning)	■ Can react with foods if untreated ■ Susceptible to corrosion ■ Not suitable for microwaves
Aluminium	■ Lightweight ■ Pliable ■ Can be produced as foil ■ Can produce film coating when vaporised onto paper or plastic ■ Can be used as a laminate ■ Easily shaped, e.g. ring pulls ■ Can be printed ■ Easy to recycle	■ Expensive to produce ■ Not suitable for microwaves ■ Can be crushed if thin

Table 8.20 continued

Material	Advantages	Disadvantages
Plastics 	■ Lightweight ■ Cheap to produce ■ Strong ■ Rigid or flexible ■ Can be coloured ■ Made clear or opaque ■ Can be printed ■ Can be made inert to avoid tainting foods ■ Fluid resistant ■ Easily shaped ■ Can be heat resistant	■ From non-sustainable sources ■ Some plastics are difficult to recycle ■ Can leach toxins into food

Advertising

Package design is the job of a graphic design team. The colours used in a design, type of font and images all help to promote the product, give it character and influence the consumer to buy.

Consumers buying ready-prepared frozen or chilled meals expect the container to act as the cooking dish. Packaging of such meals reflects this. Advertising and cooking information is printed on a paperboard outer, the product itself is sealed in a plastic or ovenable paperboard cooking dish (see *Materials used in packaging*).

Fig 8.31 Packaging suitable for a ready meal

Materials used in packaging

The most commonly used plastics are:

■ PVC (polyvinyl chloride) – plastic bottles and containers, plastic bags (not suitable for carbonated liquids)
■ PET (polythene terephthalate) – bottles for carbonated drinks, oils, coating for ovenable paperboard

■ LDPE (low-density polyethylene) – used for Tetra Pak liquid containers
■ CPET (crystalline polythene terephthalate) – withstands high temperatures, used for ready meals
■ Polystyrene – yoghurt pots and meat trays
■ Polypropylene – sweet and snack wrappings

Specialist packaging

Materials can be combined to create specialist packaging. Some examples include:

■ Ovenable paperboard is board coated with PET enabling it to withstand temperatures ranging from -40°C to 230°C; it is used for ready meals.
■ Tetra Pak is made from layers of LDPE, paperboard and aluminium. It creates a rigid, strong medium and is used in the production of containers for liquids.

Fig 8.32 Specialist packaging

- **Gualapack** is becoming a very popular form of packaging. It is made from a combination of plastics with aluminium. It is used to create pouch-like containers for a wide variety of products including foods such as cereals, drinks, soups and ready meals.

MAP and EMAP

Gases can be introduced to preserve freshness and colour. Modified atmosphere packaging (MAP) involves sealing foods in a plastic container or bag which has first been flushed with a combination of gases. The gases used are oxygen, nitrogen and carbon dioxide.

Equilibrium modified atmosphere packaging (EMAP) is a variation on MAP which has been developed as a result of consumers' desire for ever more ready-prepared foods; it is particularly effective for fruit and vegetables.

These packaging systems help to prevent oxidation and bacterial growth, which prolongs the shelf life of foods. Once opened and exposed to the air the products will deteriorate in the normal way. Most of the packaged fresh food bought today is packaged in this way. Examples include meat, fish, ready-to-cook meals, cheese, nuts, salad leaves, prepared vegetables and fruit.

1 What is the difference between primary and secondary packaging?
2 Why is food packaged?
3 Choose four types of packaging mentioned in this chapter. Suggest food products packaged using each method and give reasons why you think that method has been chosen.
4 Consider the legal requirements of labelling; how is it helpful to the consumer?
5 Why is the design of food packaging and labelling important for:
   (a) The manufacturer?
   (b) The consumer?

- **Compare the packaging used for takeaway Indian food and pizzas. Explain why you think they have been chosen and the advantages or disadvantages of the packing materials chosen.**
- **Describe the variety of packaging materials and methods used for soft drinks and juices. Explain the advantages and disadvantages of each and suggest reasons why they were chosen.**

Learning outcomes

By the end of this section you should have developed a knowledge and understanding of:

- The selection and use of common papers and boards
- The properties of papers and boards
- Common paper terms
- Modelling and display materials
- Smart materials

Introduction

There is a wide range of graphical areas within this chapter, some you will be familiar with already, so basic information has been provided to clarify principles and gather further evidence through the tasks set. In other areas, more detail has been provided, especially in relation to some industrial processes that will have a direct impact on your choice of graphical product and the design and development of that product. For example: can it be manufactured on a specific type of material within budget? Are there environmental implications?

Attention to detail and a clear understanding of the processes involved will lead to a high-quality, justifiable graphic product.

What is a graphic product?

A graphic product should represent the design and development of a 3D product that is driven by 2D interaction. It should represent the rigour of the core of the syllabus but also include many of the areas that will be discussed in this chapter.

Examples include point of sale and focused architectural modelling, which may include an interior. Do not just focus on 2D work such as leaflets, posters and websites: these should form part of a larger group of products that promote an item. To justify them, they could be complimented by pop-ups or promotional items that include the use of smart materials, which may form part of a larger campaign.

Selection and use of common papers and boards

You will already have some experience of working with different types of paper and board. Table 9.1 gives you an overview of the terms and properties of these.

Table 9.1 Common papers and boards

Type	Weight	Description
Newspaper	48 gsm	Inexpensive, off-white paper; it is used commercially; normally printed onto using the four-colour process
Layout paper	55 gsm	Highly recommended for design and technology sketching and quick marker rendering; work from the reverse of the pad and overlay the paper, allowing you to trace and refine sketches; don't forget to place a sheet in between the pages when you render; inexpensive
Tracing paper	60 gsm	Expensive; useful for specific line drawings and templates
Marker paper	70 gsm	Expensive; as in the name, very useful for marker rendering and applying markers to the reverse of the sheet for special effects; colour is enhanced with the use of this paper and the marker is easier to control
Photocopying paper	80 gsm	Relatively inexpensive; almost as good as marker paper for rendering; this is a bleached paper and is easily available in A4 and A3
Cartridge paper	120 gsm	Expensive, off-white paper; heavier than marker paper; again very nice to work with; will take watercolour as well as markers; choice is down to budget and personal preference
Corrugated card		An inexpensive card; useful for prototyping ideas quickly; used for protective packaging especially in the retail industry
Mounting board	1000 gsm	A two-part mounting board; very high quality; the top layer is normally supplied in a range of colours; ideal for mounting work and making models either by hand or with a laser cutter; can be expensive
White-lined Chipboard		A more cost-effective version of solid white board, with bleached white board on one side and unbleached, low-cost board on the reverse; ideal for products such as cereal packets
Solid white board		Manufactured from bleached card; easy to print on; gives the customer a feeling of confidence, especially when buying luxury goods
Foam board		Foam-centred card with bleached card on either side; available in a variety of thicknesses; the professionals choice for model making as it is both light weight and strong
Foil laminates		Used within the food industry to provide a hygienic surface for fast food and takeaways; the laminate can be used on most of the above boards; available in gold and silver

The selection of paper or board will depend heavily on its intended use. You must evaluate the properties of the paper/board before you begin work. For example, if you are folding work you must be aware of the direction of the grain of the paper, or, if you intend to use a laser cutter to prototype, double check that the board does not give off toxic fumes.

Common paper terms

Paper can be supplied in many different sizes. You will be used to working with A2, A3 and A4, however commercial printers tend to work with slightly different page sizes; in this instance you will hear many printers refer to SRA. This is slightly larger (oversize), which allows the printer to manufacture work that can be printed edge to edge, an important point to refer to when developing your own design work. Some desk-top inkjet printers will take oversize A3, allowing you to follow exactly the same process to proof work.

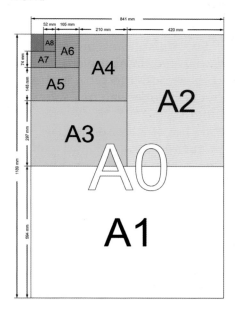

Fig 9.1 Paper sizes A0 to A8

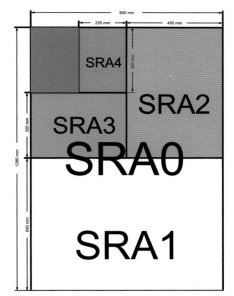

Fig 9.2 Paper sizes – SRA

Other important terms include:

■ gsm – grams per square metre; this represents the weight of the paper. Letterheads and compliment slips tend to be printed on 80 to 120 gsm, whereas a business card will be printed on 300 gsm board
■ Coated paper/stock – this will include papers such as gloss, matt and silk; each of these reacts with the ink slightly differently, providing different surface finishes

■ Uncoated paper/stock – ideal for newspapers, inserts, books, and so on; they tend to absorb the printers ink and are less expensive than coated stock
■ Calendering – the paper is passed through highly polished rollers giving it an even, smooth finish; it is a specific requirement for printing processes such as rotogravure

> Find as many different types of paper as possible – marker, gloss, silk, photocopying paper – and run them through an inkjet printer and note the saturation of ink. Are they all appropriate for your target market?

Modelling and display materials.

A relevant example that is widely used in a school workshop and industrial environment is corruflute. It is manufactured by extruding polypropylene. This is a cost-effective material which is readily available in a wide range of colours and is easy to cut and join. Its main disadvantage is its intolerance to accuracy and should be used with caution where detail is a requirement.

Corruflute is widely used in the sign industry to screen print temporary display boards, a good example is an estate agent's '*For Sale*' sign.

Cross section of corruflute:

Corrugated Paper/ Board	Widely used in the packaging industry to protect and package many 'fragile' products, the board is made up of a fluted corrugated layer and a flat backing layer. The main disadvantage with this type of board is its difficulty to print on to using the lithographic process. A thin laminate can be included in the manufacturing process to compensate for this.
Bonded Paper	70-100gsm a very popular paper used to produce high quality letterheads and envelopes, it is a stiff and durable substrate which promotes clear and crisp printed images.

Smart materials

This is a very exciting area that is constantly being researched and developed. Below are some of the smart products that are commonplace today:

- **Thermochromic pigment** – sometimes referred to as a smart colour; this is incorporated into a special ink and changes colour with a rise or fall in temperature. Products such as kettles, thermometers and garments widely use this material; it is also available in sheet form
- **Photochromatic pigment** – this can be applied in a similar way to thermochromic pigment; the colour changes according to the light conditions. Products such as sunglasses and make-up are obvious choices for this material
- **Phosphorescent pigment** – this is a revolutionary smart material replacing the previous radioactive phosphorescent while absorbing more than ten times the amount of energy, allowing the product to glow in the dark safely. Products include watches, safety signs and instruments
- **Polymorph** – while not strictly a smart material, it is however a very quick and useful way of manufacturing small plastic products including linkages; normally supplied as plastic granules (also available in a tube) you simply soak in warm water, manipulate and then allow to dry

Section B	Plastic sheet

Learning outcomes

By the end of this section you should have developed a knowledge and understanding of:

- Modelling and display materials
- High-impact polystyrene, acrylic and Styrofoam
- Polypropylene, PVC and LDPE

Modelling and display materials

Plastic sheet offers a wide range of opportunities within graphic products. It is widely available, reasonably cost effective and easy to work with, allowing you to either rapidly prototype your ideas or to produce a professional final product using a range of manufacturing techniques such as vacuum forming, laser cutting and hot wire sculpting, or traditional manufacturing processes such as the band saw, or manually using a scalpel to model materials such as polypropylene.

You should have access to most of these processes including the software to help you achieve your final design.

High-impact polystyrene (HIPS)

High-impact polystyrene is a very popular plastic used extensively in design and technology. It is ideal for vacuum forming (moulds are normally manufactured from MDF, which can be quite time consuming if prototyping) and injection moulding. It is available in a range of colours including transparent.

Acrylic

A rigid, colourful and easy to form plastic. It is ideal to line bend and excellent for point of sale and applying vinyl graphics to. It is readily available in a wide range of thicknesses and styles, including transparent and fluorescent, however it can be brittle when worked.

Styrofoam

A versatile material, often underestimated in design and technology. It is easily worked with a hot wire cutter or scalpel, and is also very easy to shape with sandpaper. It combines well with other graphic materials like foam board. It offers a wide range of 3D opportunities, especially when sheets are stuck together using contact glue.

Polypropylene (PP)

A very versatile material that is available in sheet form as well as in a wide range of colours including fluorescent and transparent. It is easy to work with and non-toxic if used with a laser cutter. It allows nets and live hinges to be manufactured quickly and easily, is virtually tear-proof and satisfies many design solutions.

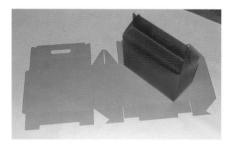

Fig 9.3 Polypropylene sheet and net

Polyvinyl chloride (PVC)

It can be vacuum formed (polystyrene is also widely used in this area, although it is not suitable for exterior applications) and is widely used in the sign and graphics industry (most vinyl is made of PVC). Do not cut this using a laser cutter! It has excellent UV stability, is easy to manufacture, available in a wide range of colours and extremely durable.

Low-density polyethylene (LDPE)

Available in a wide range of colours, it is flexible and soft, and widely used in the production of carrier bags and domestic bottles. It can be easily printed on.

> Using Styrofoam, model an interactive MP3 player. Consider in detail the function of the graphical interface. You should combine a number of materials and manufacturing techniques to present the final product.

Section C — Pop-ups and mechanical techniques

Learning outcomes

By the end of this section you should have developed a knowledge and understanding of:

- V folds: single and multiple
- M folds
- Parallelograms
- Movement using slides and moving arms – linear, rotary, oscillating and reciprocating
- Linkages, rotary to linear conversions

Pop-ups

This is always an important area when considering a range of promotional graphic products. The design process should eventually give rise to the development of complex pop-ups. You have already completed most of this section in primary school, but you now need to reinterpret this information at GCE level.

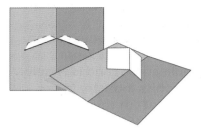

Fig 9.4 V folds

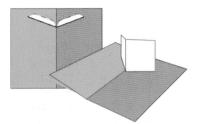

Fig 9.5 M folds

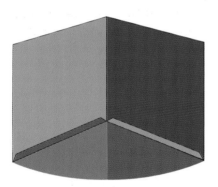

Fig 9.6 More complex folds

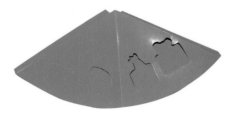

Fig 9.7 More complex folds

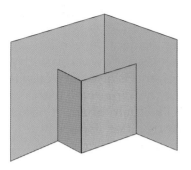

Fig 9.8 Pop-up card

V (valley) and M (mountain) folds, shown in Figure 9.4 and 9.5, are principally the same, the difference being that the V fold folds inwards and the M fold folds outwards. All measurements should be precise when constructing these folds otherwise the card will not close properly.

Figure 9.6 and 9.7 are examples of more complex solutions in relation to V folds: the base folds up when the card is closed. Die cut shapes have also been incorporated into the base.

A shareware program that allows you to develop 3D pop-up cards on screen is available to download at www.tamasoft.co.jp/craft/ popupcard_en/

> Design and make a pop-up birthday card using multiple V folds.

Parallelograms – single and multiple

This is a simple technique to manufacture a pop-up card. Always make sure that the parallelogram remains parallel to the spine of the card. The complexity of a solution can be achieved by adding to the number of parallelograms in a card and folding them inside each other.

> Design and make a pop-up Christmas card using multiple parallelograms.

Movement using slides, moving arms and linkages

There are many opportunities to control movement within graphic products. The different types of movement tend to be categorised into four different areas: linear, rotary, oscillating and reciprocating.

Linear is the most basic type of movement: an object is moved in a straight line. A good example is a drawer opening.

Fig 9.9 Linear movement

Rotary is a circular movement, such as the wheel turning on a car. Many different conversions can take place to convert rotary to linear motion.

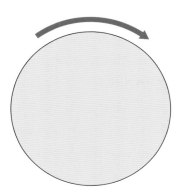

Fig 9.10 Rotary movement

Oscillating is a side-to-side movement about a pivot point; good examples include clocks and metronomes.

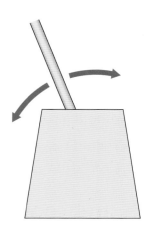

Fig 9.11 Oscillating movement

Finally, reciprocating motion is an up and down movement. A good example is a crank turning an engine or a cam controlling a pop-up toy.

> **Can you list five different types of graphic products that use motion as a promotional tool?**

Linkages

There are many opportunities to include movement within your work using slides and linkages. Linkages are made by connecting levers together with various fasteners such as nuts, bolt, paper fastenings and even folded paper. As long as you have free movement you can manufacture almost any design. There are many different examples that can be designed and manufactured to promote a graphic product; these can be quite complex to design and build. In Figure 9.12 to 9.14 is an example of a linear to rotary movement taken at three key stages. Note the coloured areas and the transfer of movement.

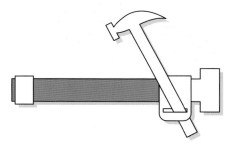

Fig 9.12 Linkages

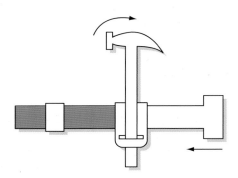

Fig 9.13 Linkages

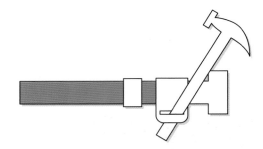

Fig 9.14 Linkages

Section D — Finishing processes

Learning outcomes

By the end of this section you should have developed a knowledge and understanding of:

- Varnishes
- Spot varnishing
- Lamination
- Embossing, foil blocking and paper coatings
- Cropping/trimming

Introduction

This is an area that should always be carefully considered. You must think about the quality of the finish of your product. Do you want to draw the consumers' attention to a specific part of the product? How will you highlight important features? Are they appropriate techniques given the design brief and the specifications?

Varnish

This is becoming a very popular and cost-effective method of promoting and presenting a graphic product. There are a number of different finishes available: gloss, matt, satin and even tinted.

The varnish can be applied as a flood coat to the entire page immediately after printing. This can enhance the colour of the product, allowing the designer to create a specific impact.

All printing presses can apply a varnish as this is essentially an ink without a pigment.

Only coated stock should be varnished to prevent the varnish soaking into the paper.

UV varnish and spot varnishing

A UV varnish will create a similar finish to a spirit-based varnish, however it is considered to be of a higher quality. It requires specific UV machinery that cures the varnish instantly.

Spot varnishing is widely used using this method, which follows a similar principle to offset lithography; that is, a plate is used to determine which parts of the product are 'spot varnished'.

Fig 9.15 UV varnishing

Fig 9.16 UV varnishing

Fig 9.17 UV varnishing

UV varnishing takes place on an independent unit as illustrated in Figure 9.15 to 9.17. The stock is calendered through a set of rollers to ensure that the print surface is even and flat, ready for the application of the UV varnish using the 'rubber coating roller' (Figure 9.17). The varnished work is then passed underneath a UV light and it is dried instantly.

The main disadvantage of this process is the negative impact it has on the environment. Many manufacturers and designers are opting for an aqueous coating, which is a more expensive method but it is water based. Unfortunately spot varnishing is unavailable with this particular method of production.

Lamination

Lamination can relate to a number of different production techniques when we consider graphic products. Simple desk-top encapsulation is probably the most familiar process (desk-top laminator) although it is important to realise the difference. Encapsulation is when the whole product is sealed, that is, a plastic film encapsulates your paper leaving a plastic border around the paper; lamination is the application of a plastic coating to the front and/or rear of your product.

There are two different materials used in this process: film or liquid. A graphic product can be flood coated with a liquid to protect it (a similar process to varnishing).

Liquid lamination is used predominantly within the sign industry to protect full-colour signs from UV rays, as well as day-to-day abrasion. The other method is to laminate the product with film using a hot or cold laminator. Products such as menus, exhibition stands, anti-graffiti posters and floor graphics use this method of protection. Once you have applied a coating to your product you cannot print over it.

Fig 9.18 Lamination

Fig 9.19 Lamination

Fig 9.20 Lamination

The film is made from polypropylene with an adhesive that melts at 115°C. The rollers apply a pressure of 8 bar as the product is fed through; this ensures that the film sticks securely to the substrate. As the final product rolls off the laminator it is fed through a series of rollers (Figure 9.20) that straighten it out ready for delivery.

Embossing/foil blocking

These two processes are invariably linked because they share the same manufacturing processes, which are also closely linked to die cutting.

Embossing effectively raises an area of the paper/card by applying pressure and heat to it. Two dies are used, which are made from brass, and foil blocking or hot foil blocking simply transfers foil (normally polyester) onto the substrate under pressure. This process has become very popular again and is used on items such as wedding stationery and holograms.

Fig 9.21 Embossing

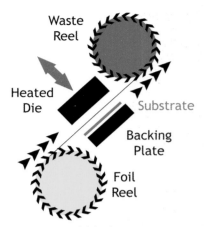

Fig 9.22 Foil blocking

Cropping and trimming

Once the 'job' has been printed it needs to be trimmed (cropped) to size. The designer would have included trim/crop marks on the piece of work. It is important to note that if the brief requires the 'job' to be printed edge-to-edge, then a 'bleed' must be included as illustrated in point 2 of Figure 9.23. All work is printed onto SRA size and the bleed is generally between 3 and 5 mm. Once it has been trimmed you are left with an edge-to-edge print. Note the guillotine; these can be pre-programmed to cut reams of paper to the correct size.

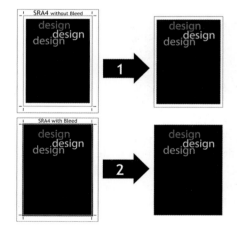

Fig 9.23 Cropping and trimming

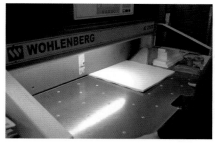

Fig 9.24 Cropping and trimming

Fig 9.25 Cropping and trimming

Fig 9.26 Cropping and trimming

Fig 9.27 Cropping and trimming

The example in Figure 9.28 shows crop marks, registration marks and colour bars in place ready for the printing process. The registration marks allow the operator of the printing press to check that all four (or more) colours line up on the final print sheet; the colour bars allow the printer to check that the press is printing the correct amount of colour. To do this they use a densitometer; the ink can either be controlled manually or electronically.

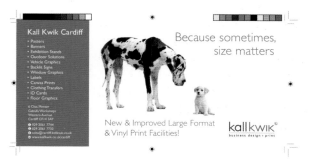

Fig 9.28 Kallkwik – crop marks etc.

Fig 9.29

Gather different types of media that have incorporated bleed into the final design. How does this fit into the design process?

Set up an A5 sheet; design a simple magazine cover with a 6 mm bleed and print this onto A4 with crop/trim marks and trim accurately to size. How can this provide further design opportunities?

Learning outcomes

By the end of this section you should have developed a knowledge and understanding of:

- Closures – tab-lock, tuck-flap, slit-lock, postal-lock, crash base
- Nets for simple and complex products
- Press forme design and construction

Closures and nets for simple and complex products

Tab–lock

The simple CD case in Figure 9.30, normally manufactured from either card or polypropylene, uses a tab-lock closure. It is a durable solution that allows the end user to open and close the product easily. Various designs can be applied to the tab to increase security of the product.

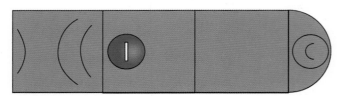

Fig 9.30 Tab-lock closure

Tuck–flap

These closures all tuck into place and require no gluing. They can be opened and closed an indefinite number of times. They are commonly found in many retail stores packaging chocolate, perfume and some mobile phones. Additional security can be achieved by applying a security sticker over the flap to monitor any tampering.

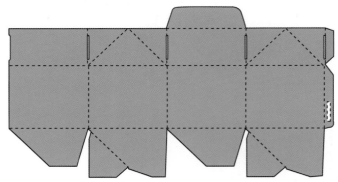

Fig 9.31 Tuck-flap closure

Slit–lock

The addition of slits in the lid provides an extra level of security/tamper proofing, allowing the retailer/customer to check the quality of the product.

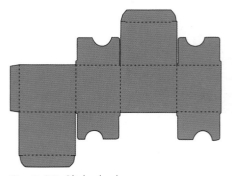

Fig 9.32 Slit-lock closure

Postal–lock

There are a number of variations that can be employed to control security with a postal-lock. It is easily detectable if opened as the tabs will crease, eventually tearing. The dagger lock is a slight variation that will completely tear when opened, making it unusable.

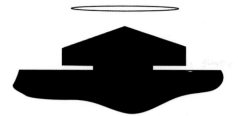

Fig 9.33 Postal-lock closure

Crash base

This type of carton is used when fast assembly is needed. They are pre-glued and folded flat for delivery. For assembly, the carton needs to be opened and the base slides into position and locks when all sides meet. These are very popular in supermarkets and fast-food restaurants. An example of a crash base package used to store a mug is shown in Figure 9.35 to 9.37.

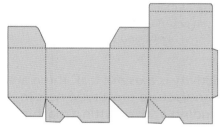

Fig 9.34 Crash base

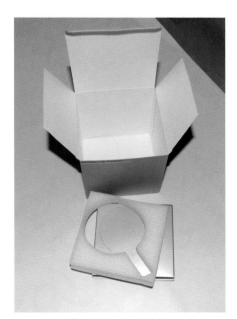

Fig 9.35 Crash base packaging to store a mug

Fig 9.36 Crash base packaging to store a mug

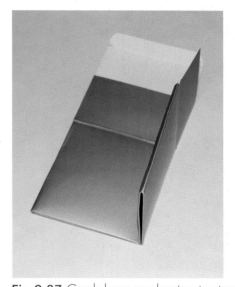

Fig 9.37 Crash base packaging to store a mug

A range of materials can be used to construct nets, Figures 9.38 to 9.43 show a few examples that use a variety of techniques. They have been manufactured using a laser cutter and printed vinyl (note that you must be careful which vinyl you use on the laser cutter as some vinyls emit toxic gases when burnt).

Fig 9.38 Nets

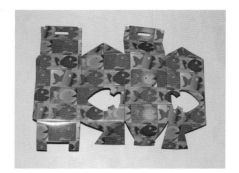

Fig 9.39 Nets

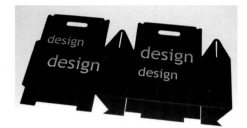

Fig 9.40 Nets

Fig 9.41 Nets

Fig 9.42 Nets

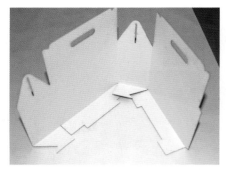

Fig 9.43 Nets

Mountboard is also a particularly useful material to use if you have access to a laser cutter as it allows images and detail to be engraved easily. This is a very popular method of prototyping for many designers.

> Can you, in principle, design and make a simple press forme. Think about similar products that fulfil a similar function, for example pastry cutters. You may also like to refer to Section F for further solutions.

While the above activity is a useful task, there are quicker ways to manufacture prototype nets in the classroom, including a vinyl cutting machine, by hand or, as already discussed, a laser cutter. Creases can easily be inserted with a rule and the blunt end of a scalpel.

| Section F | Manufacturing processes |

Learning outcomes

By the end of this section you should have developed a knowledge and understanding of:

- Image setting, plate/screen production
- Photomechanical transfer techniques
- Commercial printing processes – offset lithography, screen process printing, flexography, gravure, digital printing
- Cutting and forming processes – die cutting, folding, perforating, laser cutting
- Paper and board manufacture and production

Commercial printing process

You need to be aware of five main manufacturing processes in relation to graphic products:

- Offset lithography (rotary press)
- Flexography (rotary press)
- Gravure (rotary press)
- Screen process printing
- Digital printing

You are used to developing and manufacturing images in the classroom, however the industrial processes and requirements are very different.

While the two areas are still distinctly different you may well have experience of using 'industrial' software to generate digital artwork such as Adobe Photoshop, Adobe Illustrator for vector design, Adobe Indesign and/or QuarkXpress for final desk-top publishing. However, if you are required to realise your work on a large scale this needs to be set up in a required format for the commercial printer.

The choice of printing methods relies on a number of job/customer requirements:

- Cost
- Quality
- Production run
- Choice of substrate

These are just a few examples of how the printing method chosen can be affected by job/customer requirements. Can you think of another four?

Each of the methods we are going to discuss in this chapter has particular advantages and disadvantages associated with the above.

All of the processes use similar manufacturing techniques (rotary press) and all will use combinations of the same type of printing inks. However, each process will place certain restrictions on the designer and it is important to evaluate this when you begin to develop your work and to be able to justify your chosen methods.

Printing inks

When preparing work for full-colour print you will need to convert your image to CMYK so that plates can be made for it. Most graphics software seems to default and import as RGB (red, green and blue); this saves memory and allows the machine to process the image quickly. You will notice a difference when you convert from RGB to CMYK; in most cases the image will appear darker.

This is where many problems begin for designers, if you commit unchecked work to print you may end up with just a black square where your RGB image was, or your CMYK image maybe dull or too small (because of incorrect resolution).

> Image **resolution** is a major consideration. If you are preparing a range of complimentary graphic products that includes 3D products as well as websites and commercial print-based work, you may well be flicking between a wide range of resolutions. It is important to remember that you cannot upscale an image for print – that is, if you have used an image from the internet, as a rule of thumb this cannot be used for printing as most resolutions in this format are 72 dpi (dots per inch). The general accepted resolution is 300 dpi at the required size for commercial print work.

CMYK = cyan (C), magenta (M), yellow (Y) and black (K, key)

Fig 9.44 CMYK colours

The inks are normally applied in this order (CMYK), although this is not always necessary. The alternative method of printing to a budget or at the request of the customer is to print a spot colour. This is sometimes useful if the product has up to three spot colours, although two spot colours is the norm (you can apply these as tints as well). You still have to make up separate plates but these can be run effectively on smaller presses, ideal for letterheads, business cards and invitations. This method is extremely cost effective.

> Look at different types of packaging and letterheads. What type of ink was used on them. Is the quality any different?

One last point to be aware of is colour matching. Your screen will always display a different colour to the final print.

When producing work it is important to use a colour matching swatch. These are made by various companies but they allow you to be specific, especially if you want a colour to print exactly right (on the commercial press). You can also match it to the swatch and the colour library found in your software.

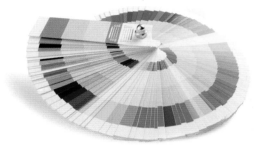

Fig 9.45 Colour swatch

> Using the same image, print off an RGB version and a CMYK version. Compare them to the screen image.

Offset lithography

This method is based on the repulsion of oil and water. The ink is attracted to the image that is etched on to the printing plate. This then repels the water, which keeps the rest of the plate clean.

Pre-press

Artwork is normally supplied to the printer in a digital format by the designer; the preferred method today is a PDF file. This sidesteps many issues associated with font copyright and supplies the artwork in a 'camera-ready' format.

If the customer requires very high-quality images, these are either taken by a professional photographer using a high-end digital camera (normally costing in excess of £10,000) or transparencies can be supplied from a 35 mm camera. These can then be scanned in at this stage using a drum scanner. This is a high-end scanner capable of reproducing the image in a digital format directly from a 35 mm transparency. The transparencies are attached to the drum, the unit is closed and the images are scanned.

Fig 9.46 Drum scanner

Accurate reproduction of work throughout the whole process is very important and viewing booths are situated throughout the print shop at key locations during the manufacturing process. The light is balanced to a specific standard that allows the designers and printers to view the image in the same way. The scanned transparency can be checked against the proof and the work can be adjusted to provide a metameric match.

Fig 9.47 Viewing booth

Once all the artwork has been imposed the digital files are ready for low- and high-res proofing. These are RIP'd (raster image processor): this basically turns all of the information such as fonts and vector graphics into a format that the proofers and image setters can understand.

Figure 9.48 is a double-sided, low-res proofer, which allows the printer/designer to supply a 'dummy' book proof that can be used for checking where trimming will take place.

Finally a high-res proof is supplied to the customer. This final print is calibrated to the printing press and provides an accurate representation of the final product.

Fig 9.48 Low-res proof being printed

Once the artwork has been approved by the customer it is then ready to be transferred onto the printing plates.

Plate making – image setting, plate production, photomechanical transfer techniques

Plate making has changed considerably in recent years. Before any print work can take place a set of aluminium plates has to be manufactured to transfer the images onto the paper using specific colours.

A standard full-colour brochure would need four different plates –CMYK – however, a single spot colour piece of work will require only one plate, and if a varnish or special colour such as gold were to be applied, this too would need its own plate.

While this is not an expensive process for offset lithography (the plate-making process for flexography and gravure is very different) there are a number of quality assurance and environmental implications to consider.

Historically every printer would have sent artwork off to an image setter to produce a set of films that would then be transferred onto plate. Fortunately this is no longer the case. While many printers outsource the production of plates, it is now a direct process known as CTP (computer to plate). This eliminates many of the environmental issues: chemicals are no longer needed to develop films and so on and, in most cases, the developer used to develop the plates (Figure 9.52) is completely recycled.

Plates are stored flat and in different sizes depending on the size of the printing press (Figure 9.49).

The plate is fed to the image setter and is wrapped around an imaging cylinder. This is spun at 3.5 K RPM. The lasers are fixed and burn into the plate in three-inch swathes producing the latent image. The plates are developed and stacked.

All plates are checked before any print run. They are then gummed to prevent any oxidisation: this simply washes off when the plates are attached to the press. Finally the plates are put into a folder with both the low- and high-res proofs.

Fig 9.49 Plate making

Fig 9.50 Offset plates

Fig 9.51 Plate making

Fig 9.52 Plate making

Fig 9.53 Plate making

Fig 9.54 Plate making

Photomechanical transfer is a largely redundant process within offset lithography. Research the process. Can you find any local printers who still use it?

Sheet-fed and web-fed printing press (offset)

Printing presses come in many different sizes. Some are sheet fed (Figure 9.55), some are web fed (Figure 9.56), and some are web-to-sheet fed, that is the web is cut to the required paper size by the press before entering the manufacturing process.

A major consideration for you as the designer is quality, time taken and cost of the finished product. We will discuss a number of alternative printing methods throughout this chapter but it is vital to consider appropriate manufacturing processes and techniques.

A number of examples of offset lithography presses are shown in Figure 9.57 to 9.59, ranging from:

- A single-colour press (Figure 9.57) – ideal for printing letterheads and business cards. You could potentially run full colour on this machine however the press would need to be cleaned in between plates, making it completely uneconomical in terms of time and money
- A five-colour press (Figure 9.58) – many printers use four-colour presses but this has the advantage of running that extra special colour such as gold or a varnish
- Finally a ten-colour press (Figure 9.59) – this substantially increases speed and press capability, easily allowing duplex, varnishing and specialist colours to be applied. Note the use of ink tanks: because of the size of these presses manual application of inks is impossible

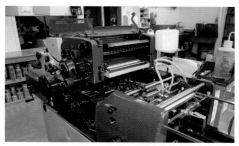
Fig 9.57 Printer

Fig 9.58 Ink

Fig 9.59 10 colour press

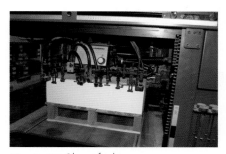
Fig 9.55 Sheet-fed press

Fig 9.56 Web-fed press

Fig 9.60 Ink tanks

Try to find examples of printer products with speciality inks and varnishes. How have they been used? Does it make a difference?

Fig 9.61 Plate change

Offset litho in action. Note the plates being fitted in Figure 9.61; you can follow the sequence in Figures 9.62 to 9.67. Note the inks on the rubber cylinders: they transfer the image through pressure from the plate directly onto the paper; also, note the sequence.

Fig 9.62 Black plate

Fig 9.63 Cyan plate

Fig 9.64 Magenta plate

Fig 9.65 Speed printing

Fig 9.66 Printer's eye glass

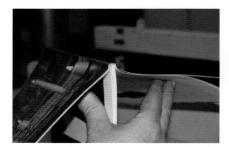

Fig 9.67 Perfect binding

This is a very quick process with many presses now having built-in ovens and finishing lines attached directly to them.

With a magnifying glass, check the quality of print on different products. What are the differences between offset litho, flexography and rotogravure?

Finishing line

Once you have manufactured your product you will need to consider finishing. This does not just include trimming but also how the sheets of paper will be joined together to form a book, which will usually be perfect bound (Figure 9.67), or a magazine, which will be run along a stitching line that staples the product (Figure 9.68), or perhaps just a four-page leaflet that will be a simple fold and trim.

Many printers have their own purpose-built finishing lines, allowing the whole process to take place on the same site.

> There are many different types of joining techniques. Can you identify two more that you could use?

Fig 9.68 Finishing line

Recycling

This is an important point as you will be expected to consider this when you answer examination questions. Large printers try to recycle everything they use, from direct to plate, making the chemicals used for developing plates redundant, through to recycling all the off cuts. This is monitored very carefully.

Fig 9.69 Recycling

Letterpress

The traditional letterpress is now a largely redundant method of commercial printing although it is used by many hobbyists and specialist craft manufacturers who print high-quality books or specific pieces of artwork.

Fig 9.70 Print blocks

Fig 9.71 Letterpress

Fig 9.72 Moveable type

Fig 9.73 Label printer

This is a very simple process that can be used in the classroom if it is appropriate for your project, but consider all types of printing technologies before investing too much time in this area.

The letterpress works on the principle of direct impression. Moveable raised letters (see Figures 9.70 and 9.72) are inked up and then transferred by pressure onto a receptive substrate such as paper. The print quality is very high although you will be restricted by colour and design.

The rotary letterpress is still used today for some graphic products. Flexography and lithography technology has been incorporated into the development of these machines and they are ideal as an alternative solution to flexography for some products such as adhesive labels, books and posters.

Flexography

In principle this follows a similar method of production to offset litho. The main difference is the production of the plates: like letterpress these are relief plates, but the images are fixed (unmoveable type) on photopolymer plates. These plates are developed by exposing them to actinic radiation. As in screen printing they are then washed, removing the unprocessed areas of the plate with solvent, and then finally dried by further exposure to radiation. The plate is then ready for the printing press.

Because the plate is flexible it can be attached to the plate cylinder. Ink is transferred from the fountain roller to the anilox roller, which has very fine, delicate dimples on the surface that transfer the ink to the plate and then to the substrate by an impression cylinder.

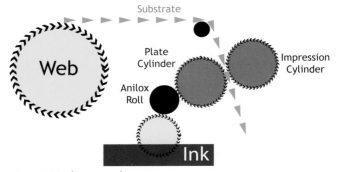

Fig 9.74 Flexography

Flexography can be used to print a wide range of substrates, such as labels, cardboard boxes, plastic bags, and pharmaceuticals as well as standard paper stock, due in some part to the quick-drying inks used. However, a disadvantage is its ability to print acceptable half tones, and this needs to be considered at design concept stage.

Halftone refers to the printing technique that simulates a gradient or continuous tone. Various sized dots are printed equally apart and, when viewed by the human eye (Figure 9.76) at distance, they appear to be a continuous tone.

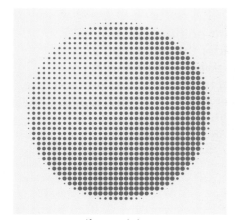

Fig 9.75 Halftone globe

Fig 9.76 Halftone viewed by human eye

Some examples of the wide range of products printed by flexography are shown in Figures 9.77 to 9.80.

Fig 9.79 Cinema tickets

Fig 9.77 Open box

Fig 9.80 Pizza box

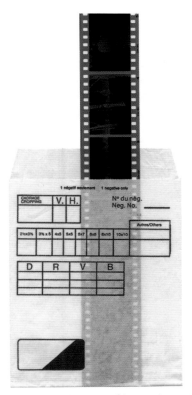

Fig 9.78 Negative film and envelope

Gravure (rotogravure)

Rotogravure is generally the choice of production for print runs of 500,000 or more. Based on the rotary process, it is a very simple process that contains few component parts: an impression roller, a doctor blade, an engraved cylinder and an ink fountain.

The engraved cylinder is normally laser etched producing small holes, or 'wells', at different depths that hold the ink in place. The deeper the 'well', the darker the colour.

Ink application is very different to the other rotary processes. Instead of applying varying sized dots, gravure applies a continuous tone, thus producing vibrant colours on various substrates. Its main disadvantage is that text is also printed this way, sometimes leading to fuzzy edges.

Rotogravure is used to produce glossy magazines, packaging, stamps and wallpapers.

The quality of print is extremely high, with excellent multicoloured fills on a wide variety of substrates including polyethylene and polypropylene.

Fig 9.81 Money

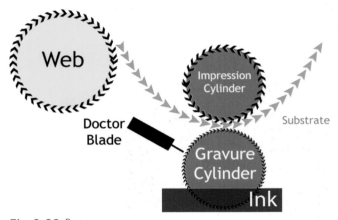

Fig 9.82 Rotagravure

Fig 9.83 Vector Christmas stamps collection

Note the simplicity of the process. The cylinder is wiped clean by the doctor blade and the substrate is pressed onto it by the impression cylinder. A separate gravure cylinder is required for each colour.

Screen printing process

A very popular and accessible industrial process for all students. Screen printing has the advantage of being a relatively simple and cost-effective process that is able to be used with most substrates. The main disadvantage is the need to manufacture separate screens for different colours; this can lead to an increase in production costs and the time it takes to manufacture the final product.

1. Artwork is originated on a suitable computer package
2. If required, colour separations are produced
3. The screen is prepared by coating it with a light-sensitive solution
4. The screen is exposed to UV light with the artwork underneath it
5. The screen is then washed to remove the emulsion
6. The screen is then fitted and placed on top of the substrate
7. Ink is pushed through the screen with a squeegee
8. The screen is carefully removed and the ink is left to dry
9. The process is then repeated for further colour application

Digital printing

This has been a revolution to the printing industry and, as technology progresses, so does the speed and efficiency of these machines.

Fig 9.84 Digital printing

Digital printers look very similar to a photocopier but they produce full-colour work of equivalent quality to offset litho and in less time, providing print on demand. There is no need to produce separate plates for each colour; the machine simply interfaces via a RIP.

Most machines are able to print on a variety of substrates, allowing short runs of leaflets and business cards with lamination at very competitive prices. There are restrictions on the weight of paper/board.

Digital printers use toner to apply the colour in a similar way to a laser printer; the colour then 'sits' on top of the paper as opposed to being absorbed into the paper.

Fig 9.85 Digital printing

Many of these machines have a built in finishing unit allowing 'jobs' to be folded and stapled ready for final trimming (Figure 9.86).

Fig 9.86 Digital printing

The main advantage that digital printing has over conventional presses is its ability to print every page differently if required; designs can be personalised for clients/customers at the touch of a button and at no extra charge. Most print shops charge per click, that is, front and then reverse (duplex), making it a very economical process for both the seasoned professional as well as the amateur.

Laser cutter

An ideal piece of machinery that is very similar in function and capability to those found in industry. Like vinyl cutters these machines are driven by vector-based software packages such as Adobe Illustrator and Corel Draw. they will cut through almost any material except metal and certain plastics.

The opportunities for bespoke modelling and the time these machines can save when you are working up an idea can make the classroom process professional by its very nature. They are ideal for making up nets quickly, scoring and perforating card and cutting through Perspex to make accurate 3D models. This can all be done by hand with a ruler and craft knife but this process can give you that extra freedom.

Fig 9.87 Laser cutting

Fig 9.88 Laser cutting

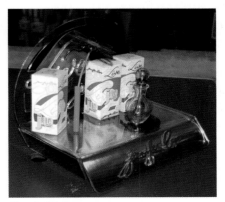

Fig 9.89 Laser cutting

Fig 9.90 Laser cutting

Fig 9.91 Die cutting process

Die cutting

This is a common manufacturing process used with many forms of graphic products, from simple folders and invitations through to complex pop-ups and nets.

Fig 9.92 Die cutting process

The process

You can actually manufacture a die cutting machine in school (sometimes know as a press forme). It is a relatively manual process though which requires a high level of skill and patience.

The printer/designer normally gives the die cutter a paper template (see Fig 9.91). This is then normally manufactured from 18 mm MDF using a jig saw ready for the 'cut rules/knives' to be inserted. These are supplied in straight lengths and are measured in points (3 points is approximately equal to 1 mm). Note the way in which the die has been planned, with a number of holes drilled to prevent the shape from simply dropping out, but close enough together to maintain an even cut and fold when the cutter is being used. Foam is placed around the cutting rules/knives. The cut height is higher than the reverse of the die: this prevents the cut rule/knife from being pushed out of the die.

Fig 9.93 Die cutting process

Fig 9.94 Die cutting process

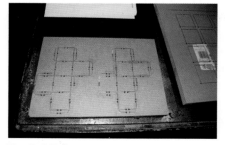

Fig 9.95 Die cutting process

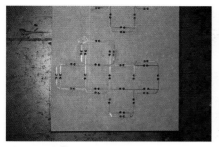

Fig 9.96 Die cutting process

Fig 9.97 Die cutting process

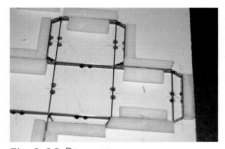

Fig 9.98 Die cutting process

Cut rules/knives can be easily shaped, although it is important to consider their limitations as well when you design your graphic product.

Manufacture

Die cutting machines are available in a wide range of shapes and sizes, but essentially they all perform the same task. Below are two examples of die cutting machines.

Rotary machine

This is suitable for high-volume work. As you can see a matrix is attached to the cutting knives when the die is first placed into the machine. The die is slowly fed through the machine and the matrix attaches itself to the cylinder; this protects the blades while the machine is cutting. Expect to see half a million impressions from a well-maintained die.

Fig 9.99 Die cutting – rotary machine

Fig 9.100 Die cutting – rotary machine

Fig 9.101 Die cutting – rotary machine

Fig 9.102 Die cutting – rotary machine

Flat bed machine

This is very similar in principle to the foil blocking machine discussed earlier. It is ideal for shorter runs and smaller products such as wedding invitations and bespoke business cards. The die is mounted and the card is pressed onto the cutting and creasing knives, leaving a clean-cut profile of the final product.

Fig 9.103 Die cutting – flat bed machine

Fig 9.104 Die cutting – flat bed machine

Fig 9.105 Die cutting – flat bed machine

Paper and board manufacture and production

Commercial paper making is one of the largest manufacturing processes on Fourdrinier machines.

Fig 9.106 Paper factory

1 Wood pulp is made into slurry with water and starch
2 The slurry enters the headbox where it is mixed and deposited onto a fine mesh (up to 200 m per minute)
3 The fibres touch, forming a web; the water drains away through the mesh
4 The paper web then passes through a series of pressing rollers
5 Moisture is removed by absorption as the web is carried on thick felts
6 The web then passes through a series of steam-heated rollers
7 Sizing agents are added to the web
8 The web is then passed through polished calendar rollers, which improve the final finish
9 The paper is rolled and stored with a moisture content of approximately six per cent

Fig 9.107 Roll of paper

Vinyl cutting

This is the predominant manufacturing process in the sign industry today and allows the designer to directly control and manufacture a graphic product using a CAD/CAM machine (the vinyl cutter) and a consumable product (vinyl).

Initially vinyl cutting was used for bespoke work, as the process is/was very labour intensive, making it expensive and restrictive. Technology has progressed to such an extent that it is no longer prohibitive to experiment and develop new market opportunities.

This is one of the few industrial processes that is very close to the classroom experience. Design and construction of graphical work follows a similar process and all computer work is carried out using a vector-based drawing package such as Corel Draw, Adobe Illustrator or more specialist sign making packages such as Signlab or Gerber Omega.

Essentially the CAD/CAM machine cuts through the vinyl, which is loaded into the machine, with a very small blade that is allowed to rotate freely in the machine. It allows very intricate work to be cut.

Hopefully you will have used one of these machines in school, but what you may not be aware of is the variety of machines and consumables that are available to the designer/manufacturer.

There are three main methods of graphic product production using vinyl:

- Standard/Solid vinyl cutting
- Digital Print and cut
- Thermal Print and cut

Each area can use the same type of vinyl but each area needs a different type of CAD/CAM machine to produce the final piece of work, a potentially expensive process.

Fig 9.108 Full colour, external, UV-resistant vinyl sign produced using thermal print and cut

Vinyl

There are many different types of vinyl and we will discuss the four most popular types. However there are specialist products available for certain markets such as an 'ultra-destructible' vinyl which, once applied, is almost impossible to remove. It is ideal for the security market/tagging

or a 'void' product which is used in the electronics industry to secure cases and hard drives. The list is becoming endless as manufacturers develop holographic vinyls that complement transparent inks.

Fig 9.109 Rolls of multicoloured vinyl

Fig 9.110 Vinyl rolls

Vinyl rolls are supplied in a variety of sizes and lengths. They are normally supplied in either 10, 25 or 50 m rolls with a width of 380, 610 or 1220 mm. As you can see from Figures 9.110 and 9.111, these are solid colours, and if a design is multicoloured then the final product has to be built up using the available colours. This is not normally a problems unless the sign/product has to Pantone match existing colours.

Note the application tape; this is normally supplied in a 100 m roll and is supplied in either standard or high-tack format; application of your vinyl is impossible without it.

Other areas include specific promotional items such as floor graphics, which you will probably have seen in your local supermarket; as you can imagine, these have to be durable as well as colourful.

When choosing vinyl the designer must be aware of the environment and its purpose.

Monomeric calendered vinyl

These vinyls are very cost effective. They are available in most of the popular colours and are mainly suitable for indoor use and limited outdoor use with an average durability of two to three years. They can be quite brittle and are not recommended for small, intricate design work.

Polymeric calendered vinyl

These are available in a wide range of colours and are suitable for most external signage with an average life of five to seven years. They tend to be the main choice for professional designers/sign makers. This product is easy to weed and significantly reduces blade wear. This is a more flexible substrate allowing conformability over a wide range of surfaces, making intricate, detailed work possible.

Cast vinyl

Although it is very expensive in comparison it is certainly worth including here. This vinyl meets all the technical requirements of most design briefs: it is chemical and fuel resistant, can be used in any environment, it conforms easily over rivets and can also be Pantone colour matched.

Digital print vinyl

This is generally supplied in a roll width of 720 mm or 1220 mm and meets the standards of the vinyls above. There is a wide range of other media available as well including banner and canvas material. With the development of new inks for these machines, digital print work can be carried out on most standard vinyls.

Fig 9.111 Digital print and cut system – virtually maintenance free, these machines can use anything from four colour cartridges to eight

Design and manufacture

Depending on the specification of the brief, the designer draws the outline of the product he or she wishes to cut from vinyl. If the job needs more detail, for example colour, then this can be added at this stage and the design will normally be digitally printed and cut using the same machine.

This works in a similar way to your desk-top printer but on a larger scale. Most printers are solvent based so suitable extraction is required when they are being used.

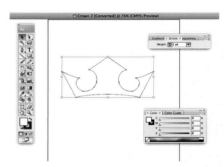

Fig 9.112 Crown screen shot

Thermal print and cut system

The other alternative method of colour production using vinyl as the substrate is to thermally print the design on to the vinyl and then cut. Figures 9.113 to 9.115 outline the process.

Fig 9.113 These machines are quite small and use colour foils to transfer the colour onto the vinyl. These machines were and are still very expensive to buy, originally costing in excess of £40,000 when they were first released

Fig 9.114 The foil and vinyl are loaded into the machine. This vinyl is supplied with sprocket holes that allow the vinyl to be fed through the machine in a straight line and at a set speed

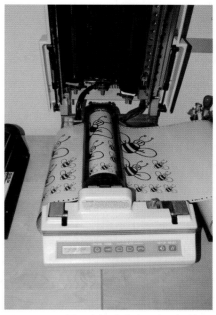

Fig 9.115 The black from the foil is transferred onto the vinyl using heat. The advantage of this process is that the images dry instantly. Unfortunately it is labour intensive, especially when the images are flooded with colour

Even though this is a complete print and cut solution, the vinyl needs to be cut separately and this involves placing it into a vinyl cutter (note the sprockets) and lining the image up to the cutter using a 'bomb site', which accurately aligns the cutting knife with the image.

Fig 9.116 Thermal print and cut system

Once the vinyl has been cut, the excess needs to be 'weeded' out. Note the way that the images have been imposed to make sure that all the vinyl is used. At this stage it is very important to make sure that the vinyl is peeled away cleanly and does not pick up any of the remaining substrate.

Fig 9.117 Thermal print and cut system

Once the excess vinyl has been removed, application tape can now be applied. Note the use of a squeegee: you must make sure that all air bubbles are removed and that the tape is securely attached to the substrate. You are now ready to apply it to your product.

Choose a photograph of your favourite product. Scan it, download it, then drop it into a 'tracing' package such as Coral Trace or Adobe Illustrator CS3. Print out the original bitmap image and the new vector image. What are the differences?

Fig 9.118 Thermal print and cut system

Primarily you will be working with **vector graphics** within vinyl design and laser cutting. This is because the technology needs to interpret and cut the lines that you have drawn.

Have you noticed that photographs pixelate when you enlarge them? This is because they are **bitmap graphics**. They have the advantage of including a lot of detail (as you would expect from a photograph) but they cannot be resized or used as a cutting path.

A vector graphic is made up of individual lines and objects: the most obvious example is clipart. They have the advantage of being able to be resized without the loss of drawn detail, but they lack 'real life' detail.

design

Fig 9.119 Design bitmap

design

Fig 9.120 Design vector

Section G — Holography

Learning outcomes

By the end of this section you should have developed a knowledge and understanding of:

- The use of holograms within graphic products for functional, aesthetic and security applications

Holographic images provide unique opportunities for designers of graphic products. They are also a reasonably cost effective method of manufacture (once the machinery has been purchased).

Holography can be a very interesting way of decorating products: think of toothpaste boxes and wrapping paper. Many manufacturers now use this type of laminate to brand their products.

Manufacturers of security products and luxury goods have also embraced this technology, using it to prevent fraud, counterfeiting and to provide authenticity of the product.

Holograms are very flexible and can be applied to most substrates. They cannot be reproduced using conventional printing techniques, making it almost impossible to copy.

Fig 9.123 Holographic paper

Fig 9.121 Credit cards

Identify five different graphic products that use holograms. Why are they used in this way?

Fig 9.122 Euro bank note security mark

CHAPTER 10

Manufacturing

Further coverage of many topics in this chapter will be found in Chapter 5 Core knowledge.

Learning outcomes

By the end of this section you should have developed a knowledge and understanding of:

- One-off, batch, high-volume production systems
- Modular/cell production systems
- Just-in-time manufacture
- Repetitive flow
- Continual flow
- In-line assembly
- Cell production
- Automated production
- Robotics
- Bought-in parts and components, standardised parts
- The implications of these industrial production processes/procedures
- Appropriate manufacturing methods that take into account the properties of different materials
- The effects of the manufacturing process on the properties and structure of materials

Introduction

Processes used in manufacturing are quite diverse depending upon the complexity of the industry studied. Processes observed could vary from a manual system where single, one-off items are being produced to the use of fully automated high-volume production systems.

Modern industrial methods are moving towards fully automated systems, however there are many examples of combinations of manual and automated production systems to produce items as a single product and in volume. This section explores a variety of manufacturing processes that you may see in an industrial situation.

One-off, batch, high-volume production systems

One-off (also known as job, or custom) production is where a single item may be needed. This method is suitable for producing individual items (often to a particular customer's requirements) but at a high cost. In such cases the designer, who could also be the manufacturer, would work closely with the customer to produce an item that meets an agreed specification. A high level of skill is required. Examples of products produced using this method include single items of furniture, a wedding cake or a ball gown.

Fig 10.1 A one-off dress

Larger quantities of items are produced using batch production. This method involves the same item being made repeatedly over a period of time, such as 1000 loaves of bread made in a few hours daily in a bakery.

The manufacturing system used can include elements of line and one-off production. Batches produced can be increased or decreased according to demand. As this method produces a lot of items, the individual unit cost starts to reduce when compared to one-off items.

Evidence of batch production can be seen in smaller industries where continuous production may prove to be costly or when the firm does not know what the demand for a particular product will be. The use of computer-controlled machinery and equipment is common, and this enables quick changes of computer programs between batches of different products or components.

Fig 10.2 Batch production: orange truffles

The continuous production of standardised items over a long period of time is known as mass production or high-volume production. This method makes use of dedicated machines with trained workers carrying out specific tasks as the manufactured item moves along an assembly line.

The use of such a system was first credited to Henry Ford at the beginning of the 20th century when production of his Model T Ford began. This method of production was quickly adapted and modified by other companies. It has evolved over the years but the appeal to factory owners remains the same: the manufacture of large quantities of a product at a lower cost.

To enable this lower unit cost, investment in machinery and automated production equipment is necessary, and these initial costs can be very high. Only if the sale of large numbers of the item can be guaranteed can this be justified. Economies of scale are possible, including bulk buying of materials. Although a relatively low skill level is required to operate machinery, highly skilled labour is crucial for the design, programming, and management of the production lines and equipment.

Fig 10.3 Mass production

Just-in-time manufacture

Just-in-time manufacture (JIT) is a method used by a lot of firms. The philosophy behind it is that the right part will be available in the right place on the production line at the right time. The implementation of such a system makes the manufacture of the product more efficient; however, in order for it to be effective, detailed production planning is necessary. The success of JIT production depends upon many carefully controlled systems including JIT delivery of parts to the factory.

The car manufacturer Nissan relies on JIT systems and therefore the location of their site in the north east of England had to be carefully considered so that the delivery of parts from suppliers would not be delayed. The late arrival of supplies would prevent the items arriving to the appropriate point of the production line at the set time and this would have a heavy financial effect on the company as the production line would have to be slowed down or, at worst, stopped.

Through the use of JIT delivery the manufacturing company becomes more financially smart as parts are no longer stockpiled onsite due to items arriving as they are needed, and therefore a quicker financial return is made on their investment as delivered parts are used immediately.

Processing

In the industrial situation the basic aim of the company is to turn raw materials into a product that a customer would like to buy. Raw materials are delivered to the factory and, after a specified number of processes have been carried out, a final product is manufactured that can then be packaged and delivered to a retail outlet.

How the product is manufactured will depend upon the complexity of the item and the environment in which it is to be produced.

In-line production

In-line production is used where the products move continuously along the assembly line with processes being carried out or parts added in sequence. Such a system is used mainly for the mass production of items such as cars, dishwashers and televisions.

The items are progressively assembled as they flow along a production line. The process is repeated constantly during the working day with the assembly line only stopping in the event of a breakdown. Work carried out during the identified manufacturing stages could be done using a trained workforce, automation, or a combination of the two.

In-line production plants are very expensive to set up; however, as they are mainly used to produce huge quantities of items, the individual product cost is reduced a great deal compared to batch and one-off methods.

Fig 10.4 A production line used in cheese manufacturing

Automated production

Automated production is carried out using machines that have been pre-programmed to work within set parameters. The role of the operator in these instances is an observer who will check that the machine is performing as required.

Samples will be taken at identified intervals and tested. If any discrepancies are found the machine will be stopped, the fault will be corrected and then production continued.

In car assembly plants the car body shell arrives at a welding station on the assembly line. At that point a series of sensors detect the specific body type and robots then place and weld into position the appropriate body panels.

Continual flow production

Continual flow production is used to manufacture products that are going to be produced over a long period of time. Such methods are used when refining oil or in other chemical industries. Continual flow production plants usually make use of automated monitoring and control systems as part of the quality assurance process.

Fig 10.5 An assembly line

Repetitive Manufacturing

In repetitive manufacturing products remain unchanged over a long period of time and are not manufactured in individually defined lots. Instead, a total quantity is produced over a certain time at a certain rate. This type of production will be used to produce food items such as chocolate bars or high demand drink products.

Cell production

A group of machines or a group of workers and machines that carry out tasks together in order to produce components or a complete product are known as production cells. People who work in this arrangement will be required to work as a team and may carry out several tasks in producing the final item. In the mass production of electrical goods, such as washing machines, robot cells may be used to weld, assemble and spray particular parts.

Robotics

Computer-controlled devices known as robots are used in industry to carry out material lifting, handling and placing components. They are also used to carry out repetitive tasks that workers may find tedious or operations that could be classed as hazardous such as working with corrosive liquids or processes that may cause fumes that may have an effect on workers health.

Industrial robots are built to copy human movements; they are expensive to produce but these costs reduce after continued, prolonged use. Robots offer many advantages to the workplace including repetition of tasks, greater accuracy in carrying out tasks avoiding the possibility of human error, reduction in labour costs, and they can work continuously. Robots are adaptable as they can be reprogrammed to carry out new tasks once an operation is completed.

The introduction of robots into the workplace may have had an effect on the number of workers employed but some human input is still necessary in order to carry out routine maintenance, quality control checks of work and for programming.

Fig 10.6 Robots in action

Fig 10.7 A robot used in industry

Bought-in parts and standardised parts

Production of complex items such as electrical goods and cars are not entirely carried out in one location. Several industries may contribute to the production of a variety of parts and then these items are brought together as the final item is assembled. Parts such as motors, seals, glass doors and wheels that are needed in the production of a washing machine may be manufactured, to order, by another supplier.

Standard components and parts are used extensively. These are usually manufactured to comply with international standards for size and specification to ensure interchangeability and to enable repair and maintenance of products as required during their life. They are readily available from suppliers and include bolts and screws, electronic and electrical connectors, batteries, buttons and pop-fasteners, paper and ring binder mechanisms, cardboard, roof tiles and building blocks.

By using bought-in components and standard parts, the manufacturer can reduce the amount of processes that are needed on the production line and this will reduce the number of employees and the range of skills or expertise required from the workforce. It will be the responsibility of the suppliers to meet deadlines and to guarantee the quality of the part produced.

> Find three standard components or parts. Carry out research to determine their precise specification. Put together a short presentation for your group explaining the role of these parts in successful named products or systems. Include an explanation of the benefits to the manufacturer of using these standard parts in the products or systems you have chosen.

The effects of manufacturing on the properties of materials

Materials used during the manufacturing process must be carefully selected as certain materials will react differently in a variety of situations. Material properties must be considered when a product is at the planning stage. Materials may appear to be appropriate and, from an aesthetics point of view, fit the purpose. However, when in use they may prove to be brittle or wear out quickly. Similarly, during the manufacturing stages processes such as pressing or stamping may affect the structure of the material, and procedures such as annealing may be needed to return work-hardened materials back to their original state in order to avoid stress fractures during use.

Section B — Manufacturing, production and planning

Learning outcomes

By the end of this section you should have developed a knowledge and understanding of:

- Preparation and processing of materials
- Assembly stages during production
- Sequencing and timings of manufacturing stages
- Production planning
- Costing

Introduction

Manufacturing in any company uses various resources including raw materials, people and processing equipment. There are relationships between resources, for example the most suitable process may depend upon the materials used, or the availability of a skilled workforce. For this reason manufacturing resources must be considered in the design of manufactured products to optimise production efficiency.

Resources of the appropriate quality and quantity must be available and at the appropriate time. Time itself may be considered a resource when comparing alternative processes or sequences of manufacturing stages.

Think about the manufacture of products you have designed and made, and others you have studied. Identify materials and processes used in their manufacture and possible alternatives.

Preparation and processing of materials

Materials must be sourced to appropriate quality standards, in the right quantity and at the right time for production needs. Many materials and components can be bought pre-prepared, from electronic components ready bandoliered for use in automated printed circuit board population, to chopped fresh onions used in manufacturing ready meals.

Fig 10.8 Custom bandoliered electronic components

Using pre-prepared components can lead to savings in storage, handling, processing and other costs (for example, in the case of the onions, in hygiene measures and waste disposal). It is important that manufacturers consider all potential savings when deciding whether to carry out material preparation in-house.

Whole processing steps may be eliminated by selecting specific materials, such as pre-coated sheet materials, or by processing materials prior to manufacture.

Fig 10.9 Component preparation: spraying chair parts

Assembly stages during production

Manufacturing production stages typically include preparation, processing, assembly and packaging. The number of assembly stages required varies with the product and scale of production. Assembly may be simply mixing components. In the case of complex products, there may be many sub-assembly stages before the final product takes shape. An example is a car, where the engines are assembled in one factory and then transported to another for assembly into the car body.

Sequencing and timings of manufacturing stages

Some stages of production can be carried out concurrently, for example preparing different sub-assemblies; others must be in a set order.

Before detailed production planning can take place, it is essential to consider timings and sequences of manufacturing activities in detail. Graphical methods are used in many organisations and computer software aids complex production analysis. Gantt charts are used in building and construction projects. Critical path analysis is used to identify key stages and critical points to aid project management and ensure they keep to schedule.

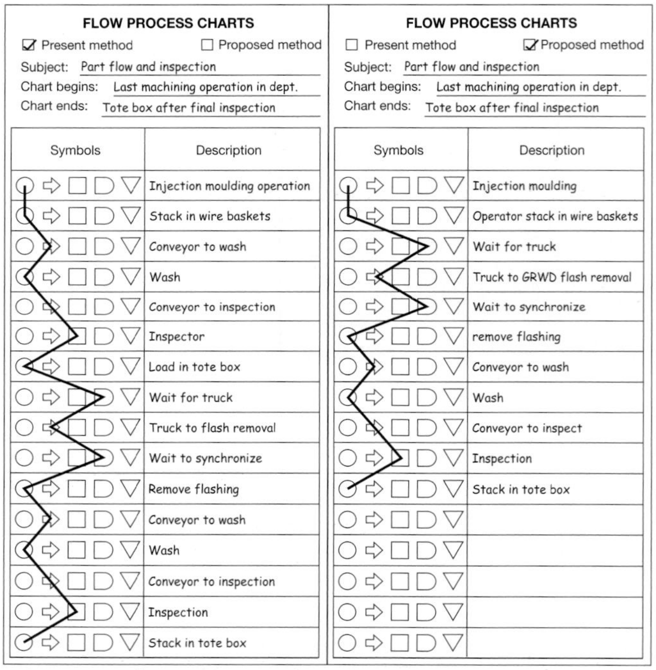

FLOW PROCESS CHARTS		FLOW PROCESS CHARTS	
☑ Present method ☐ Proposed method		☐ Present method ☑ Proposed method	
Subject: Part flow and inspection		Subject: Part flow and inspection	
Chart begins: Last machining operation in dept.		Chart begins: Last machining operation in dept.	
Chart ends: Tote box after final inspection		Chart ends: Tote box after final inspection	
Symbols	Description	Symbols	Description
○⇨□D▽	Injection moulding operation	○⇨□D▽	Injection moulding
○⇨□D▽	Stack in wire baskets	○⇨□D▽	Operator stack in wire baskets
○⇨□D▽	Conveyor to wash	○⇨□D▽	Wait for truck
○⇨□D▽	Wash	○⇨□D▽	Truck to GRWD flash removal
○⇨□D▽	Conveyor to inspection	○⇨□D▽	Wait to synchronize
○⇨□D▽	Inspector	○⇨□D▽	remove flashing
○⇨□D▽	Load in tote box	○⇨□D▽	Conveyor to wash
○⇨□D▽	Wait for truck	○⇨□D▽	Wash
○⇨□D▽	Truck to flash removal	○⇨□D▽	Conveyor to inspect
○⇨□D▽	Wait to synchronize	○⇨□D▽	Inspection
○⇨□D▽	Remove flashing	○⇨□D▽	Stack in tote box
○⇨□D▽	Conveyor to wash	○⇨□D▽	
○⇨□D▽	Wash	○⇨□D▽	
○⇨□D▽	Conveyor to inspection	○⇨□D▽	
○⇨□D▽	Inspection	○⇨□D▽	
○⇨□D▽	Stack in tote box	○⇨□D▽	

Fig 10.10 Flow process chart

The chart shows operations carried out, their type and timings. By examining idle time the sequence of operations may be improved. Network analysis is another graphical tool for determining the optimum sequence of stages.

Production planning

To ensure all resources are used effectively and efficiently to meet customer requirements, thorough planning is essential. Production planning sets out timings and sequences of manufacturing activities in detail. Production plans vary between sectors, but all include consideration of:

- Manufacturing specification requirements
- Materials and components
- Human resources
- Tools and equipment
- Services
- Processes
- Timings
- Quality assurance

Production plans need to be reviewed as opportunities and resources become available to ensure efficiency is optimised.

Costing

The aim of any business is to survive and, if possible, to make a profit. In general, any new product should contribute to the overall profitability of the company. Break-even analysis (a comparison of expected total costs and gross profit) determines the production level above which a profit will be made. If projected sales are not above this break-even point, it is unlikely that the new product will go into production.

Minimising production costs without compromising product quality is a key aim of production planning. Decisions made on preparation and processing of materials; assembly stages and the sequencing and timings of manufacturing stages (see above) are all underpinned by consideration of costs.

Section C — Manufacturing quality control

Learning outcomes

By the end of this section you should have developed a knowledge and understanding of:

- Total quality management (TQM)
- Quality assurance
- Quality control
- Recording and use of data, tolerances, fit, finish and performance

All of the material in this section is covered in Chapter 5: Core knowledge.

Section D — Manufacturing sectors

Learning outcomes

By the end of this section you should have developed a knowledge and understanding of:

- Why manufacturing sectors exist
- The advantages of belonging to a sector
- The relationships between sectors
- The modern technologies and materials used in sectors

Introduction

Manufacturing businesses in a sector make similar or related products. Manufacturing sectors and sub-sectors exist for every type of product, from aeroplanes (aerospace) to bread (food).

Why manufacturing sectors exist

One reason why manufacturing sectors are designated is to facilitate statistical analysis. In this way it is possible to identify which sectors are growing and developing and which may be showing early signs of decline. This may trigger external action, for example in the form of funding to support expansion, or even import restrictions and tariffs to give home-produced products a competitive edge. Another reason is that grouping similar companies together gives them a collective identity.

The advantages of belonging to a sector

Companies in the same manufacturing sector are likely to use the same type of processes and need employees with similar skills and experience. Though they may be competitors in the market, there are advantages to be gained from working together and sharing methods and research. Events such as trade fairs, conferences and exhibitions target specific sectors for this reason, and allow suppliers of goods and services aimed at the sector to promote themselves to potential customers.

Fig 10.11 A trade fair check in

External organisations such as government departments, training providers and trade unions provide services and support for individual manufacturing sectors and sub-sectors.

Fig 10.12 A mobile phone

The relationships between sectors

Manufacturing sectors and sub-sectors are convenient groupings of companies. Some products may seem to 'belong' to more than one sector. An example is mobile phones with claims to the electronics, communications and, increasingly, the computer sectors.

The modern technologies and materials used in sectors

Innovation in manufacturing processes and materials enable new products to be introduced and existing products to be improved or manufactured more efficiently or sustainably. While some technologies, such as computer-integrated manufacturing, are used across a range of manufacturing sectors, some are more sector specific. For example, freeze drying is primarily used in food and pharmaceuticals. Similarly, materials such as biodegradable polymers are used in packaging (for example Biopol shampoo bottles). Shape-memory alloys are mainly used in engineering and electrical products.

Section E Manufacturing systems and control

Learning outcomes

By the end of this section you should have developed a knowledge and understanding of:

- Monitoring, testing and tracking during production
- Computer-integrated manufacturing (CIM), programmable logic controllers (PLCs), robotics, automation and embedded systems in industrial appliances

The availability of computers has had a considerable effect on manufacturing production. The development of PLCs and robotics, microchip technology and computer-controlled handling devices has increased the accuracy and flexibility of the manufacturing process. Computer-integrated manufacturing systems have developed from these technologies, making manufacturing more efficient.

Monitoring, testing and tracking during production

Production systems must be monitored to ensure they are working as required and to avoid wasted production or breakdowns. Simple mechanical systems are physically checked regularly, but modern computer-controlled systems frequently have sensors and in-built checking systems.

Fig 10.13 A CNC woodworking machine with conveyor and monitor

CIM, PLCs, robotics, automation and embedded systems in industrial appliances

CIM is the integration of the product development and manufacturing engineering functions through the use of IT, which has had a significant impact on manufacturing management in general.

Most of the manufacturing automation systems implemented today make use of computer technology. Although automation preceded computer technology, it is computer power that made automation flexible and more effective in applications other than mass production.

PLCs are computers designed specifically for industrial use in process automation. They are built to operate in extreme temperatures, and to resist vibration, impact and electrical noise. PLCs allow multiple inputs and outputs through sensors and actuators of various types and were initially developed to allow rapid and economical modification of automated manufacturing systems, for example when a new model of car was introduced.

Since robotic systems were first introduced in the mid-twentieth century, their applications and versatility have increased dramatically. Most robots have many degrees of freedom. In manufacturing applications, they can be used for assembly work, processes such as painting and welding, and for material handling. More recent robots are equipped with sensory feedback. Through vision and tactile sensing, they can check production quality and self-reset as necessary.

A key advantage of robots is that they can be used for repetitive, monotonous, mundane tasks that need precision. They can also be used in hazardous environments not suitable for human operators.

Many modern industrial appliances designed to carry out specific tasks are controlled by embedded systems in the form of pre-programmed microchips. These include control systems on freezing tunnels, and bottling lines. Their output functions may depend on inputs from operators or sensors, but they are limited.

Fig 10.14 A bottling plant

Learning outcomes

By the end of this section you should have developed a knowledge and understanding of:

- Common processes for working with materials – drilling, sawing, shaping, abrading
- Processes used to manufacture products from wood – laminating, bending, routing/profiling, turning
- Processes used to manufacture products from metal – milling, turning, casting, modifying characteristics using heat, pressing and stamping – see Chapter 7 Engineering
- Processes used to manufacture products from plastic – compression moulding, injection moulding, vacuum forming, rotational moulding, extrusion and blow moulding
- Processes, materials and components used to manufacture products from differing materials
- Processes used in assembling and joining similar and dissimilar materials
- The design of simple jigs, presses and moulds – see Chapter 7 Engineering
- Joining methods using fittings, adhesives, heat and common joints

Common processes for working with materials

Drilling

Drilling can be carried out by machine or manually and is generally the best method of producing through or blind holes in a workpiece.

When manual operations are used a hand-powered drill is used to provide the action. Portable cordless power drills are a popular tool today, with rechargeable batteries, and consequently few operations are now truly carried out using hand power.

The drill is probably one of the most widely used machining processes and most workshops contain at least one pillar drill or bench-mounted machine drill to provide power and accuracy. Multi-head and multi-function drilling machines can be found in manufacturing industries.

Bench drills or pillar drills are found in most workshops. The work table can be moved up and down the vertical column and is clamped at the selected height.

Fig 10.15 Typical bench-mounted drill

The drill is normally located in a three-jaw chuck, which is rotated by the drive system. Fig 10.16 shows the belt drive. The chuck is moved up and down by a feed handle that drives the rotating spindle via a rack and pinion mechanism.

When drilling holes in a material a number of factors should be considered, including:

- Material being drilled
- Hole size
- Hole quality
- Rotation speed/feed speed required
- Depth of hole

Fig 10.16 The belt drive system

- Through or blind hole
- Need for coolant
- Capacity of drilling machine
- Method of work holding: hand held, vice, clamped
- Orientation of drill (horizontal, vertical drilling, angled)
- Swarf control

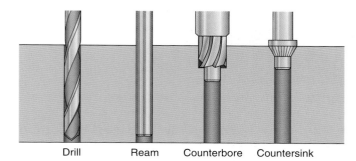

Drill   Ream   Counterbore   Countersink

Fig 10.17 Drilling holes

Holes can be drilled, but also reamed – a sizing process by which an already drilled hole is slightly enlarged to a desired size. They may also be counterbored or countersunk, which both involve the enlarging of one end of a hole to accommodate a bolt head or screw head so that it will be below or flush with the work surface.

For accuracy and safety, the drilling speed is important; for example, aluminium can be drilled at a faster speed than mild steel. Tables giving the correct drilling speeds can be found in many engineering reference books.

Common drill bits

Twist drills are probably the most common drilling tools. They can be used on timber, metal, plastics and similar materials, although HSS (high speed steel) twist bits should be used for drilling

metals. Twist drills are normally available in sizes 0.8 to 25 mm. They are designed for drilling relatively small holes.

Fig 10.18 Twist drill

Spur point bits are also known as wood or dowel bits. The bit leaves a clean-sided hole. Spur point bits should only be used for drilling wood and are for relatively small sized holes.

Fig 10.19 Spur point bit

Flat wood bits are for power drill use only. The centre point locates the bit and the flat steel on either side cuts away the wood. These bits can be used to drill fairly large holes. Sizes range between 8 and 32 mm.

Fig 10.20 Flat wood bit

Hole saws have interchangeable toothed cutting rings and can be used to cut small or large holes in thin sheet metals as well as wood or plastic. They are best used in a power drill at low speed as the blade saws its way through the material.

Fig 10.21 Hole saw

Forstner bits are used to form holes with a flat bottom, such as for kitchen cupboard hinges. These are best used in a pillar drill. If used freehand, the positioning is difficult to control as there is only a very small central point.

Fig 10.22 Forstner bit

Wood auger bits are ideal when drilling deep holes in wood or thick manmade boards. Generally, auger bits should only be used in a hand brace but versions are available for portable power drills. The bit will cut a clean and deep hole.

Fig 10.23 Wood auger bit

Expansive bits are similar to auger bits, but the distance of the cutting edge from the screw point of the drill can be adjusted.

Adjustable tank cutters are used to cut large diameter holes in thin sheet materials.

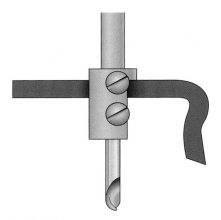

Fig 10.24

Safety procedures when drilling

Drilling takes place in many different practical situations. In some, such as using a battery-powered hand drill to cut a hole in a car body panel, the work is large enough and has enough mass to stay still while being worked on. When a smaller workpiece is to be drilled however, it must be held securely – both to assure accurate placement of the hole, and to prevent it binding to the drill bit and spinning around with the rotation of the drill. This can obviously be a safety hazard, especially if the work is a piece of sheet metal with sharp edges.

There are several methods of work holding available, from clamping the work in a vice to be drilled with a hand drill, to fixing it in a machine vice attached to the table of a pillar drill, or clamping it to the bed or a work bench with G-cramps if the work is larger. As with any process

of this nature, appropriate guarding and PPE (personal protective equipment) should be used.

Sawing

A wide range of saws are available, all varying in size and with different sized teeth that are set according to the material to be cut. Saws tend to be divided into groups known as back saws, frame saws and hand saws.

- **Back saws** – tenon saws are used for general cutting out in wood; dovetail saws are a smaller version of the tenon saw used for finer, more accurate work. The back of the saw limits the depth of the cut

Fig 10.25 Tenon saw (top), dovetail saw (bottom)

- **Frame saws** – coping saws are used for cutting curves in thin sectioned wood; hacksaw are used for cutting out metal; junior hacksaws are used to cut thin sections of metal and tubes; piercing saws make use of a fine blade and are used for delicate work including cutting out sections by silversmiths or jewellers

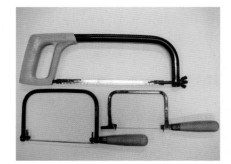

Fig 10.26 Frame saws

- **Hand saws** – cross cut saws are used on large sections of wood when cutting across the grain; panel saws are used for cutting out panels in large sheets of wood; rip saws are used to cut or rip down the grain on large sections of wood

Fig 10.27

The width of the cut that the saw makes is known as the 'kerf' and this must be wider than the blade itself to avoid it sticking when the operation is carried out. To create the necessary gap the teeth are 'set' by turning them to the left or right. Smaller saw blades tend to be set in a wave form.

A general rule to ensure successful cutting of a material, avoiding jumping or sticking, is that at least three teeth should be in contact with the material at any one time. The number of teeth per inch has to be taken into consideration when selecting a saw: finer toothed saws tend to be used for metals.

> With a magnifying glass, look closely at the shape and 'set' of the teeth of a range of saw blades. Look at both hand-held saws and machine saws. Can you see a difference between those designed to cut metals and those designed to cut wood?

Machine saws

Scroll saws are fixed saws that are useful for cutting intricate shapes on thin sheet materials.

Fig 10.28 Scroll saw

Jig saws are portable power tools that can be used to cut around curved shapes in sheet material. The blade reciprocates and cuts through the wood.

Fig 10.29 Jig saw

Power hacksaws are also available for cutting through bars of metal and plastics.

Bandsaws are used to cut curves and other shapes in wood, metal and plastics. Intricate shapes can be cut with narrow blades.

Fig 10.30 Bandsaw

Fig 10.31 Circular saw

Circular saws are used for cutting large sheets and straight edges.

It is unlikely that you will use a bandsaw or circular saw in the school workshop. They are however used by trained teachers and technicians to cut and prepare wood, and are widely used in the manufacturing industry. As with any machine, training, appropriate guarding, extraction and PPE should be used.

Snips and shears can be used to cut thin metal sheet and soft plastics. These may be hand held or bench mounted. When bench mounted the lever greatly increases the force that can be applied.

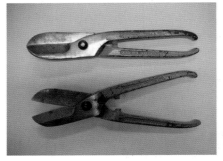

Fig 10.32 Tin snips

Fig 10.33 Bench shears/guillotine/notcher

Shaping

Shaping of materials can be carried out with tools such as files, rasps and surforms.

Files are available in a variety of shapes, length and grade of cut. The shape of file – flat, round, half round, three square and knife – can be selected according to the profile of material

being cut. A swiss or needle file is a smaller version of the engineering file and is used for more delicate work.

The selection of cut will depend upon the amount of material to be removed or the surface finish required. Rough cut and bastard cut are used to remove material quickly. The use of these files is followed by a second cut file and finally smooth or dead smooth cut files could be used to produce a surface ready for finishing with emery cloth.

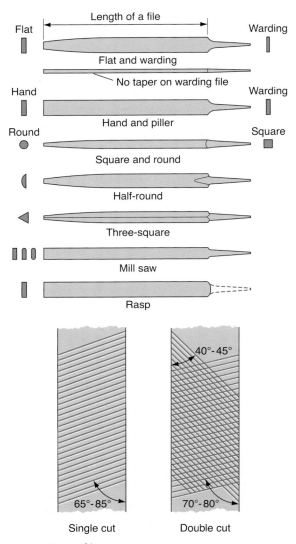

Fig 10.34 Some file types

Filing is normally carried out by pushing the length of the work, which is called cross filing; surface finishing is done by draw filing.

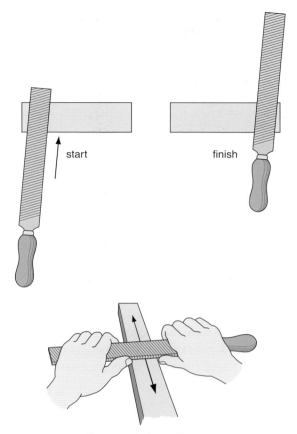

start　　　　　finish

Fig 10.35 Cross filing and draw filing

Rasps are similar to files but they have coarser teeth and are more suitable for use on wood.

Surform tools are available in a range of shapes and sizes with replaceable blades. Due to surforms having a range of blades all with different cutting edges they can be used on a several material groups including hardwoods, softwoods, soft metals, nylon, acrylic and plastic laminates.

Fig 10.36 Surforms

Abrading

Material can be removed from woods, metals and plastics by the use of abrasives. Disc and belt sanders are commonly used. Abrasion is the action of wearing away a surface by friction, and can be a means of shaping a solid form.

An abrasive is any substance that wears down a surface by rubbing against it. Abrasives are available in many forms, including powders, compounds, papers, wheels or disks, brushes, belts, and more.

Grinding of metals can be carried out using a disc or angle grinder, surface grinder or off-hand grinder. All the grinders make use of discs that have been made from abrasive grit that has been bonded together. The size of the abrasive grit used determines the grade or coarseness of the discs.

Fig 10.37 A bench mounted grinding machine

Glasspaper is an abrasive sheet that can be used manually to smooth wooden surfaces. Emery cloth can be used on metals; silicon carbide paper can be used on plastics. Aluminium oxide is another grit commonly used in abrasive sheets.

Disc sanding machines use abrasive sheets mounted on a backing disc.

Fig 10.38 A disc sander

Polishing or buffing machines use abrasive compounds such as tripoli applied to cloth mops.

Fig 10.39 A polishing machine

Belt sanders use a continuous abrasive band over two rollers and workpieces are pressed against it. These machines are used for cleaning up and finishing work. You may also have used portable power sanders in your workshop.

Fig 10.40 A belt sander

Processes used to manufacture products from wood

Laminating

Thin sheets of wood known as veneers can be joined together and cramped onto formers to form shapes that may not be associated with natural timber. Items such as salad servers, curved stool seats or chair backs can be created in this way. Large curved laminated beams are created for use in buildings, and are sometimes referred to as 'glulam' structures.

By producing an item through the process of laminating its strength is increased greatly. Plywood is manufactured by laminating an odd number of thin layers of wood veneers together, each layer glued at 90 degrees to the previous layer. An odd number of veneers is used so that the grain on the outer surfaces runs in the same direction. By creating the material in this way a

very strong product is achieved, especially when compared to natural timber of an equivalent thickness.

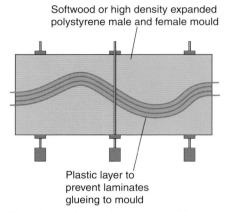

Fig 10.41 A salad server mould

A simple male and female mould can be used to form a salad server. The contact surfaces must be high quality. A plastic film can be used to protect the product and moulds from glue spillage.

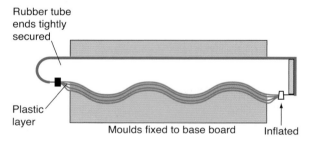

Fig 10.42 Wine rack shelf

A sealed, pressurised rubber tube system is preferred for more complex shapes.

Safety tip
Use commercially available equipment. If you manufacture your own, ensure that it has been correctly checked and tested.

Larger structures, such as table supports, could be produced using a male former secured to a base. The laminates are glued and held using clamps with protective blocks. The clamps would be tightened in sequence.

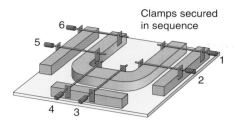

Fig 10.43 Laminating larger structures

A steel strap could be utilised for larger batches.

Vacuum press systems are commonly used in commercial furniture manufacture and are increasingly used in schools and colleges. Large moulds can be built quickly from expanded polystyrene, with a thin laminate glued to its surface to ensure a high-quality finish and strengthen the mould. The laminates are glued, placed on the mould and put inside the heavy-duty clear PVC bag. Air is withdrawn from the bag using a vacuum generator. Atmospheric pressure (up to 8250 kg per square metre) ensures uniform and even compression.

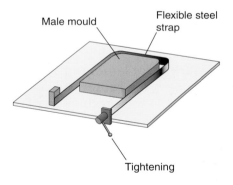

Fig 10.44 Using a steel band

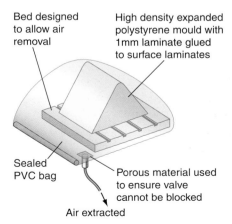

Fig 10.45 A vacuum press systems

Pneumatic and hydraulic clamping systems are also used in the industrial context.

Laminated products that will be used in wet or damp conditions require a waterproof adhesive.

Bending

Due to its grain structure, trying to bend wood can create a range of different problems and cause weaknesses in its structure. A traditional method of shaping wood is kerfing. This involves saw cuts being made close together on the inside surface where a bend is required, giving space for compression once the wood is bent.

Steam applied under controlled conditions to a piece of wood will make it soft, semi-plastic and compressible. In this state the wood can be clamped in a jig or former and held in place until it dries out. The amount of time needed for steam bending varies according to the timber selected.

Fig 10.46 Steam bent products

Kiln-dried wood is often unsuitable for steam bending because the lignin in the wood is set during the hot, dry kiln process. Air-dried (seasoned) wood such as ash, elm, cherry, hickory and oak can be steam bent. The samples must be straight grained and free from defect.

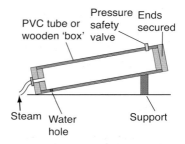

Fig 10.47 A simple steam chamber for bending wood

The sample is usually steamed for approximately three minutes for each millimetre of thickness. The steel clamp former (Fig 10.44) could be used to support the work while it takes on the desired shape.

Safety tip

A steam chamber can be purchased commercially or can be built by technically capable adults. Ensure that the system used to provide steam is constantly topped up and that a safety valve is fitted to the chamber.

Another method of holding steam-bent work is shown in Fig 10.48.

Various types of wood-based manufactured boards are now available for creating curved shapes in furniture and buildings. These include flexible plywoods and MDF called 'Bendy Ply' and 'Bendy MDF'.

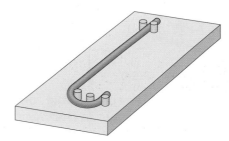

Simple dowel bending jig for steam bending a walking stick

Fig 10.48 Holding steam-bent work

Carry out an internet search using 'Bendy Ply', 'Bendy MDF', and 'curved shapes in wood.' Be amazed at the wide variety of different materials and technologies now available.

Routing/profiling

Routing can be carried out using portable electric routers and also by CNC routers/CAM. The routing process allows a variety of shapes and sections to be created on the surface and edges of strips of wood. Examples of routing can be seen when one looks at items around the home such as edges that have been applied to tables and cabinets. In the kitchen, door and drawer panels may have been manufactured using a router.

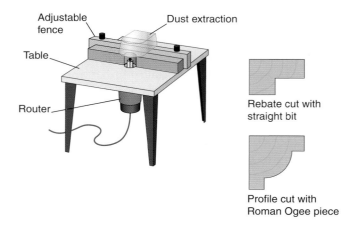

Fig 10.49

For larger pieces of work or for larger batches requiring the same process, the router can be secured to a table to enable the production of moulded shapes. A wide variety of cutters can be used to generate different profiles.

Fig 10.50 Spindle moulding machine

Dedicated spindle moulding machines are used when very large batches are required or more complex profiles are needed.

CNC routers are used for cutting and shaping wood. Shapes are first created using 3D CAD software.

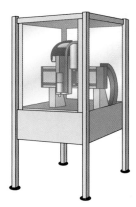

Fig 10.51 A typical example of a CNC router used in schools and colleges

Turning

Wood can be turned using a wood lathe in two ways:

- Using a faceplate and holding it on the lathe headstock. This method is used to create items such as bowls, plates and dishes
- Between centres when slim, longer items such as turned table legs, candlesticks or stair rails can be produced. Wood used in this process needs to be held between the headstock using a driving dog centre and supported with a fixed or revolving centre in the tailstock

Although almost any wood can be turned, those particularly suitable for turned work include ash, beech, cherry, elm, sycamore and teak.

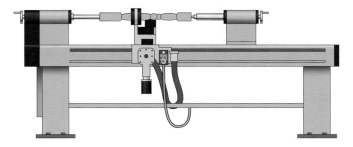

Fig 10.52 Automatic copy lathe

Furniture manufacture and the construction industry require large batches of turned components, for example legs, banister rails and so on. Copy lathes can be used that can either

manually or automatically trace the profile of the required shape and save, store and repeat it. Alternatively, CNC machines that use a profile generated by CAD software can be used.

Processes used to manufacture products from metal

This material is covered in Chapter 7 Engineering.

Processes used to manufacture products from plastic

Compression moulding

Compression moulding is a process that is mainly used to process thermosetting plastics. Examples of common products made by this process include light switches, plug sockets and tops for jars and bottles.

Fig 10.53 A product made by compression moulding

The process involves inserting a pre-measured amount of plastic into a closed mould and subjecting it to heat and pressure until it takes the shape of the mould cavity and solidifies.

- The mould in this process is made up of two parts and is usually made of highly polished, high-carbon steel. Each part of the mould has a heater plate attached to it
- The measured plastic granules (known as charge) are placed into the bottom part of the mould; the second mould is attached to a hydraulic ram, which helps to apply the pressure once the mould is closed

- Once the mould is closed the plastic is heated until it reaches the required temperature. The pressure is applied and the plastic takes up the form of the mould. The heaters are turned off and the job allowed to cool before the mould is opened. The base resin, being a thermosetting material, cures and hardens, cross linking the plastic
- The part is then ejected and removed

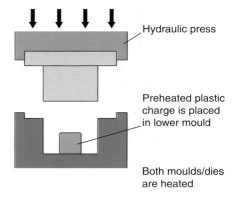

Fig 10.54 The compression moulding process

The cycle time for compression moulding is very long compared to a process such as injection moulding but it has its advantages. These include low capital cost: the tooling and equipment is simpler and cheaper. There is no need for sprues or runners, which reduces waste. However there are limitations upon the size and complexity of products that can be manufactured.

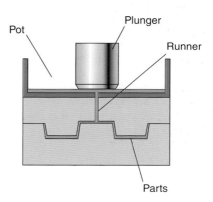

Fig 10.55 The transfer moulding process

Transfer moulding is a modified version of compression moulding and is aimed at increasing productivity by accelerating the production rate. The process involves placing the charge in an open, separate pot. A hydraulic ram then squeezes the soft plastic into a mould where it sets into the defined shape. It means that sprues and runners are needed. The surfaces of the sprues and runners are kept at high temperatures (approximately 150°C) to promote curing of the thermoset. The entire shot including the sprues is then ejected.

Injection moulding

Injection moulding is the most commonly used method for mass production of thermoplastics, due to high production rates, dimensional accuracy and quality of finish.

Polythene bowls, nylon gear wheels and polystyrene casings can be manufactured using the injection moulding process. Products can be manufactured with features such as finger grips, surface texture and other intricate details such as fastening clips and reinforcing ribs/webs as an integral part of the moulding.

Injection moulding machines are similar in some aspects to extruders, except that where an extruder continually forces material through a die, an injection moulding machine uses force to apply the required amount of plastic into a mould to produce the specified product.

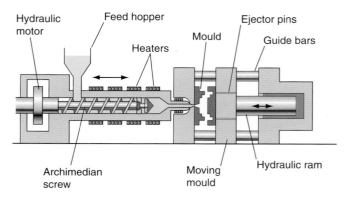

Fig 10.56 A typical injection moulding machine

- Plastic granules are placed into the machine through a hopper
- The injection moulding machine consists of a heated barrel equipped with an Archimedean screw (driven by a hydraulic or electric motor), which feeds the molten polymer into a

temperature-controlled split mould via a channel system of gates and runners

■ The screw melts (plasticises) the thermoplastic pellets to the point of melting and becoming liquid, and also acts as a ram during the injection phase

■ The liquid plastic is injected (forced through a nozzle) into a mould tool that defines the shape of the moulded part

■ Water is pumped around the mould chambers to rapidly cool the moulding

■ The mould opens and the part is ejected by ejector pins. Once removed from the machine the item will need to be finished or 'fettled' by removing the sprues and any flashing.

The pressure of injection is high and depends on the material being processed. Tools (moulds) tend to be manufactured from tool steel or aluminium alloys. The mould may have more than one cavity.

The high costs associated with tool manufacture means that injection moulding lends itself to high-volume manufacture; this is the main limitation of the process. Once a designer has decided to injection mould a product, they then need to produce a design that favours this process. Some of the considerations would be:

■ Make the wall thickness of a product uniform and to a minimum without compromising strength. It is better to use ribs than increase thickness; the thicker the walls of a product the longer the product cycle and the higher its cost

■ Provide generous fillet radii, avoiding sharp corners where possible

■ Avoid using blind holes as more complicated moulds will be needed. Holes are produced using core pins

■ A draft angle of at least one degree is needed so that the product can be ejected easily. Undercuts should also be avoided as they require sectional moulds that can be withdrawn after moulding to release the component

■ Moulded components will have visible gates – this is the point at which the plastic is injected into the mould, and sometimes you will notice surface marks created by ejector pins

Injection moulded components can be complex, as can be seen in the injection moulded parts shown in Figures 10.57 and 10.58.

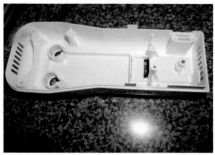

Fig 10.57 The casing for a food mixer

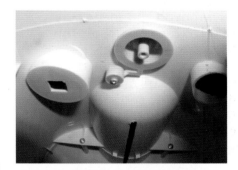

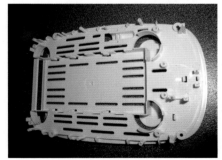

Fig 10.58 The casing for a toaster

Vacuum forming

Vacuum forming is a process that is commonly used to make simple trays or containers. In industry, products produced can vary from polystyrene food trays used in supermarkets to acrylic sinks and baths. Products are manufactured from thermoplastic sheets by a sequence of heating, forming, cooling and trimming.

The main stages in the process are outlined below:

- Before carrying out the vacuum forming process, a quality mould, identical to the finished product, has to be produced
- The mould is placed on the bed or platen of the machine. The platen is then lowered and a piece of thermoplastic sheet is clamped into position onto an air-tight gasket
- The heater is switched on until the plastic becomes soft. Once the plastic is soft the platen is raised and the mould pushes into the plastic
- The heat is then removed and the vacuum pump switched on to remove the air. The plastic is forced against the mould by atmospheric pressure. Where a deep draw is required a top 'plug' may be used to push material into the mould during the forming process
- The material is allowed to cool. The cooling process may be shortened with blown air or even a fine water spray
- The component is then released from the mould by introducing a small air pressure
- After moulding, any mould finishing may be performed (trimming, cutting, drilling, polishing, decorating and so on)

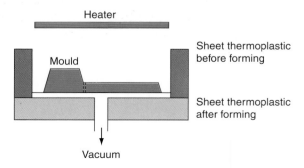

Fig 10.59 The vacuum forming process

Moulds are usually made of aluminium because of its high thermal conductivity. For low-volume production, wood-based materials or plaster of Paris are often used. Draft angles are needed for easy removal. The main limitation on the shape of the product is that is should contain no holes, but they can be machined at a later stage.

Very small air extraction holes are required in the corners of complex moulds to aid removal of the air. While 'male' moulds for the plastic to form around are much easier to produce, 'female' moulds, for the plastic to form inside, ensure a more constant wall thickness of plastic in the finished component.

Webbing, where the plastic creases as it stretches around or into the mould, can be a problem.

The limitations of the process in terms of thickness of material, the depth of moulding, and the thorough and even heating of the plastic must be observed.

More sophisticated machines and moulds are used for continuous automated production of high-volume items like yoghurt pots, disposable cups and sandwich packs.

Rotational moulding

Rotational moulding differs from other plastic processing methods in that the heating, melting, shaping and cooling stages all occur after the plastic is placed in the mould, and no external pressure is applied during forming. Hollow-shaped products such as, traffic cones, litter bins and plumbing components (for example underground drainage pipe connectors) can be produced using rotational moulding.

Rotational moulding normally uses thermoplastics but thermosets can also be used. The process has a relatively long cycle time, but it enables complex parts to be moulded with low-cost machinery and tooling. The advantage of this is that large products can be produced economically with relatively low mould costs.

- A pre-determined amount of polymer powder is placed in the mould. With the powder loaded, the mould is closed, locked and loaded on to the arm of the machine and into the oven

Once inside the oven, the mould is rotated around two axes, tumbling the powder. As the mould becomes hotter the powder melts and sticks to the inner walls of the mould. The plastic gradually builds up an even coating over the entire inner surface of the mould

When all the plastic has been used to form the required shape, the heater is switched off and the mould cooled either by air, water or a combination of both. The plastic solidifies to the desired shape of the mould with a uniform wall thickness

When the plastic has cooled sufficiently to retain its shape and be easily handled, the mould is opened and the product removed. At this point powder can once again be placed in the mould and the cycle repeated.

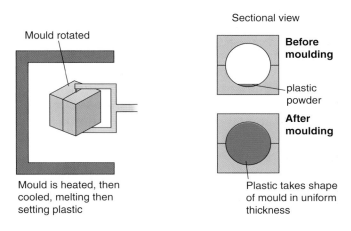

Fig 10.60 The rotational moulding process

Extrusion

Extrusion is a continuous process used to produce both solid and hollow products that have a constant cross section. Polythene pipes and nylon curtain rails are produced by extrusion, and uPVC window and door frames are extruded in a semi-automated process to meet the high-volume demand in the construction industry.

The thermoplastic is heated and extruded out of a die, a continuous process capable of forming an endless product that is cooled by spraying water and then cut to the desired length. The shape of the die used during this process will determine the cross section of the extrusion.

The speed of the process varies according to the cross section being manufactured: thicker sections are extruded slowly as time is required for the initial heating and subsequent cooling of the larger quantities of material involved.

Uniform wall thicknesses are important to produce a straight extrusion.

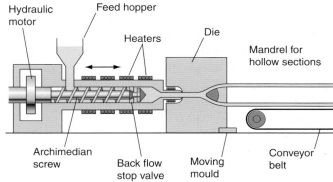

Fig 10.61 The extrusion process

Plastic granules or powder are loaded into a hopper on the extrusion machine

They are fed from the hopper into the heating chamber by a rotating screw. As the screw turns the granules pass through the chamber, and as the temperature increases the plastic becomes molten

The screw continues to turn compacting the plastic and forcing it to leave the chamber through a pre-shaped die in a continuous stream

As the plastic emerges from the die, it is hot and will lose its shape unless it is cooled fairly quickly. Water baths or jets of air tend be used for this purpose as the item is supported on a moving belt

Blow moulding

Blow moulding is a process in which a softened thermoplastic tube inflates against mould walls and hardens by cooling, typically to form a hollow vessel or container. Mass produced drinks bottles as well as other hollow items, such as large water containers, can be produced using blow moulding.

Blow moulding is fast and efficient. The hollow products manufactured this way usually have thin walls, but range in shape and size.

Although there are different versions of the blow moulding process, for example injection blow moulding and extrusion blow moulding, they basically involve blowing a tubular shape (parison) of heated plastic within the cavity of a split mould. Air is injected into the parison, which expands to the shape of the mould in a fairly uniform thickness.

Injection blow moulding is used for the production of hollow objects in large quantities. The main applications are bottles, jars and other containers. The injection blow moulding process produces bottles of superior visual and dimensional quality compared to extrusion blow moulding. The process produces them fully finished with no flash.

- The process requires a split mould to be accurately produced prior to the production system being put into action
- A parison is produced by injection moulding or extrusion and may consist of a fully formed bottle/jar neck with a thick tube of plastic attached; it is transferred to the blow mould while still hot
- The process commences with the mould open and a hollow length of heated thermoplastic placed between the two parts of the mould
- The mould is then closed and compressed air is blown inside the hollow tube, forcing it against the walls of the mould. As the walls of the mould are cold, this causes the plastic to cool and therefore harden quickly
- Once the plastic is cool the mould can be opened and the product removed; excess material is trimmed off and the product is ready for use

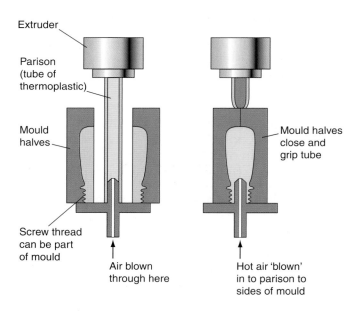

Fig 10.62 The blow moulding process

Plastic domes that may used as food covers can also be produced by blow moulding in a similar way to that already described in vacuum forming. A shaped ring will be clamped over the thermoplastic on the vacuum forming machine. The plastic will be heated as described in the vacuum forming section but once it reaches the required temperature, instead of removing air out and causing a vacuum, the motor is reversed and air is blown into the chamber causing the heated plastic to rise; a shape will be produced according the clamped ring positioned on the surface.

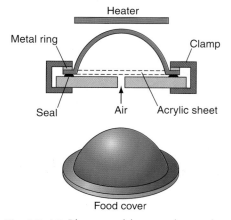

Fig 10.63 Blow moulding a plastic dome

Collect a range of products and components manufactured in plastics. In your group, try to identify the most likely plastics and processes used to produce the components. Write down the reasons for your decisions.

1 Many inkjet printer components are manufactured by extrusion. Describe in detail the process of extrusion moulding. Use sketches where appropriate.
2 The switch on a hair dryer is compression moulded. Describe in detail the process of compression moulding. Use sketches where appropriate.
3 Give two reasons why vacuum forming is an appropriate process for the production of a car dashboard. Describe in detail the process of vacuum forming a car dashboard including the most important features of the mould. Use sketches where appropriate.
4 Bottle screw tops are manufactured by injection moulding. Describe the production cycle for injection moulding a screw-on bottle top. Use sketches where appropriate.
5 The process of injection moulding has many similarities to that of die casting of metals. What are the main differences and similarities?

Processes, materials and components used to manufacture products from differing materials

When selecting manufacturing methods, care must be taken that materials and components being processed are not adversely affected. In the food sector, combining ingredients by cooking often changes their composition.

Processes used in assembling and joining similar and dissimilar materials

Many of the joining methods covered in the next sub-section can be used to join dissimilar materials. There are a range of technologies that are specifically designed to join dissimilar materials.

Inserts used in plastic mouldings provide features such as threaded anchorages and electrical terminals. Metallic inserts can be added at the moulding stage or by pressing into moulded or drilled holes. Alternatively, an interference fit joins two parts by friction. They are usually brought together by pressure, or expansion or contraction of one part, using heat, cooling or shape memory properties. Other mechanical methods include crimping, stitching, bending and using fixings such as screws rivets and bolts.

Increasingly, adhesives are used for joining dissimilar materials in manufacturing. Some are developed for a specific use to meet set performance criteria. The use of adhesives involves additional processing, for example preparing surfaces to be joined and time waiting for setting or curing. Increasingly adhesive tapes are used, reducing this 'idle' time.

In the electronics sector, surface mount adhesives (SMAs) hold chips on the bottom side of printed circuit boards during the wave solder or solder reflow processes. These are formulated to be dispensed from syringes or applied by stencil.

For automated production, additional or modified joining methods may be used. For example using: double-sided tape or sleeves to hold components in place in sub-assemblies, fusible bonding of textiles to aid handling, and tapered (dog-nose) bolts to reduce precision needed for their correct location.

The design of simple jigs, presses and moulds

This material is covered in Chapter 7 Engineering.

Joining methods using fittings, adhesives, heat and common joints

Most products are produced as an assembly of a number of different components. To complete the product the components must be joined together in some way either as part of the manufacturing process or by the end user. The method used to join components will depend on many factors including:

- The materials to be joined and whether they are similar or different to each other
- The strength required from the joint
- How permanent the join must be – whether the joint needs to be dismantled for maintenance or adjustment
- Manufacturing considerations such as the speed of assembly or the skill of the workforce or end user

It is impossible to provide details of all the different methods here so what follows is an overview of some of the more important ones.

Fasteners

Mechanical fastenings are important in the production of many (if not the majority) of products. They allow similar and dissimilar materials to be joined and many allow the joint to be disassembled for maintenance and/or adjustment.

Screws – self-tapping threads

Self-tapping screws are those that cut their own thread and require only a plain hole, called a pilot hole (or sometimes no hole at all) before use. They are characterised by a fairly coarse open spiral thread with a tapering diameter – sometimes reaching a sharp point. The most common self-tapping screws are wood screws. Wood is a material in which it is impossible to produce a threaded hole ready for a machine screw. Self-tapping screws are also available for

other materials, especially metals and plastics, but the angle of the thread and point are not the same as those for wood. Although a self-tapping screw will be thought of as a simple component, they are in fact quite sophisticated pieces of engineering in their own right.

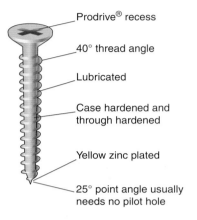

Prodrive® recess

40° thread angle

Lubricated

Case hardened and through hardened

Yellow zinc plated

25° point angle usually needs no pilot hole

Fig 10.64 A typical wood screw

Screws – machine screws

Machine-threaded screws rely on the hole into which they will be inserted having a thread already formed internally. Machine screws can be identified by their evenly pitched thread and constant (non-tapering) diameter. It is essential that the thread of the screw and hole correspond exactly: historically there were many

Fig 10.65 A typical machine screw

different types of thread but in recent years ISO metric screw threads have become the manufacturing standard. Diameters typically range from 2 mm to 20 mm, with the thread denoted M2 to M20.

Screws – different head patterns

Over the years many different drive systems have been introduced to supplement the familiar slotted and Pozidriv® screws. In some cases these have been developed to allow greater torque to be applied to the screw and in others they are intended to prevent the end user dismantling the product. Examples of these are shown below.

Fig 10.66 Screws are available with a number of different types of heads to suit different applications

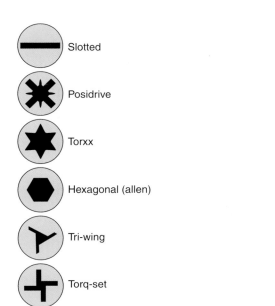

Slotted

Posidrive

Torxx

Hexagonal (allen)

Tri-wing

Torq-set

Fig 10.67 Some of the more common drive systems used on consumer products

Fig 10.68 A couple of the wide range of anti-tamper screw heads and sockets available to manufacturers

Nuts, bolts and washers

In some respects a bolt is very similar to a machine screw. The difference is that a screw is threaded over its entire length whereas a bolt is threaded at the end and has a plain shank running between the thread and the head of the bolt.

A bolt is inserted into a hole that is slightly larger in diameter than itself, and the screw threads at the end of the bolt are secured into a nut or threaded component at the far end of the hole. Bolts are available with a similar range of head types to the machine screws shown in Figure 10.66.

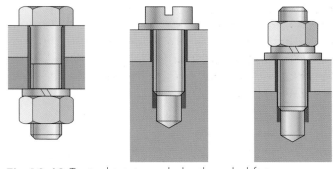

Fig 10.69 Typical joints made by threaded fixings

There are several different types of nuts available for use with bolts. The most common is a plain hexagonal nut but other types are frequently used, for example locking nuts that prevent accidental loosening of the joint, and wing nuts that allow the joint to be tightened without the need for tools.

Fig 10.70 From left to right: plain hexagonal nut, nylon locking nut and wing nut

Washers are an essential part of a bolted joint. Several different types of washer are available for different applications. They help to prevent damage to the components being joined, and some are developed to prevent loosening of the connection. Most common are plain and spring washers.

Fig 10.71 From left to right: plain washer and spring washer

Rivets

Riveted joints are usually associated with joining flat sheet metal components, but the principle can be applied to most resistant materials. The stages involved in making a basic riveted joint are shown in Figure 10.72.

Fig 10.72 A simple riveted joint

In practice a complete riveted joint may be made up of many individual rivets to give the required strength and the diameter of the rivet can be from 3 mm to 50 mm.

A significant drawback to the standard riveted joint is that access to both sides of the joint is needed. This is overcome with the use of pop (or blind) rivets. These tend to be relatively small in diameter, such as that shown in Figure 10.73.

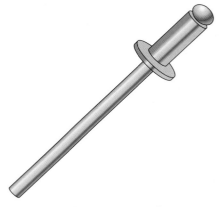

Fig 10.73 Blind rivets usually consist of a hollow aluminium alloy rivet (rather like a top hat) and a steel pin. They are passed through the hole in the components as shown in Figure 10.75

Fig 10.74 A rivet gun

A special tool known as a rivet gun is used to pull the steel pin through the rivet, distorting the alloy stem of the rivet so that it clamps against the back (or blind) side of the components. When sufficient force is exerted the pin fractures leaving the aluminium rivet in place.

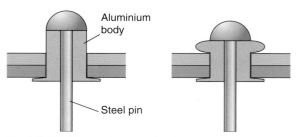

Aluminium body

Steel pin

Fig 10.75 Stages in pop riveting

This technique allows the connection to be made very rapidly. It is used widely in the aerospace industry to attach external panels to the wing and fuselage of aircraft. In manufacturing situations such as this the hand-operated rivet gun is likely to be replaced with an automated, or semi-automated pneumatically operated machine.

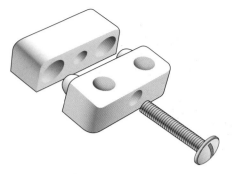

Fig 10.77 Block connectors – one part of the connector is screwed into each of the components to be connected; they are then assembled using the screw

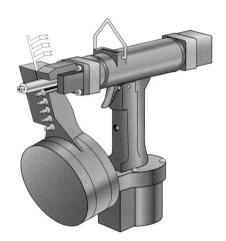

Fig 10.76 A semi-automatic pop-riveting gun

Make a study of the use of mechanical fasteners in products. Look around your school, college or home. Find as many different types, different materials and different finishes as you can. Using the internet as a resource for images, produce a poster for your DT area.

Fig 10.78 Barrel and screw fittings – they produce a strong joint commonly used in bed frames but also in a wide range of products in similar and dissimilar materials; they are often used in conjunction with dowels, which may or may not be assembled with adhesive

Special-purpose fastenings and fittings

Knock-down (KD) fittings

KD fitting is a generic term for a range of mechanical joining systems developed for the manufacture of flat-packed furniture. They are manufactured in huge numbers, which results in extremely low unit costs. They allow furniture to be assembled at home without the need for specialist equipment or skill that would be the case for traditional furniture.

Some of the more common examples are shown in Figures 10.77 to 10.80.

Fig 10.79 CAM lock fittings – used extensively in flat-packed furniture; they are quick and simple to assemble and like, barrel and screw fittings, are often used together with dowel joints

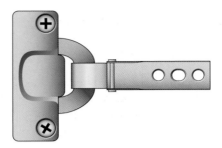

Fig 10.80 Cabinet hinges – ideal for hanging doors to cupboard carcasses made from man-made boards. They fasten onto the faces (rather than the fragile edges) of the boards and provide easy adjustment for correct door alignment

Although all KD fittings are simple to use, their success relies on the amazing accuracy of modern manufacturing methods that utilise CNC machinery to pre-drill holes ready to receive the fittings. If KD fittings are to be used in the school workshop, jigs and templates are invaluable to position holes as accurately as possible.

Examine a range of products in different materials. Identify the fasteners and fittings used in their assembly and make sketches and notes of how they have been used. Explain the benefits they bring to the product and the advantages to the manufacturer and user.

EXAMINER'S TIPS

You will be expected to be able to suggest ways in which a product may be assembled and to be able to sketch and explain the way that any fasteners or fitting would be used.

Adhesives

In some situations adhesives offer significant advantages for designers/manufacturers and the user of products:

- They can produce very discrete joints with little visible evidence of how the joint is secured

- As part of a manufacturing process they can be made using unskilled labour
- They make permanent joints that, when correctly made, will not become loose under normal design loading
- They can produce very high-strength joints

A demonstration of an advantage of a glued joint compared to a joint made with mechanical fastening can be made using paper, paper glue (in stick form) and staples.

Join two pieces of paper by overlapping their edges. For one sample use the glue to make the join. For a second sample use a number of staples. Which joint has the highest tensile strength (i.e. when the pieces of paper are tugged apart)?

This demonstration shows the effect of stress concentration – which causes the paper to rip around the location of the fasteners. This is why the wings of some high-performance aircraft are bonded to the fuselage with adhesive rather than relying on more traditional mechanical fastenings.

Key Points

'Tack' or 'grab' is the immediate stickiness of glue when it used. This is quite a different property to the ultimate strength the joint made with the glue will develop. Do not think that because a glue does not feel 'sticky' that it will not work well when used correctly.

Adhesive technology is developing very rapidly and is closely related to advances in polymer chemistry, with many modern adhesives being a form of thermoplastic or thermosetting plastic.

Drying adhesives

The standard woodworkers' white glue – polyvinyl acetate (PVA) – is a rubbery synthetic polymer usually sold as an emulsion in water. It works very well with porous and semi-porous materials to which it bonds by penetrating the surface. As the water evaporates the flexible polymer is left behind attaching the two (or more) surfaces together. As it relies on the evaporation of the solvent (in this case water) it is known as a 'drying adhesive'.

To work really effectively pressure must be applied to the joint, usually through clamps or other devices that can be attached temporarily. Drying time is 30 to 60 minutes under normal workshop conditions.

In industry, the drying time of PVA adhesive is drastically reduced by the application of radio frequency (RF) energy to the assembled joint using a tool that looks a bit like a domestic iron. The glue reacts to the RF energy and gets hot. The heat is uniform throughout the joint and is almost instantaneous. As a guide, 1 kilowatt of RF power will cure 60,000 mm^2 (an area equivalent to the size of an A4 piece of paper) of glue per minute.

Hot melt adhesives

Hot melt adhesives are a type of thermoplastic (which can include PVA) that become a viscous liquid at fairly low temperatures. They are applied to the joint while hot and the joint is closed immediately. The main advantage of this type of adhesive is the speed of curing, but they seldom result in a high-strength joint. Typically they are used commercially in packaging applications where speed rather than strength is required.

Reactive adhesives

Polyurethane resin (PU) is a multipurpose adhesive that works with almost any porous material. It is formulated to begin the curing process when the resin comes into contact with moisture (normally contained naturally in the materials to be joined). This is known as a 'reactive adhesive' and usually these have a shorter setting period than the drying adhesives. Another well-known example of a reactive adhesive is cyanoacrylate or Superglue.

It is possible for reactive adhesives to respond to other curing conditions such as UV light. Curing times of a matter of seconds can be achieved, which has obvious benefits to manufacturers.

Other reactive adhesives include epoxy resins (a thermosetting plastic). These are usually supplied in two parts that are mixed to initiate the curing process. Epoxy resins form a strong bond with the surface layer of the materials to be joined and so work well with both porous and non-porous materials. They are useful because they

can join materials that are quite dissimilar to each other and a joint made under ideal conditions will have great strength.

Other synthetic resins used as adhesives include urea formaldehyde polymer, sold as a high-quality waterproof wood glue under the well-known trade names of ResinMite (formerly Cascamite) and Extramite.

Silicone adhesives

Silicone adhesives and silicone sealants are based on tough silicone elastomeric technology. Silicone adhesives have a high degree of flexibility and very high temperature resistance (some are flexible between -40°C and 1200°C) when compared to other adhesives, but they lack the strength of other epoxy or acrylic resins.

The most common silicone adhesives and silicone sealants are room temperature vulcanizing (RTV) forms, which cure through reaction with moisture in the air and give off acetic acid fumes during curing.

Pressure-sensitive adhesive (PSAs)

PSAs are very common in everyday life in the form of adhesive tapes such as masking tape or sticky tape. Industrial versions of these tapes exist and are widely used in the automotive and similar industries to permanently secure items of trim and so on, and to temporarily attach labelling such as barcodes to items during assembly and retail.

Solvent adhesives

Solvent adhesives are a niche group of adhesives used solely to join some plastic materials, particularly those containing styrene. They work by dissolving the surface of the materials to be joined (which must be the same or very similar in composition) into a semi-liquid state. The surfaces then flow together until the solvent evaporates, leaving the components permanently joined.

Contact adhesives

Contact adhesives are either based on natural or synthetic rubbers in a solvent solution. This is applied to both surfaces to be joined and allowed to dry. During this process the rubber forms a bond with the surface of the material and, when the two items are brought together, the faces of the adhesive bond rapidly to each other. This joins the two materials together quickly and without the need for lengthy clamping or support. However, the strength of the joint is not terribly high and the system is most effective where the area to be glued is large compared to the forces to be resisted. They are used extensively for fixing laminates such as Formica to flat surfaces.

> In a small group, explore the strength of adhesives. Use a range of different adhesives and join a range of different materials, following the manufacturer's instructions closely (for example, allowing time for the glues to fully set). Devise and carry out some simple tests (safely and sensibly!) and record your findings.

Methods of joining using heat

These techniques are used principally to join metals.

Soldering

Two or more metal components are joined together by melting and flowing a metal filler material into the joint. The metal filler material has a low melting point (around 400°C) so that the process does not involve melting the metals to be joined.

Traditionally an alloy of lead was used as the filler but concerns about the health and safety implications of the fumes produced during heating has resulted in lead-free alternatives becoming more common.

The molten metal forms an alloy with the surface material of the pieces to be joined and so acts very much like an adhesive. For the bond to be successful the metals to be joined must be extremely clean so, in addition to clean working conditions, a chemical known as a 'flux' is used while making the joint to clean the surfaces and prevent oxidisation of the molten metal. The technique does not work for all metals as it

depends on the crystalline structure of the materials and their ability to form a suitable alloy with the filler material.

It is most commonly used for electrical connections (because of its low electrical resistance) and can be automated for the production of printed circuit boards, although it is better known as a craft process completed using a small electric soldering iron.

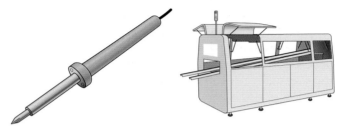

Fig 10.81 The opposite ends of the scale of soldering equipment: on the left a typical light-duty soldering iron, and on the right an industrial scale 'wave soldering machine' used to produce printed circuit boards

Brazing

In principle brazing is very similar to soldering but takes place at a higher temperature (>450°C) by using a different filler material. The filler materials are typically either an alloy of silver (when the technique is sometimes called silver brazing or silver soldering) or brass or bronze-based alloys.

As with soldering a flux is applied to the joint before heating to clean the surfaces of the components to be joined to allow the filler metal to bond effectively with their surface layers.

In most cases a gas torch is used to provide the heat needed for brazing. Butane, propane and oxyacetylene can all be used successfully.

> **Key Points**
>
> It is important to remember that for both soldering and brazing, the metals to be joined do not melt, and that usually the temperatures involved in the process are considerably lower than the melting points of the metals that are joined.

Welding

Welding is usually associated with the joining of metals but in principle can be applied to many thermoplastics. The materials are joined by causing coalescence: by melting them and allowing the molten material to mingle and then cool to become solid once more.

A filler material is usually added to the pool of molten material (the weld puddle) to produce the weld. This produces an extremely strong permanent joint. It is essential that the materials to be joined and filler material are similar.

In principle many metals can be joined by welding but, because of their metallurgical structure, some are extremely difficult to weld. For example mild steel is considered one of the most easily welded metals yet stainless steel, which contains a higher level of carbon, is considerably more difficult to weld successfully.

Many different sources of energy can be used to weld:

- Friction welding uses mechanical energy (for example a rapidly rotating component pressed against a stationary component)
- Gas welding usually uses a mixture of oxygen and acetylene
- Electricity is used in arc-welding (MMA), metal-inert-gas (MIG) and tungsten-inert-gas (TIG) welding and resistance (spot) welding
- Ultrasonic energy (sound waves) are commonly used in the welding of plastics

While the metal is molten during welding it is likely to absorb impurities from the atmosphere that will weaken the material one it solidifies. In gas welding this is prevented by increasing the flow of oxygen to the flame, which effectively excludes the ambient atmosphere.

In MMA welding the molten metal is protected by a shield of inert gas produced by the

vaporisation of a powdered coating around the welding rod. In MIG and TIG welding the shielding gas is introduced directly from bottled sources.

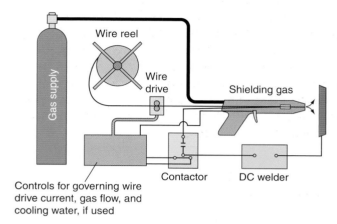

Fig 10.83 MIG welding equipment in diagrammatic form showing the principle of operation

Fig 10.84 A typical workshop MIG welding set

The higher temperatures involved in welding (approximately 3000°C is typical) can cause significant problems of distortion. For some joints, where ultimate strength is not necessary, this problem is reduced or eliminated by spot welding, that is, making the weld in isolated spots rather than a continuous seam, which generates much more heat. Specialist equipment has been developed to do this.

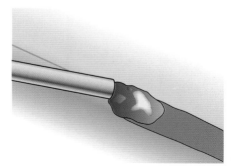

Fig 10.82 An MMA welding rod: note the dull grey surface which is a compressed powder that vaporises to create a shielding gas

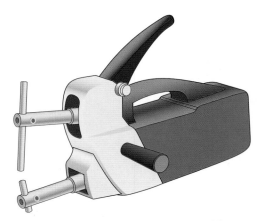

Fig 10.85 A small hand-held spot welding machine

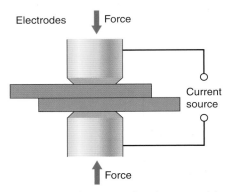

Fig 10.86 The principle of spot welding – the current source is usually a mains-fed transformer producing low voltage (1–2 volts) but extremely high currents

Although welding is a highly skilled trade it is common for the process to be automated in manufacturing processes such as car body assembly.

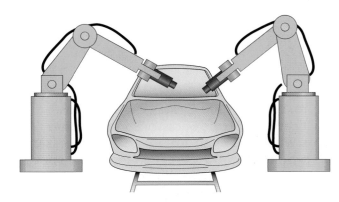

Fig 10.87 Robotic spot welding machines on a production line for motor vehicle bodies

Joints

In addition to the techniques of securing components during assembly in the previous sections, the properties of some materials mean that it is necessary to create specific mechanical joints within the components to be joined. This is particularly true of timber where the internal structure means that any joint that relies on contact of the end grain is unlikely to be successful.

Common timber joints that have evolved over many centuries include those shown in Figures 10.88 to 10.94.

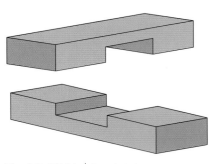

Fig 10.88 Halving joint

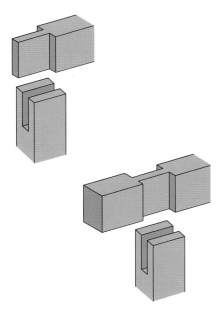

Fig 10.89 Bridle joint

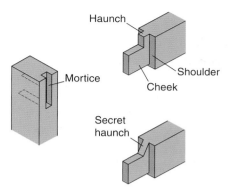

Fig 10.90 Mortice and tennon joint

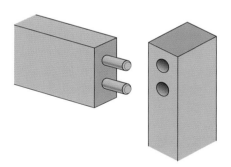

Fig 10.91 Dovetail joints

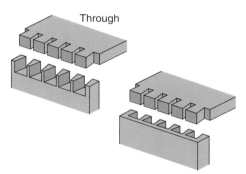

Fig 10.92 Dowel joint

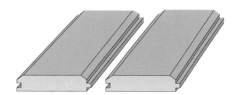

Fig 10.93 Tongue and groove

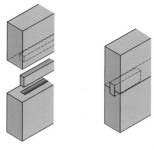

Fig 10.94 Tongue and groove with loose tongue

More recently, joints have been developed to take advantage of advances in power tool design and industrial manufacturing processes. Examples of these are shown in Figures 10.95 to 10.101.

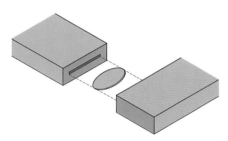

Fig 10.95 A biscuit joint – sometimes called a flat dowel joint

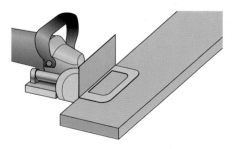

Fig 10.96

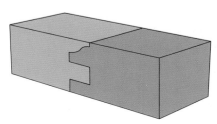

Fig 10.97 A matched profile joint produced by the router cutter shown in Figure 10.98

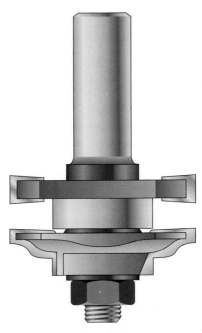

Fig 10.98 Router cutter to produce profile joints

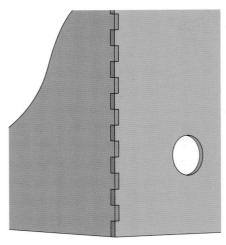

Fig 10.100 Finger joints

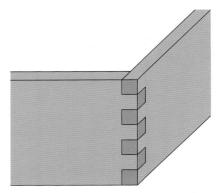

Fig 10.99 Finger joints are particularly suited to mass production. Accurate machining allows simple assembly and produces a strong joint that uses no additional fittings

> **Find four different wood joints used in furniture and household items. Sketch the joints in situ and state why that joint is suitable for that particular product or situation.**

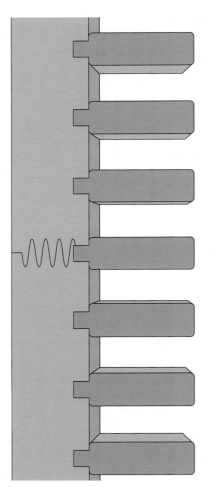

Fig 10.101 A scarf joint – a type of finger joint – is used by manufacturers to produce long lengths of timber for production. It minimises waste because short lengths that would otherwise be unusable can be joined to make usable lengths of material

CHAPTER 11

Resistant materials

Section A   Metals

Learning outcomes

By the end of this section you should have developed a knowledge and understanding of:

- The selection and use of common ferrous and non-ferrous metals
- Mild steel, high-carbon steels, stainless steels, cast iron, zinc, copper
- Brasses and aluminium alloys and tin alloys
- The properties of metal and metal products – strength, toughness, ductility and malleability, weight, durability and thermal and electrical conductivity, in terms of suitability for specific consumer products
- Up-to-date developments of new and smart metal alloys and their potential application

All of the material in this section is covered in Chapter 7: Engineering.

Section B   Plastics

Learning outcomes

By the end of this section you should have developed a knowledge and understanding of:

- The selection and use of common thermoplastics and thermosetting plastics: polystyrene, polyethylene, acrylic, polypropylene, PVC, ABS and PET, phenol resins, phenol formaldehyde, melamine formaldehyde, urea formaldehyde, epoxy resins
- The selection and use of common composite materials: Kevlar®, carbon fibre
- The properties of plastics: hardness, brittleness, tensile strength, plasticity, compressive strength, sheer strength, strength-to-weight ratio, chemical resistance, elasticity, stiffness and impact resistance
- Up-to-date developments of new and smart plastic materials and their potential applications

All of the material in this section is covered in Chapter 7: Engineering.

Learning outcomes

By the end of this section you should have developed a knowledge and understanding of:

- The selection and use of the following hardwoods and softwoods: beech, ash, oak, jelutong, sycamore, maple, teak, cedar, pine
- The selection and use of the following manufactured boards: plywood, laminated boards, chip and compressed boards
- The properties of wood and wood products: hardness, flexibility, tensile strength, compressive strength, sheer strength, strength-to-weight ratio, chemical resistance, elasticity, stiffness and impact resistance
- Up-to-date developments of new forms of wood products and their potential applications

Natural timbers

Hardwoods and softwoods

Natural timbers are classified into two groups known as hardwoods and softwoods. The group a particular timber belongs to has nothing to do with whether it is hard or soft to the touch. It is a biological classification that is related to the way in which the tree produces its seed. Although not strictly accurate, for the purposes of a design technologist, it is reasonable to say that hardwoods come from broad-leaved (deciduous) trees and softwoods from coniferous (cone bearing or evergreen) trees.

Selecting timbers

There are many factors to consider when choosing a species of timber for a particular application. The factors that need to be considered vary enormously from one product to another. For example, in some cases appearance may be the most important consideration, but in others a physical property, such as strength or stiffness, may be most significant. Typically designers will consider:

- Short-term physical properties: strength (in bending, in tension, in compression, in shear), stiffness, density, hardness
- Long-term physical properties: stability (how much the timber expands and contracts, warps or twists as moisture levels in the atmosphere change), suitability for outdoor conditions, resistance to chemicals and so on
- Working properties: ease of machining, ability to be glued, ease of finishing
- Aesthetic properties: colour, grain pattern, lustre and surface blemishes
- Commercial factors: cost, availability of sizes, consistency, wastage.

> Think about the following products: a dining chair, a window frame, a kitchen worktop, a toy. Make a list of the factors that would be important when selecting a timber for their manufacture and place them in order of importance.

The physical properties of natural timber

In common with all other resistant materials the physical properties of timber can be measured in terms of strength (in bending, in tension, in compression, in shear), stiffness and elasticity, density, impact resistance and resistance to chemicals.

However, because it is a natural material (rather than being man-made) many of the physical properties listed are influenced by natural features in the timber such as localised variations in grain pattern, natural growth defects, drying defects, moisture content or biological decay.

Growing timber as part of a tree is composed of elongated, hollow, water-filled cells (approximately 1 mm long in hardwoods and 5 mm long in softwood) 'glued' together in bundles rather like drinking straws. The rate at which the cells form depends on the growing conditions for the tree, and it is this that gives rise to the annular rings on the cross section of a tree trunk corresponding to summer and winter seasons. Tropical hardwoods have little if any evidence of seasonal growth because the weather conditions are more uniform.

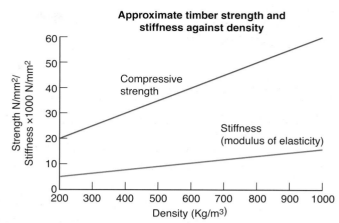

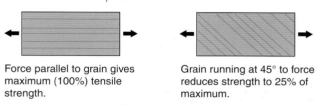

Fig 11.2 Typical values for compressive strength and modulus of elasticity

Force parallel to grain gives maximum (100%) tensile strength.

Grain running at 45° to force reduces strength to 25% of maximum.

Fig 11.3 Grain and direction of timber

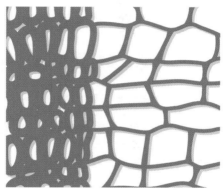

Fig 11.1 A magnified cross section of timber. It is easy to see the large (rapid growth) and the small (slow growth) cells; the smaller cells produce wood that is more dense

In general, the strength and stiffness of timber is proportional to the density of the timber. Typical values for the compressive strength and modulus of elasticity (found from bending tests) are shown in Figure 11.2. You should also consult the material selection charts referred to in Chapter 7 Engineering.

The clearly defined grain structure of timber means that, unlike many other materials, timber is not isotropic (that is, its properties vary in different directions because of the direction of the grain).

The grain and pattern of growth means that the way in which a tree is converted from logs to usable boards has a significant impact on the way the way the timber will behave. This is because the cross section of an individual board will depend greatly on how it is cut from the circular cross section of the log – and the cross section of a piece of timber will define how it will respond to changing moisture content in the atmosphere throughout its life. This is known as timber 'movement', and can result in warping, twisting and other undesirable effects.

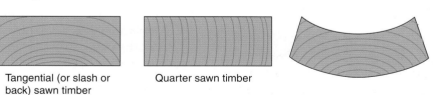

Tangential (or slash or back) sawn timber

Quarter sawn timber

Fig 11.4 Tangential-sawn timber tends to warp as moisture levels in the atmosphere change; although quarter-sawn timber expands and contracts it remains rectangular in section

> Examine products that are made from solid timber. Look at the 'end grain' and try to identify how the wood has been cut from the natural log. Can you find examples of natural timber affected by 'movement'?

For commercial purposes timber is graded so that users who require timber with specific minimum strengths can order appropriate supplies. Visual grading and grading by machine testing are used and both are equally valid.

Visual grading assesses the size, frequency and positions of defects in the timber, such as knots and sloping grain. Visual grading of softwoods is carried out to BS 4978.

Machine grading measures the resistance of the timber member to flexing, which gives a measure of the strength of the piece. Machine grading is carried out to the requirements of BS EN 519.

Common hardwoods

There are hundreds of different types of hardwoods with a wide range of appearances and physical properties. You will be expected to know the main characteristics and typical applications of the timbers set out in Table 11.1.

Table 11.1 Common hardwoods

Timber	Origin	Description	Physical characteristics	Typical uses
Oak	Europe and North America	European oak is typically yellow brown in colour. American oak is either darker in colour (red oak) or lighter (white oak). All usually have attractive grain patterns	Strong and tough. Resistance to outdoor conditions varies greatly with species – European oak is generally very durable whereas American white oak is not. Density lies between 720 and 790 kg/m^3	Heavy structural use, cladding, exterior joinery, interior joinery, furniture, flooring
Beech	Europe, especially central Europe and the UK	The wood is typically straight grained with a fine, even texture. Pink/pale red, reddish brown (after steaming), white/cream	Beech is stronger than oak in bending strength, stiffness and shear by some 20 per cent, and considerably stronger in resistance to impact loads. Density is around 720 kg/m^3	Interior joinery, furniture, workbenches, mallets, kitchen ware, chopping boards

Table 11.1 continued

Timber	Origin	Description	Physical characteristics	Typical uses
Ash	Europe and North America	There are several different varieties of ash. Typically straight grained the colour varies from dark brown (black ash) to very pale (white ash)	Very tough and flexible. Density varies considerable with exact species from 560 kg/m³ (black ash) to 710 kg/m³ (European ash)	Interior joinery, sports goods and handles for tools where the ability to absorb shock loading is important. Used for laminating to produce furniture
Jelutong	Malaysia	Usually pale in colour and very fine grained with little pattern visible	Rather soft, weak and brittle. It is extremely easy to work and can be finished easily. Density is around 470 kg/m³	Modelling, small household products (boxes/CD racks and so on), patterns for sand casting
Sycamore/maple	Europe (sycamore) North America (maple)	White/pale cream. Subtle grain pattern. Fine grain allows very crisp, detailed work.	Similar in strength to oak. Density is around 630 kg/m3	Furniture, interior joinery, craftwork. Rock maple is used for flooring in sports halls
Teak	Burma, Indian peninsula	Yellow/brown in colour. Straight grained with a rough, oily texture. Very expensive; other varieties (for example iroko) are used as alternatives	A little less resistant to impact than oak but slightly stronger and stiffer. Very durable – it resists outdoor conditions well because of the high content of resin. Density about 660 kg/m³	Furniture and joinery (both interior and exterior), favoured for outdoor chairs and benches

EXAMINER'S TIPS

Your knowledge of the different species of timber listed in the table needs to be specific so that you can justify the use of a named species of timber in a given situation by referring to its particular properties.

Common softwoods

There are many different species of tree that are grown commercially to produce softwoods. In general coniferous trees grow more quickly than deciduous trees, which means that softwoods are usually less expensive than hardwoods. It also means that commercial production of softwood timber is sustainable.

Many of the species are very similar and are grouped together for the purposes of timber classification. You need to know the main characteristics and typical applications of the timbers set out in Table 11.2.

Table 11.2 Common softwoods

Timber	Origin	Description	Physical characteristics	Typical uses
Cedar (western red)	North America	Colour of new timber varies greatly from white to reddish-brown, but all become silver-grey over time when exposed to air	Soft and not very strong. It is very durable and tends to shrink and expand less than other timbers in changing atmospheres. Density is around 390 to 500 kg/m³	Cladding, high-quality garden sheds and summer houses, roof and wall (exterior) cladding, especially in North America
Cedar (of Lebanon)	Morocco/Algeria	Straight grained but with many disturbances. Notable for pungent cedar odour	Quite soft and not very strong. Density is around 580 kg/m³	Cedar of Lebanon is traditionally used to line drawers because of its pleasant perfume
Pine (several varieties available, also known as redwood and/or deal)	U.K, northern Europe, Russia	White/creamy coloured timber. Straight grained but sometimes marked by knots	Reasonably strong and stiff but the quality varies because of the wide and varied distribution. Density is around 510 kg/m³	Exterior joinery, interior joinery, Heavy structural use (for example roof trusses and timber-framed buildings). Can be treated for exterior applications such as fences and garden sheds

> Find products that are made from timber. Identify the specific timber from which they are made and relate the properties of the timber to the attributes required for the product.

Man-made boards

Wood-based boards are categorised into two main groups:

- Laminated boards (including plywoods)
- Compressed boards (including chipboard and MDF)

Man-made boards offer a number of useful advantages to designers and manufacturers compared to natural timbers:

- They available in large sheets – 1220 by 2440 mm is a standard size, although considerably larger panels can be obtained
- They are relatively stable – they do not shrink or expand, warp or twist as much as natural timbers
- Their properties are consistent across the whole board
- Their properties are consistent from one board to another
- Many are available pre-finished using laminated plastic sheet (melamine), foil (PVC) or natural veneer surfaces
- They may be treated with flame-retardant chemicals

For many products these advantages are so overwhelming that man-made boards have effectively eliminated all alternatives. In particular, because of their consistency and stability, they are so suitable for mass-produced furniture such as kitchen cupboards, that an extensive range of assembly fittings (usually referred to as knock-down (KD) fittings) have been developed.

> Examine a piece of flat-packed furniture. Try to identify the joints used in its construction and any other hardware components used, such as hinges and drawer fittings. Make sketches of the fittings and try to explain how they work.

Man-made boards also offer some environmental benefits because they can be made from lower grade timbers and so avoid much of the wastage associated with manufacturing with natural timbers.

Laminated boards

Laminated boards are produced by gluing layers together in a well-organised structure. The adhesive used is usually a synthetic resin that sets rapidly during manufacture using heat and pressure.

Plywood

Probably the most well known and widely used laminated board. Thin layers (or veneers) of wood are laid one on top of another with the grain direction at right angles in successive layers. There are always an odd number of layers so that the grain on the two outer faces of the board is parallel. Alternating the direction of the grain through the thickness means that the board has similar structural properties in all

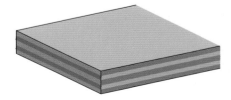

Fig 11.5

directions, unlike the natural timber from which it is made. The number of layers depends upon the thickness of the board, which for most commercial applications lies between 4 mm (three plies) and 27 mm (19 plies).

There are many different types of plywood, suitable for different applications. Most are strong, stiff and resist impact well. The exact type of plywood produced depends on the timber that is cut to make the veneers and the adhesive used to glue the layers together.

Typical adhesives used are urea formaldehyde (UF), phenol formaldehyde (PF) and melamine urea formaldehyde (MUF). UF produces a board suitable for interior use but boards using PF and MUF can be suitable for exterior conditions depending on the species of timber used for the veneer.

Many different types of timber, both hard and softwoods, are used to manufacture plywood. Some of the more common are birch, beech and spruce.

Grading systems are used to indicate the quality of the board. Some are based on the surface appearance (for example BS EN 635) and others are based on the structural and other physical properties (for example BS EN 636).

Specialist plywoods

Birch plywood is also available for craft use and this can be as little as 1 to 1.5 mm thick, allowing curved surfaces to be produced easily. Also suitable for curved surfaces is 'flexiply' or 'bendy ply', which is available in different thicknesses including 5 mm and 8 mm, and manufactured from veneers of very open-grained timber species, which produces a relatively flexible board.

Blockboard and laminboard

These are similar to plywood in that there are layers of timber through the thickness of the board, but in each case the centre of the board is made from parallel strips of timber running perpendicular to the surfaces of the board.

In blockboard the core is made from sawn timber battens approximately 25 to 30 mm wide; in laminboard the core is made from vertically placed veneers, usually made from the same veneer as the facing surfaces.

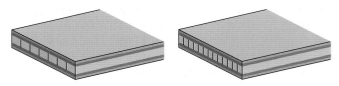

Fig 11.6 Sections showing blockboard (left) and laminboard (right)

This construction means that the boards are stiffer and stronger parallel to the core strips than perpendicular to them. For this reason blockboard and laminboard are usually used for load-bearing shelves and similar applications where the direction of loading is constant.

Compressed boards

Compressed boards are made from chips, strands or particles of timber that are glued together, as the name implies, under a compressive force to produce large, flat sheets.

There are several different types of compressed boards, two of the most widely used of which are chipboard and MDF/HDF.

Chipboard

Chipboard is a cost-effective panel product made from small (2 to 3 mm) chips of timber often from the waste by-product of another production process. Compared to plywood it is rather weak and flexible and has very poor durability. Nevertheless it is commonly used for flat-packed furniture where it is pre-finished with a surface veneer or coating that disguises its humble origins and gives the panel some durability in situations where moisture may be present, such as kitchen cupboards or worktops.

Medium- and high-density fibreboards (MDF and HDF)

In some respects MDF/HDF is similar to chipboard but it is manufactured from much finer fibres of wood rather than chips. As the name implies HDF is more dense than MDF (approximately 800 kg/m^3 compared to 600 kg/m^3), giving a harder surface, greater stability and an ability to be shaped more precisely. The finer fibres mean that the face surfaces of these boards are considerably smoother than that of chipboard and this allows a high-quality painted finish to be obtained with little preparation. There are significant health and safety issues about MDF/HDF because of the release of fine fibres and particles of formaldehyde (adhesive) into the atmosphere during any working of the material.

Up–to–date developments

Although timber is one of the most traditional of all materials used to manufacture artefacts and products, modern developments in manufacturing methods allow designers to use wood creatively.

Increasingly 'engineered wood products' allow innovation in commercial-scale projects. One example of this is flexible sheet materials used to produce curved shapes – especially in furniture.

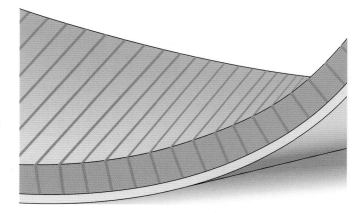

Fig 11.7 A flexible board used for curved surfaces

Another development in the manufacture of sheet material is cement-bonded particleboard for use in the building industry. This overcomes problems of poor durability associated with most compressed boards, so cement-bonded boards can be used for exterior as well as interior applications. By replacing the formaldehyde bonding adhesive of standard compressed boards it is also claimed to reduce the health and safety risks they pose.

A major area of innovation is the use of lamination techniques where strips of solid timber are glued together with their grains running parallel to produce structural members. These combine great strength with the opportunity to produce curved forms with the aesthetic properties of solid timber. This technique, known commercially as 'glulam' (glued lamination), has been used to produced some spectacular structures.

See www.vebjorn-sand.com/leonardo.html for an application of glulam techniques.

Fig 11.8 Dining chair in sycamore by Tim Rinaldi – note the delicate joint between the legs and the seat, which has no under frame. Metal-doweled joints fixed with a two-part epoxy resin adhesive give sufficient strength in a design that would be impossible using traditional wood joints

Section D — Surface finishes

Learning outcomes

By the end of this section you should have developed a knowledge and understanding of:

- The nature and suitability of surface finishes and coatings across a range of products relating to decoration, resistance to decay and wear, absorption and aesthetic qualities
- Methods of preparing surfaces to accept finishes
- Finishes for metal – paints, dip coating, varnishes, lacquering and electroplating, galvanising, plastic coating
- Finishes for plastic – edge polishing, chemical finishing
- Finishes for wood – varnishing, waxing, oiling, stains, polishing, interior and exterior finishes, chemical preservatives, pressure impregnation, lipping

Introduction

The purpose of applying a finish to any material is to:

- Protect against wear, dirt, damage, corrosion and decay
- Enable easy maintenance by dusting, washing or polishing
- Change the appearance of the end product

Fig 11.9 Materials exposed to harsh elements outdoors – protection against corrosion and decay is very important

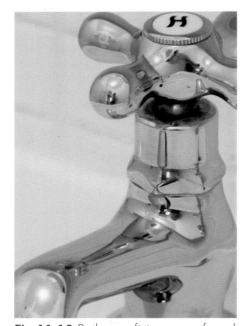

Fig 11.10 Bathroom fittings are often chrome plated to improve appearance and durability; the surface finish also protects against corrosion in the hot steamy environment

Fig 11.11 Wooden garden furniture is regularly oiled to improve durability and improve appearance

Any designer will need to consider surface finish. The first stage of any finishing process is preparation:

- **Wood** surfaces should be made smooth using a smoothing plane and/or glass paper; dents and scratches will look worse if painted over.
- **Metal** surfaces need to be degreased and cleaned with white spirit before painting or applying a plastic coating
- **Plastics** are generally self finishing but edge finishing/polishing is often necessary; textures are often used in plastic moulded products to achieve different surface finishes

Finishes for wood

Varnishes

Polyurethane varnish and lacquers give an attractive and hardwearing finish. Available as clear or colour, a matt, satin or high-gloss finish, this gives a very tough surface finish that it is both heat and water proof, and will stand up to hard knocks. It can be applied with a brush or spray.

French polishing is one of the oldest forms of finishing wood, using a methylated spirit-based 'shellac' giving a beautiful surface finish. However, it does require great skill and time to apply the polish.

Friction polish is a type of shellac that can be applied to turned wood products while they are revolved in the wood lathe.

Wax/polish

Beeswax polish is often used in interior furniture to achieve a natural-looking finish. Before the wax is applied, usually with a cloth, the surface should be sealed with a cellulose sealer to help make the finish last longer. Wax finishing will not stand up to heat and many liquids may stain or mark the final surface.

Oil

When a natural appearance is required, one method is to apply oil to the wood surface. This highlights the wood's own colour and grain, making it water-resistant with a non-shine finish:

- Olive oil – this can be used as a finish when the wood is going to come into contact with food (for example salad servers)
- Danish or linseed oil – this can be used on most woods; it needs two to four coats for the best protection
- Teak oil – as the name suggests this oil is ideal for such woods as teak and iroko; it is based on linseed oil with additives such as silicone to give a harder wearing surface

Most oil finishes are applied with a cloth and worked across the grain. The surface is wiped clean with a second cloth and left to dry before a second coat is applied. Hard rubbing with a soft cloth can build up a sheen over a period of time.

Specialist hard wax oils can be used on floors; the benefit of these over varnish is that the wax can be reapplied without the need for re-sanding and preparing the wood.

Fig 11.12

Fig 11.13

Stain

Stain is used to colour and enhance the grain. Stains can be matched to wood colours and types, or they can be used to add colour. Stains alone do not usually protect the wood.

Paint

Paint provides colour and protection. Acrylic paints are water-based and are generally not waterproof. Oil-based paints are more expensive but are waterproof and tougher. Polyurethane paint is particularly hard wearing.

Paint provides a bright-coloured finish and a degree of protection. Softwoods tends to be painted as hardwoods cover less well and look more interesting when left naturally. Several different types are available from flat non-shine (matt) through satin to very high gloss. Non-toxic paints are available for children's toys and furniture.

Laminated plastics

When a wood surface is to be used constantly and subjected to hard wear, for example a kitchen worktop, many finishes are unable to cope. This problem is overcome by adding a laminated plastic sheet to the surface (for example Formica or melamine).

Protecting wood outdoors

One of the major difficulties in using wood in outdoor applications is its ability to absorb water, causing the wood to expand, degrade and rot. Wood can also be subject to attack by bacteria and fungi if they are able to gain access.

One approach to protecting timber is, therefore, to provide a barrier between the wood and its environment, for example using an exterior yacht varnish or a teak oil. The other is to impregnate the internal structure of the wood with a preservative, for which an effective method is pressure treatment. Once treated by this method the timber can last 50 years in all weather conditions. Creosote (tar oil) is still used commercially, but for general domestic use it has been replaced by ranges of high-performance preservatives in a range of colours.

The effectiveness of any preservative depends on the penetration. Brushing or spraying is not recommended because it only gives superficial surface protection to the timber. They are, however, useful where timber needs to be treated in situ and can extend life of a product by several years if applied regularly.

Creosote

Creosote is a heavy-duty preservative that is toxic to most fungi and insects. It is highly water repellent and can be used for ground contact. Its water repellent properties give it excellent weathering characteristics. Creosote-treated wood usually has an odour, but it is an excellent preservative for applications such as bridges, telephone poles, railway sleepers and marine piles.

Fig 11.14 Wooden train rails

Immersion treatments require a suitably dimensioned bath of preservative in which wood can be fully immersed. Hot and cold bath treatment in open tanks is a more controlled method of immersion treatment and consists of immersing the timber in the bath, raising the temperature to about 85°C then allowing the preservative to cool, or transferring the timber to a separate cold preservative bath until cool.

Pressure impregnation

Pressure impregnation is widely used commercially. The wood is placed in a cylinder which is then sealed. A vacuum is applied and the cylinder flooded with preservative. The pressure in the cylinder is raised and held until the timber refuses to absorb further preservative. A treatment cycle can vary from one-and-a-half to five hours, depending on the wood species.

Tanalising involves impregnation with Tanalith E, an environmentally friendly wood preservative that is applied in a vacuum pressure timber impregnation plant. When impregnated into the timber the chemicals become chemically fixed into the timber and cannot be removed.

Lipping

Solid wood lippings of a variety of shapes and sizes can be adhesive bonded to the edges of manufactured boards for strength, for example to a door to apply fittings such as hinges, or to furniture components requiring a decorative wood edge.

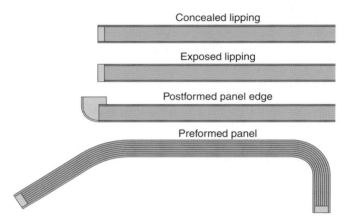

Fig 11.15 Lipping

Powder coating

The rapid growth of wood powder coating is a result of technological advances and manufacturing expertise. Finishes can have the quality of a grand piano with a smooth high gloss, or simulate solid surface granite in a smooth or textured appearance.

Powder coating on manufactured boards is more durable than traditional paint. It is impact, chip, temperature and stain resistant. Plus it's well suited to hot, wet conditions like changing rooms and saunas because it won't fade in sunlight or warp in humidity. Because it's a uniform, seamless coating, wood powder coating won't peel or delaminate.

Fig 11.16 Powder coating wood requires substantial investment in state-of-the-art technology and expertise

Powder coating involves the use of an electrostatic charge, both positive and negative, which offers optimum adhesion in the bond and significant strength.

The process of powder coating metal has been around for many years. Because powder coating on wood is applied using a 5 to 8 mil (127–203.2 micrometos) thickness, it creates more hardwearing finish for longer product life. It is now commonly used in office furniture.

Fig 11.17 Unique colours and textures give great design appeal, such as hammertones, wrinkled finishes, granites and so on

Finishes for metal

Like wood, metals can be finished with paints, varnishes and lacquers. Some metals require little surface finish, for example aluminium and

copper form an oxide coating when used outdoors; in both cases this coating helps to protect the material from further corrosion.

Anodising

Aluminium is often anodised to form a thicker oxidised layer encouraged by electrolysis; this coating can be coloured for decoration.

The product is placed in a solution of sulphuric acid, sodium sulphate and water. The product is used as the anode and lead plates as the cathode. When current is applied, a thin coating of oxide forms on the anode. Anodising can be used to ensure components have a consistent finish.

Fig 11.18 Anodised aluminium window frames

Metals can also be dip coated, electroplated or galvanised. All of these processes (as well as painting, lacquering and so on) involve applying a layer of a second material.

Electroplating involves the coating of an object with a thin layer of metal by use of electricity. The metals commonly used are gold, silver, chromium, tin, nickel and zinc. The object to be plated is usually a different metal, but can be the same metal or a non-metal, such as a plastic part for a car.

Fig 11.19 Jewellery may be gold or silver plated

Electroplating

Electroplating usually takes place in a tank of solution containing the metal to be deposited on the work. This metal is in a dissolved form called ions. An ion is an atom that has lost or gained one or more electrons and is thus electrically charged. You cannot see ions, but the solution may show a certain colour; a nickel solution, for example, is typically emerald green.

The object to be plated is negatively charged and attracts the positive metal ions, which then coat the object to be plated and regain their lost electrons to become metal once again.

A familiar example of this process is the experiment often performed in chemistry lessons in which a key is plated with copper.

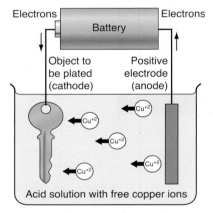

Fig 11.20 Positively charged copper ions are free in the solution, but are being attracted by the negatively charged key. As the ions contact the key, they regain their lost electrons and become copper metal and stick to the key wherever they touch it. This is the basic process of electroplating, and all forms of it work the same way

Galvanizing

Galvanizing is a process in which steel is immersed in a molten bath of zinc to provide the steel with a metallurgical bond between the zinc and the steel, creating a barrier of protection from the elements.

Fig 11.21 Hot dip galvanizing

The process consists of three steps: surface preparation, galvanizing and final inspection. Molten zinc will not react or bond properly to the steel if it is not perfectly clean. The cleaning of materials for galvanizing consists of three solutions and rinses. Caustic is a hot alkali solution that removes surface dirt, paints, grease and oil. It is followed by a hot sulphuric acid bath, more commonly referred to as the pickle. Pickling removes mill scale and rust that has formed ant the surface. The final step in the cleaning process is the flux. The flux tank is contains aqueous zinc ammonium chloride. The flux prevents further oxides from forming on the steel surface prior to galvanizing. Between each step in the cleaning process material gets a quick rinse in water to prevent cross-contamination of the different solutions.

Dip coating/plastic coating

Metal, glass or other plastics can be plastic coated. The coatings are generally applied using either plastisol (a liquid vinyl) or powder (fluidised bed).

Plastisol dip coating is a thermal process. Metal parts are preheated, dipped and then heated. During the dip, heat in the parts gel the surrounding plastic material. The hotter the parts and the longer the dip, the thicker the gelled coating. During the 'cure', the plastisol fuses. Using controlled ovens, dip speeds and times, a range of wall thicknesses are achievable.

A fluidised bed is a common way of coating a product in plastic. Plastic powder is fluidised by gently blowing air into the bottom of the tank. Before the powder is fluidised it is about as penetrable as fine beach sand. Once the air is switched on, the bed rises about 30 per cent and

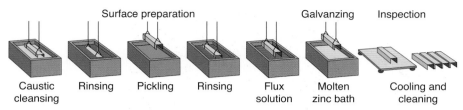

Fig 11.22 A typical galvanising line

the plastic powder looks like it is boiling at the surface.

Metal is preheated to a temperature to suit the thickness of the material and dipped into the fluidised bath. The plastic melts onto the hot metal and coats the surface. After the part is withdrawn, the residual heat in the metal fuses the coating and creates a smooth surface. For this to happen, the metal must be heated to a sufficiently high temperature to be able to retain enough heat after fluidising. Otherwise the part is returned to the oven for a very short time.

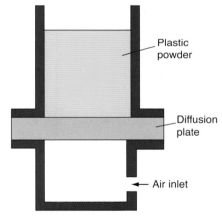

Fig 11.23

Fluid bed coatings are relatively thin. Thicker coatings need multiple dips or plastisol coating. However fluidised bed coatings have an advantage of being uniform in thickness. Many plastics are on the market in powder form, for example vinyl, nylon and polyethylene, and therefore can be applied as a fluidised coating.

1 Discuss the factors that a manufacturer must consider before deciding on the most appropriate surface finish for metal components.
2 Discuss the factors to be considered by a manufacturer when deciding on the most appropriate surface treatment for wooden products.
3 Name two appropriate surface finishes that could be applied to the external panels of domestic kitchen appliances. Give four reasons why a surface finish is necessary on metal panels used in domestic kitchen appliances.
4 Give two methods of protecting mild steel other than electroplating.
5 Describe, using notes and sketches, the process of electroplating motorcar parts.

Section E Manufacturing processes

Learning outcomes

By the end of this section you should have developed a knowledge and understanding of:

- Common processes for working with resistant materials – drilling, sawing, shaping, abrading (**see Chapter 10 Manufacturing**)
- Processes used to manufacture products from metal – milling, turning, casting, modifying characteristics using heat, pressing and stamping (**see Chapter 7 Engineering**)
- Processes used to manufacture products from plastic – compression moulding, injection moulding, vacuum forming, rotational moulding, extrusion and blow moulding; (**see Chapter 10 Manufacturing**)
- Processes used to manufacture products from wood – laminating, bending, routing/profiling; (**see Chapter 10 Manufacturing**)
- The design of simple jigs, presses and moulds (**see Chapter 7 Engineering**)
- Joining methods using fittings, adhesives, heat and common joints (**see Chapter 10 Manufacturing**)

Learning outcomes

By the end of this section you should have developed a knowledge and understanding of:

- Input devices
- Processing devices and techniques
- Output devices
- Power sources
- Structures
- Mechanisms
- Pneumatic systems

Introduction

Control systems and industrial manufacturing systems have historically used a combination of electrical, mechanical and pneumatic devices. Electronic systems are now more prevalent due to lower initial costs and the inbuilt flexibility that they can offer. Whether it is in automotive technology or domestic products, rapid changes have occurred in the last 20 years and are still happening today. This section reviews some of the devices that you will use in solving design problems and examples of their application.

Electronic input devices

Input devices are the interface between the outside world and your circuit. They can be divided into those that require user input and those that operate automatically. A number of the devices make use of a potential divider circuit to provide a voltage that changes proportionally to the condition being measured. There are some input devices that are integrated circuits and these provide a signal that can be used directly in the processing part of a circuit. The most common input devices are described below.

Potential divider

Two resistors in series connected across a power supply will have a voltage at the centre proportional to the value of the resistors. The voltage can be calculated using the formula:

$$V_{out} = \frac{R_2}{R_1 + R_2} \times V_{supply}$$

The resulting voltage can be used to operate a transistor circuit or as an input to a comparator. Figure 12.1 shows the arrangement. Remember, if $R_1 = R_2$ the supply voltage is halved.

Example: $R_1 = 2.7$ K

$$V_{out} = \frac{10}{2.7 + 10} \times 9 = \frac{10}{12.7} \times 9 = \textbf{7.09 V}$$

$R_2 = 10$ K
$V_{supply} = +9$ V

The voltage between V_{out} and positive is 9 minus 7.09 = **1.91 V**

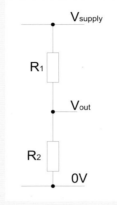

Fig 12.1 Circuit diagram

Light-dependent resistor

The light-dependent resistor (LDR) is available in two common types: the larger style being Ø13 and the miniature LDR Ø5. Both types have a dark resistance of 1 M which falls as the light level increases. LDRs are normally used in a potential divider configuration. Figure 12.2 shows both types.

Fig 12.2 LDR

PIN photodiode

This device is useful for detecting light in the infrared (IR) spectrum. When connected in reverse bias the PIN diode will start to conduct as IR light falls on the active surface. To connect in reverse bias the cathode of the diode is connected to the supply anode. The device is available as a discrete component and is also found in slotted opto switches and reflective opto switches.

Fig 12.3 PIN code

Phototransistor

This is an NPN transistor with a small light-collecting window connected to its base. The device can be connected in either common emitter or common collector mode. By adjusting the load resistor the phototransistor can act either as an amplifier or as a switch. A 10 K resistor will ensure that the device operates in switch mode. This means that the output will be either high or low, never unconnected.

Some phototransistors will include a base connection as shown in Figure 12.4, other types have only two legs, an emitter and a collector.

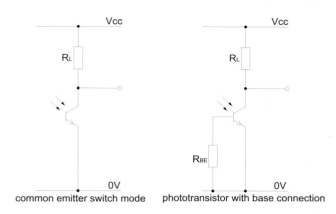

Fig 12.4 Circuit diagram

common emitter switch mode   phototransistor with base connection

Fig 12.5 Phototransistor

Fig 12.6 Slotted opto

Fig 12.7 Reflective opto

Slotted and reflective opto switch

This slotted switch can be used for event counting. If a wheel with a cut-out is rotated inside the slot, the rotation of the wheel can be sensed. Lower-cost versions will need additional components to give a digital signal but the device is available with a Schmitt output giving a digital signal that can be used directly in a logic circuit.

The reflective switch works in exactly the same way but relies on a good reflective surface to return the light emitting diode (LED) output to the sensor. The version shown in Figure 12.8 uses a photodiode as the sensor; Figure 12.9 shows a phototransistor as the sensor.

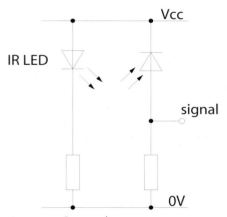

Fig 12.8 Circuit diagram

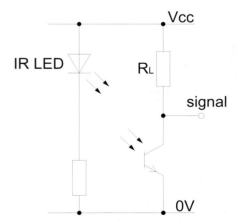

Fig 12.9 Circuit diagram

IR-emitting diode and sensor

Fig 12.10 IR pair

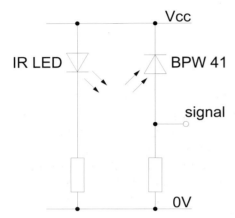

Fig 12.11 Circuit diagram

When connected as shown in Figure 12.11, the pair of components allow a signal to be transmitted between the IR LED and the input photodiode. This is how IR remote controls operate. The output LED is switched on and off in a set pattern which is then received by the photodiode. The arrangement can also be used to detect a physical blockage between outgoing and incoming signals. The photodiode is connected in reverse bias but will start to conduct when it is receiving from the IR LED. The photodiode casing is black to prevent interference from visible light.

Thermistor

Fig 12.12 Thermistor

Resistance of the thermistor changes as a result of temperature rise or fall. With the negative temperature coefficient (NTC) thermistor, a rise in temperature will result in a fall in resistance. A thermistor is useful when comparative measurement is required but more difficult to

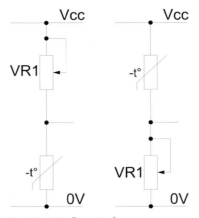

Fig 12.13 Circuit diagram

calibrate when a precise temperature is needed. They are available in a range of resistances at 25°C and are normally used in a potential divider circuit.

The operating temperature range for a thermistor can be very wide; typically -50°C to 300°C.

Figure 12.13 shows two potential divider arrangements using a thermistor; each uses a potentiometer to set the output voltage range. The diagram on the left gives a falling output voltage as the temperature increases; by placing the thermistor at the top as in the diagram on the right, the voltage rises as temperature rises.

Temperature sensing integrated circuits (ICs)

Measuring temperature in a recognised scale can be important in some applications. Devices such as the LM35 have an output voltage proportional to temperature in °C.

Other devices, such as the LM335Z and LM334Z, have an output current proportional to the absolute temperature in °K.

Fig 12.14 DS18B20, LM335 and LM334

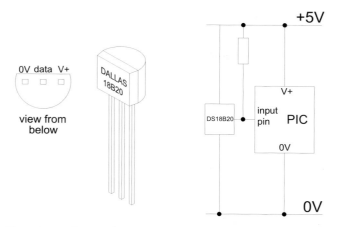

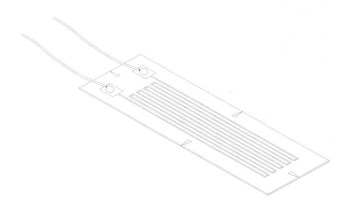

Fig 12.15 Circuit diagram

Fig 12.17 Strain gauge

One of the more useful devices to become available in recent years is the DS18B20, which is easily integrated into a PIC circuit; the output signal from the device is stored as a temperature in °C. The circuit is shown in Figure 12.15.

Microphone

You may come across three types of microphone: electret, magnetic and crystal. Preamplifier circuits for them are shown in Figure 12.16. It is also possible, though not advisable, to use a small loudspeaker as a microphone.

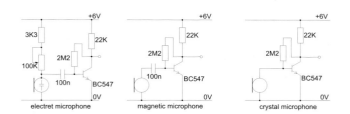

Fig 12.16 Circuit diagram

The strain gauge

This is a sensor used to measure pressure, load, torque or position.

The gauge consists of a pattern of foil laid out onto a backing sheet; alignment arrows indicate the position of attachment. The gauge can be attached to the test sample with epoxy resin

adhesive. A tensile force will cause the foil to elongate and reduce in section, causing a slight increase in resistance. Compression will give a corresponding decrease in resistance; the gauge is largely unaffected by any lateral movement. Important applications such as the aircraft industry will allow for small changes due to lateral movement.

Two configurations are shown in Figure 12.18. In the basic quarter bridge a single strain gauge

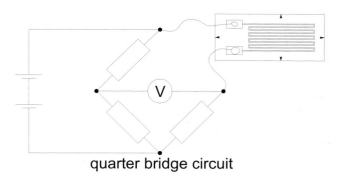

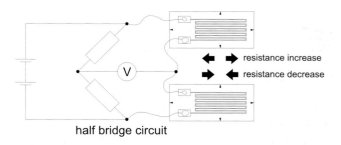

Fig 12.18 Two configurations

has been placed in a Wheatstone bridge circuit to record changes in tensile or compressive load on a material sample. The half bridge circuit is more reliable as one of the gauges can either be used as a dummy to compensate for temperature change or it can be bonded to the reverse side of the sample to record an opposing force. With an 'exciter' voltage applied to the bridge circuit, a change in the gauge resistance will cause a change in the voltmeter reading.

Pressure-sensing transducer

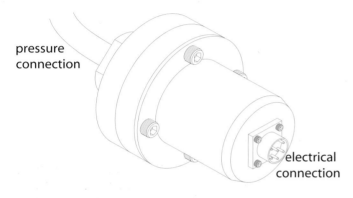

pressure connection

electrical connection

Fig 12.19 PS transducer

A variety of types are available; the transmitter shown in Figure 12.19 uses a bonded foil strain gauge as the sensor. Pressures from less than 1 bar to more than 1000 bar can be recorded accurately. The electronics in these devices will often be 'potted' in epoxy resin to ensure stability and reliability in harsh environments.

Reed switch

Fig 12.20 Reed switch

A favourite in alarm applications where it is normally encased to protect the glass body of the switch; the contacts are operated by a magnet being placed close to the switch. Reed switches are reliable because the contacts are sealed, preventing any oxidation. They are also available as reed relays for use in the output section of a circuit. The standard switch is SPST (single pole, single throw) with NO contacts but some switches have an extra NC contact which allows the switch to become SPDT (single pole, double throw) or a changeover switch.

Micro switch

Fig 12.21 Micro switch

This is a snap-action switch, patented in 1934, which still has many uses today. The switch can be operated by lever, roller or direct pressure on the switch. The double throw version makes it very suitable for use in projects as a changeover switch that can be easily mounted and gives years of trouble-free use. Micro switches are extensively used to cut power to machines when access to moving parts is required. Connection to the switch can be by spade terminal or direct soldering.

Pressure switch

The pressure switch uses a diaphragm to operate micro switch contacts when air pressure on the diaphragm increases. They are useful in positions where direct electrical connection could be hazardous, for example in a foot-operated stop switch to a machine tool.

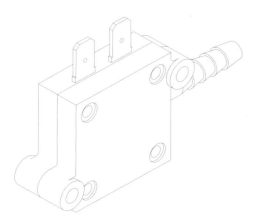

Fig 12.22 Pressure switch

Thermal switches

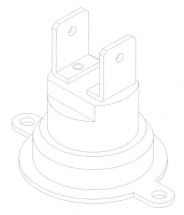

Fig 12.23 Thermal switch

As the name suggests, these switches operate as a result of a change in temperature. They can be normally closed, opening on a rise in temperature, or normally open, closing on a temperature rise. The switches are available to operate at a range of temperatures.

Tilt switch

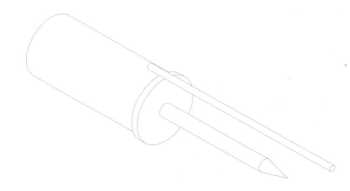

Fig 12.24 Tilt switch

Early versions of the tilt switch made use of mercury to close the contacts when the body of the switch tilted; many modern tilt switches use gold-plated ball bearings as the contactor. The angle of operation will vary but typically it is ten degrees from horizontal.

Float switches

Fig 12.25 Float switch

Designed to monitor liquid levels, these switches use a moving float with a magnet attached; the magnet operates a reed switch mounted on the body of the device.

Float switches are available for either horizontal or vertical fitting. For more advanced applications capacitive electrode devices, which use a diaphragm as the moving part of the sensor, are available.

Processing devices and techniques

Operational amplifier

The operational amplifier, or op-amp, is basically a very high gain amplifier with very high input impedance, used in the processing of analogue electronic signals. The most common device you will come across is the 741 op-amp, which has been around since the late 1960s. During that time it has been improved upon and a number of better alternatives based on FET (field-effect transistor) or MOSFET (metal-oxide semiconductor field-effect transistor) circuitry are now available. A useful general purpose version is the CA3140, which can be substituted

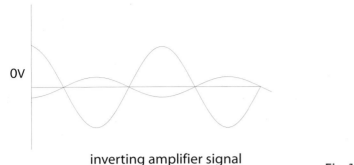

inverting amplifier signal

Fig 12.26 Inverting amp

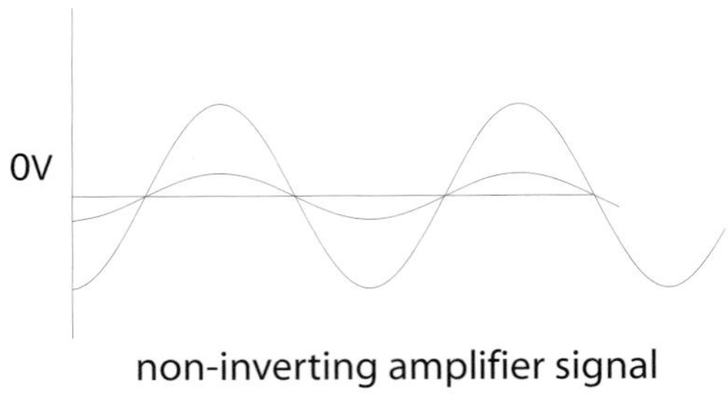

non-inverting amplifier signal

Fig 12.27 Non inverting amp

in a 741 circuit. The main uses for op-amps in project work are based around three circuits: the inverting amplifier, the non inverting amplifier and the comparator.

The circuit arrangements are shown in Figures 12.29. Amplifier circuits normally use a 3-rail power supply, for example +5 V, 0 V, and -5 V. Comparator circuits can be successfully built using a 2-rail supply. These are useful when using sensors such as the thermistor to perform a switching action at a particular temperature.

In the inverting amplifier negative feedback is provided by the feedback resistor Rf. This will feedback a proportion of the output voltage to the inverting input. The output is in opposite phase to the input.

The non-inverting amplifier has an output that is in phase with the input. Negative feedback is supplied by Rf.

In the comparator circuit shown, the op-amp compares the difference in voltage at the two inputs. If the voltage at the non-inverting input is greater than the voltage at the inverting input, the output will be close to the supply voltage. If the inverting voltage is greater, the output is close to 0 V.

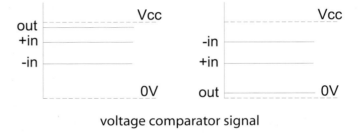

voltage comparator signal

Fig 12.28 Voltage comparator

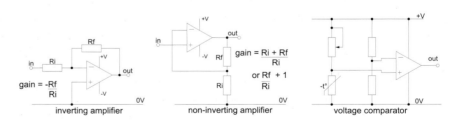

Fig 12.29 Circuit diagram

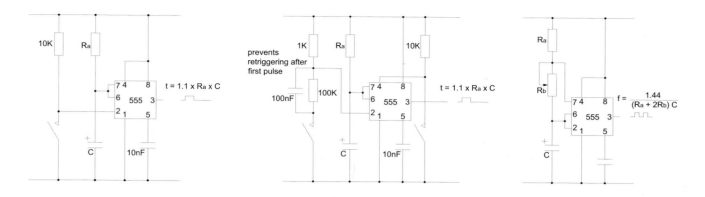

Fig 12.30 Monostable; Monostable with reset and modified trigger; Astable

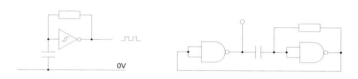

Fig 12.31 Schmitt inverter astable; NAND gate astable

Monostable and astable signals

One of the earliest ICs still in use is the 555 timer and its derivatives. This device can be set up to produce either monostable or astable signals quite easily; however, it is not the only way of producing these signals.

Dedicated ICs such as the 4047B can be used and, in the monostable configuration, offer either positive- or negative-edge triggering. Basic 555 circuits are shown in Figure 12.30.

Astable signals can be set up quickly using NOT gates and a minimum number of components, as shown in Figure 12.31. The arrangements shown are useful because they can utilise spare gates in a circuit.

For project work PIC devices offer a number of benefits when compared to dedicated ICs. In many cases they offer a reduced component count in the circuit as well as accuracy of the pulse and quick changes of pulse length or frequency.

inputs		AND	NAND	OR	NOR	XOR	NOT
0	0	0	1	0	1	0	1
0	1	0	1	1	0	1	n/a
1	0	0	1	1	0	1	n/a
1	1	1	0	1	0	0	0

(heading above table: outputs)

Fig 12.32 Truth table

Logic gates

The following truth table should help you to remember the results of applying logic in the form of two input gates. The two most important types are probably the NAND and NOR gates, as all others can be constructed from these. By doing this it is possible to reduce the number of unused gates in a logic circuit.

There are a few general facts about logic that should be remembered when choosing a logic type:

- NAND gates and NOR gates can be used to form any of the other gates
- By joining the inputs of NAND gates or NOR gates the result is a NOT gate or inverter
- To detect two logic 1 signals at the inputs, use AND logic or NAND followed by NOT
- To detect two logic 0 signals at the inputs use a NOR gate

- To detect two different logic levels, invert the logic 0 signal and use an AND gate, or invert the logic 1 signal and use a NOR gate
- When using complementary metal-oxide semiconductor (CMOS) ICs, any unused inputs should be tied to a logic level; the easiest way is to connect them to the most convenient of the power pins(7 or 14) or to any of the pins that are already tied to a logic level
- To reduce the chance of interference from any other part of the circuit, a 10 nF capacitor can be connected across the power leads of each IC

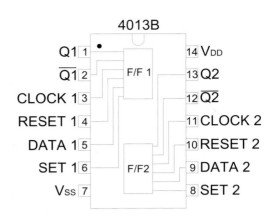

Fig **12.34** Circuit diagram

Debouncing circuits

When using logic gates it is important that any input devices produce a clean signal. Contact bounce resulting from the use of mechanical switches will become a real problem in counting circuits. Two ways of removing contact bounce are shown in Figure 12.33.

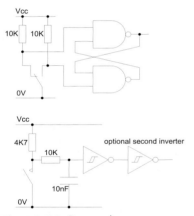

Fig **12.33** Circuit diagram

The NAND gate arrangement is a variation on the bistable circuit, which is often used as a latching device. The Schmitt trigger debouncer is normally used with a single inverter but, if the output signal is needed at the same logic level as produced by the switch, a second inverter can be added. Debouncing in a PIC circuit can be carried out if required by using software to add a short delay after an initial input pulse.

D-type flip-flop or D-type bistable

This device will allow a clock pulse to determine the point at which input data is transferred to the output. Data at the D input will not have an effect on the Q output until the next positive going edge of the clock signal. In other words, when the clock pulse is at logic 1 output Q is at the same logic level as input D. When the clock pulse is at logic 0 output Q will remain at its previous state. For this reason the device is said to have a single-bit memory.

The 4013B shown in Figure 12.34 is a dual D-type flip-flop, each one having a clock and data input and also S and R inputs, which allow it to be used as a NOR gate latch. Each output Q has a complementary or opposite output Q. The two flip-flops on the 4013 can be linked together to provide a 2-bit binary counter; this arrangement is shown in Figure 12.35.

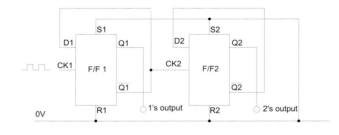

Fig **12.35** 2-bit binary counter using a 4013B

Counting circuits

As mentioned previously a counter circuit requires a clean, debounced pulse if it is to be accurate. Apart from the debouncing circuits shown in Figure 12.33 it is often necessary to shape the clock pulse of a counting device using a Schmitt trigger; these are available in quad NAND or hex inverter packages.

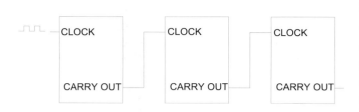

Fig 12.36 Circuit diagram

The choice of counting device will depend largely on the type of output required. The easiest ICs to use are the 4026B, which has decoded outputs for a seven-segment display, or the 4510B which is a 4-bit binary up or down counter. Both of these counters have a 10 s output, known as carry out or Cout, which can be used as the clock input for a second or subsequent counter as shown in Figure 12.36.

If counters are linked together in this way it is usual to link their reset pins so that all counters will reset together. If reset is active high, the reset pins should be connected to 0 V through a resistor; a push switch can then be used to provide the high pulse for a reset. PIC circuits can be used to drive a single display using seven outputs or they can drive a 4511B decoder using only four outputs. Additional outputs from the PIC can then be connected to lamp test, blanking and latch enable – pins 3, 4 and 5 respectively – if required.

When using an IC for the first time it is useful to

download a datasheet to provide information. A search using '4510B datasheet' as the search string will give access to a number of datasheet specialist sites.

The datasheet will have a pin diagram for the IC together with full details on its use; it is important to remember that any inputs on the diagram with either a negation circle or a bar above the label will be active when the input is low or at logic 0. Pin diagrams for the most useful devices are shown in Figure 12.37.

Additional output devices and techniques

N channel MOSFET

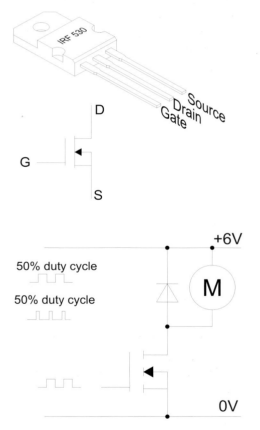

Fig 12.38

This device, which is relatively straightforward to use, allows high-current devices to be operated without the need for a relay. The symbol and pin diagram for a typical MOSFET, IRF 530, are shown in Figure 12.38. This MOSFET can be used to operate a motor using pulse-width modulation (PWM).

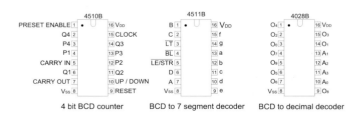

Fig 12.37 Circuit diagram

Using this technique a square wave is output to a motor rather than a continuous signal. The motor is only powered for the 'on' period of the square wave. By adjusting the proportion of 'on' time the speed of the motor can be controlled. This technique would not be possible using a relay to operate the motor; a protective diode in reverse bias is still required. The circuit is shown in Figure 12.38. Most PIC systems will allow a simple basic command to directly output a PWM signal suitable for motor control; any other method of producing an astable output with variable mark/space ratio is also suitable.

Thyristor

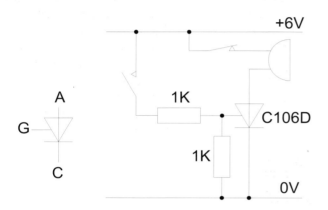

Fig 12.39 Circuit diagram

The thyristor, sometimes known as a silicon-controlled rectifier, is similar to a transistor in action. A small current to the gate allows a larger current to flow from anode to cathode. The difference in action to a transistor is apparent when the current to the gate is removed; the larger current from anode to cathode continues to flow as it has been 'latched' on. The only way to stop the flow is to break the circuit. The thyristor symbol is shown next to a typical circuit in Figure 12.39.

Output devices

The output of an electronic system is normally the part that the user is aware of. Whether it is an LED display, a motor or sound output, the principle of operating the device remains the same. If the device includes a coil it is possible for reverse emf to damage the drive circuitry when the supply voltage is cut. For this reason a protective diode in

reverse bias must be used; the 1N4001 is suitable for the majority of applications.

DC motor

The most common type is the brushed DC motor, which has carbon brushes touching the commutator as it rotates. Friction built up at this point is one of the drawbacks of this type of motor. Another failing is wear on the brushes, which eventually leads to their requiring replacement. The brushless motor is commonly used in applications where precise speed control is an issue, for example in computer disk drives and CD/DVD players. Because brushless motors produce no sparking, unlike the brushed variety they are particularly suited to harsh environments containing fuels and gases.

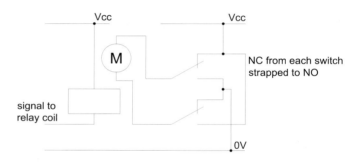

Fig 12.40 Circuit diagram

The brushed variety of motor used in the majority of school projects requiring rotary movement has two properties that can be utilised by a designer. By reversing polarity the motor direction is reversed and, if the motor shaft is driven, it will act like a dynamo or generator. Reversing of a DC motor can be carried out using a DPDT (double pole, double throw) relay as shown in the circuit in Figure 12.40.

Relay

The relay is an important output device. They are described by coil voltage, contact voltage/current and the number of contacts. When choosing a relay the first point to look at is the coil voltage; this can range from 5 V to 24 V typically. Contact rating will be given in amps for a particular voltage, for example 5 A at 24 V DC or 110 V AC. Contacts can either be a single switch, normally with double throw, or it may be up to four switches on a single relay.

When designing a PCB to hold a relay it is good practice to measure and check the spacing on the pins before placing pads; the actual relay can then be checked against a print of the circuit before manufacture. Because relays will generate reverse emf there must be a protective diode when the relay is operated by a transistor. If a Darlington array such as the ULN2803 is used, the diode is built in. Reed relays are available in either single in-line or dual in-line packages, making them very easy to design into a PCB layout.

In addition to the motor reversing circuit shown in Figure 12.40, it is possible to configure a DPDT relay to act as a latch. By connecting 0 V to one of the switch common terminals, and the NO terminal of that switch to the coil 0 V connection, the relay remains 'on' when the initial signal to the coil is no longer there. This is shown in Figure 12.41.

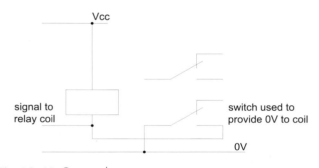

Fig 12.41 Circuit diagram

Solenoid

In many ways a solenoid is similar in terms of operation to a relay. Current draw from the solenoid will be high, particularly in the non-latching type of solenoid, which requires the coil to be powered while the solenoid pin remains operated. The latching solenoid only needs a 40 ms pulse to set or reset the pin. Many solenoids have slots and holes in the pin for attaching operating levers.

Optical devices

Many of these are based on the LED; for counting purposes the seven-segment display is available in a range of colours, sizes and light intensities. Operation of the device is similar for all though. A typical seven-segment display is

shown in Figure 12.42; it comprises seven LEDs for the numeral segments and one for the decimal point. The display will be either common anode or common cathode and it is important to order the correct type. If the display is to be driven directly from a counter IC the common cathode should be used, with a protective resistor for each segment. If, however, the counter IC cannot provide enough current to light each segment the easiest way around the problem is to pass all seven signals through a Darlington array and to use a common anode display. With larger displays, generally more than 2.5 cm, each segment will be lit by up to three LEDs in series. It is necessary to use the Darlington array option if that is the case. The pin diagram from the manufacturer will give details of the layout; it is essential to have this information before starting to lay out a PCB as pin layouts can be very different. In some cases the common connection, either anode or cathode, will have two available pins that are connected internally; only connect the most convenient on your PCB layout.

Fig 12.42 7-segment display

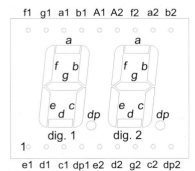

Fig 12.43 Multi-digit display

Multi-digit displays are available either as totally separate displays in the same casing as shown in Figure 12.43, or as a display where similar segments on each digit are connected. This cuts down on the tracks needed but this type of display has to be multiplexed.

Multiplexed seven-segment LED display

In order to cut down on wiring if a multiple display is being used, it is often multiplexed. This means using the same cable or tracks to carry signals for more than one digit. The technique relies on the fact that our eyes cannot detect rapid switching on and off; a display that is multiplexed is laid out as shown in Figure 12.44.

Fig 12.44 Multiplexed 7-segment LED display

The display has common pins for the segments and individual pins for each anode. This arrangement makes it possible to have any one of the three digits visible at a given time.

Fig 12.45 Liquid crystal display or LCD

Liquid crystal display

Liquid crystal displays (LCD) have a number of advantages over LED displays. They operate on a fraction of the current and have a far greater range of characters available for the display. It is possible with 'intelligent' displays to have a number of programmed messages that can be output as part of the final display. The message shown has the words 'temperature is' and 'deg. C' programmed in permanently. The number that is displayed is updated as part of a PIC program. The back of the display module in Figure 12.45 shows the battery, which is used to back up stored message data and date/real time data. This type of module is easily interfaced with a project and can give much greater flexibility when designing display outputs. The main drawback to many LCD displays is the lack of visibility in poor light. This can be overcome by providing a backlight that can be illuminated for a short time only, conserving battery power.

Piezo-electric sounder

The sound source in a piezo-electric sounder is a piezo-electric diaphragm. This is made by attaching a piezo-electric ceramic plate to a metal plate using adhesive to hold the two together. Applying a DC voltage across electrodes of the diaphragm causes distortion due to the piezo-electric effect. When an AC signal is applied to the plate, the distortion changes direction rapidly, producing sound. The sounders are available either with a built-in drive circuit or requiring an external drive circuit. The

Fig 12.46 Piezo sounder

sound produced is normally in the 2–4 kHz range. The uncased piezo transducer can be used as a trigger for either a thyristor or a 555 timer circuit.

Loudspeaker

Fig 12.47 Speakers

A loudspeaker is an electro-mechanical transducer capable of converting an electrical signal to sound. Different speaker designs are used to reproduce different sound frequencies. For low notes a woofer or sub-woofer is capable of reproducing frequencies in the region below 120 Hz. At the opposite end of the range a 'tweeter' may produce sounds of up to 20 kHz. Speakers are available that can cover the full range of sound but often at the price of added distortion. Loudspeakers are rated in terms of power that can be handled safely without distortion, impedance (normally 4, 8 or 16 ohms) and frequency response. The speakers shown in Figure 12.47 are low-cost, full-range speakers of the type used in school projects.

Stepper motor

Fig 12.48 Stepper

The stepper motor is different to a normal DC motor in that it does not have to rotate continuously. There are two type of stepper in common use: the unipolar, with four coils, and the bipolar, which has two coils. Both types are driven by a series of pulses that must be in the correct order to step the motor. If the pulses are counted, accurate positional control can be maintained. A typical stepper motor will move 7.5 degrees for each step, or 48 steps for a complete revolution of the shaft.

A drive circuit is required to power the motor coils in the correct order and there are a number of ICs that can perform this function. It is also possible to use outputs from a PIC chip to provide the pulses. Figure 12.49 shows the general layout of coils and three possible

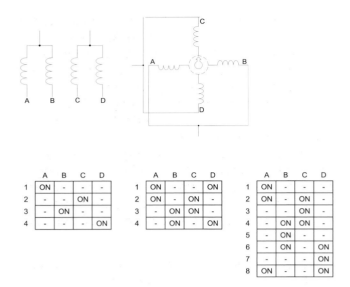

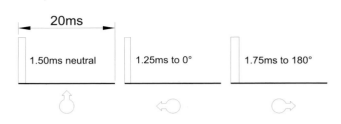

20ms

1.50ms neutral   1.25ms to 0°   1.75ms to 180°

Fig 12.51 Circuit diagram

	A	B	C	D
1	ON	-	-	-
2	-	-	ON	-
3	-	ON	-	-
4	-	-	-	ON

	A	B	C	D
1	ON	-	-	ON
2	ON	-	ON	-
3	-	ON	ON	-
4	-	ON	-	ON

	A	B	C	D
1	ON	-	-	-
2	ON	-	ON	-
3	-	-	ON	-
4	-	ON	ON	-
5	-	ON	-	-
6	-	ON	-	ON
7	-	-	-	ON
8	ON	-	-	ON

Fig 12.49 Circuit diagrams

sequences used to drive the motor. In the first sequence the coils are energised singly; to reverse the direction the sequence is reversed. The second sequence energises two adjacent coils together; this sequence will increase current draw but will also increase the torque of the motor. The third or interleaved sequence provides half steps, doubling the number of pulses needed for a single rotation.

Servo motors

Fig 12.50 Servo motor

A servo is a small, high-torque DC motor with an output shaft that can be positioned to a specific angle. The angle is determined by the length of pulse applied to the control wire; this is called pulse-coded modulation. Servos are often used in radio control applications.

Although they will only move through a preset angle when supplied new, it is possible to remove the stop from the shaft and allow the motor to travel through a complete rotation, driving it with a stream of pulses.

A typical servo is driven by a pulse every 20 ms; the length of pulse will determine how far the motor turns and in which direction. A pulse of 1.5 ms duration will put the motor shaft in the neutral position (90 degrees), a pulse shorter than this will turn the shaft anticlockwise to zero degrees, and a pulse of 1.75 ms will turn the shaft clockwise to the 180 degrees position. This is shown in Figure 12.51. To maintain torque and position the pulses must be repeated every 20 ms. A simpler way to operate the servo is by using a PIC chip, many of which can be programmed directly through a basic command to provide pulses of the correct frequency.

Power sources

Mains power in the UK is 230 V AC, plus or minus ten per cent. This type of supply has to be adapted for use in electronic systems. The first stage is to transform it to a lower voltage that is safer to use; the result will still be AC. The problem with an AC supply is that it constantly changes direction – positive becomes negative and negative becomes positive – therefore it cannot be used in circuits that can be damaged by a reverse supply.

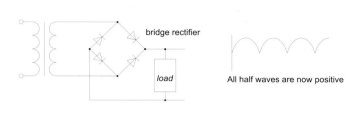

Fig 12.52 Circuit diagram

The solution is to rectify the supply to ensure that the positive terminal remains positive. Figure 12.52 shows the standard full-wave rectification circuit, which consists of four diodes that in turn block the unwanted waves and ensure that there is a permanent positive supply. The rectified wave is still not smooth enough for use in digital electronic circuits so smoothing is carried out with electrolytic capacitors.

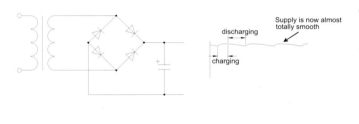

Fig 12.53 Circuit diagram

The resulting DC supply shown in Figure 12.53 is suitable for electronic circuits. Very often a small transformer is built into a fixed voltage power supply that will plug direct into a 13 A socket. If using this type of supply it is important to check that it has been smoothed before using it in a logic circuit.

Battery supplies

Batteries can be divided into two types: rechargeable and non-rechargeable. Of the rechargeable type, lead acid batteries are still

used where a large current draw is anticipated. A variety of non-spill, sealed for life lead acid batteries are available in either 6 V or 12 V rating. The older nickel cadmium (ni-cad) batteries have been largely superseded by lithium-based batteries, which have the advantage of no memory effect. The memory effect on ni-cad batteries means that if they are not fully discharged before being recharged they will tend to remember the level before recharging and, over a period of time, will hold less and less charge. For project work alkaline batteries are still frequently used, mainly because of low cost and availability.

At the smaller end of the scale, lithium coin cells are often used for memory back up in microprocessors.

When choosing batteries for a project the following points should be considered:

- Voltage
- Current draw
- Physical size
- Recharging ability
- Cost
- Disposal requirements

Voltage regulators

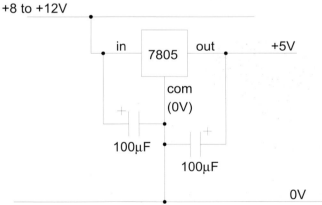

Fig 12.54 Circuit diagram

Many circuits require a precise voltage, for example those using transistor-transistor logic (TTL) or those using PIC chips. The nominal requirement for TTL is +5 V ± 0.25 V. Most PIC devices will operate with a +5 V supply, though they can use a slightly lower voltage. The simplest way of achieving this is to use a regulator of the required voltage. These devices take in a higher voltage and convert it to +5 V with very little fluctuation from that value. Figure 12.54 shows the general circuit for using a 7805 voltage regulator. The capacitors are not essential if the initial source is from a battery but will smooth out any changes or 'spikes' from a transformed supply. Regulators are available in a range of maximum current ratings, from 500 mA up to 5 A, with positive or negative output voltages. Many of the devices have overload, thermal and short circuit protection.

Zener diode

An alternative method of producing a fixed voltage is to use a zener diode. These are generally used in reverse bias and are designed to allow a current to flow when the rated value or breakdown voltage of the zener diode has been reached. When designing a regulator using a zener diode it is important to know the power rating of the diode; this allows calculation of the maximum current that can safely be passed thorough the diode. Figure 12.55 shows a circuit using a BZXC6V2 zener diode; this diode is power rated at 500 mW. The 39 R resistor will limit the current through the diode to a safe level.

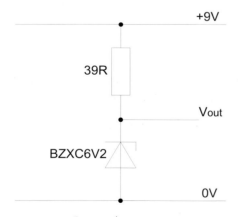

Fig 12.55 Circuit diagram

A suitable value for the resistor can be calculated in the following way:

$$\text{Safe zener current} = \frac{W}{V} = \frac{500}{6.2} = 80 \text{ mA} = 0.08 \text{ A}$$

The resistance needed to limit current through the diode is derived using Ohm's law:

Voltage $= 9 - 6.2 = 2.8$ V

Current is already calculated at 0.08 A

$$\text{Resistance is } \frac{2.8}{0.08} = 35 \text{ R}$$

The next available value from the E12 series is 39 R.

Structures

We rely on structures for our well being and everyday life. From the house we live in to the desk at school, we expect them to remain sound and not collapse. Wherever you look there are structures, man made and natural: our own skeleton is a good example of a natural one. If a structure is at rest, such as a table, it is said to be in equilibrium and hopefully will remain so even when loaded. To be in equilibrium the forces acting on the table must just balance and have no resultant. The downward force of the table in Figure 12.56 acting on the floor must be exactly balanced by the upward force from the floor.

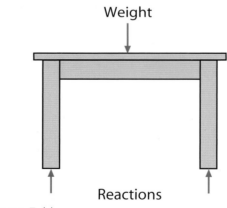

Fig 12.56 Table

Tension

If a member of a structure is in tension, it is called a tie. One way to visualise a member in tension is to see if it could be replaced by string or a similar flexible material. If the table in Figure 12.56 is analysed, the legs cannot be replaced by a flexible material so they are not in tension. The wires supporting a suspension bridge are in tension, and act to stop each end of the tie moving away from each other.

Compression

This is the opposite of tension and members of a structure in compression are known as struts. The table legs are in compression and they are acting to squash the structure. Figure 12.57 shows the bookshelf is in compression on its upper surface and tension on its lower surface. Between the two is the neutral axis.

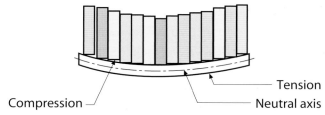

Compression — Neutral axis — Tension

Fig 12.57 Bookshelf

Bending

The bookshelf in Figure 12.57 is bending due to the weight of the books. To minimise the bending caused by the load, consideration must be given to the cross-sectional shape of the beam and where the load is to be applied. Figure 12.59a shows the cross section of some commonly used beams. Figure 12.59b shows there is a correct orientation of the beam for maximum structural rigidity, with the load being supported across the widest section of the member.

The forces acting on a beam must balance and, depending on where the loads are positioned, the calculations are made using moments. The formula for calculating the reactions at various points along the beam is given as:

clockwise turning moments (CTM) = anticlockwise turning moments (ACTM)

Torsion

If a twisting action is applied to a structure then it is in torsion; as shown in Figure 12.58; this is similar to the torque that a shaft is subjected to when a load is applied to it. When bridges are designed, consideration must be given to any torsion to which they may be subjected.

Fig 12.58 Torsion (twisting)

(a)

(b)

Load

Load

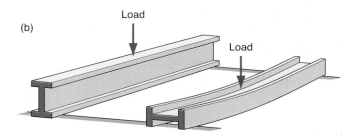

Fig 12.59 Cross-sections of beams and structural rigidity of beams

Figure 12.60 shows a see-saw balancing (see also Figure 12.81). The calculations are as follows:

moment = force (N) x distance (m)
CTM = ACTM

200 N x 1 m = 100 N x 2 m
200 Nm = 200 Nm

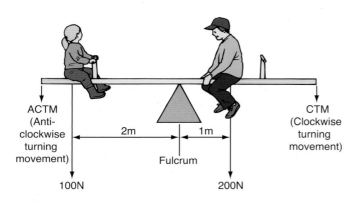

Fig 12.60 See-saw

Equilibrium of a loaded beam

A loaded beam must be in equilibrium, which means that the loading must be balanced by the reactions. The reactions are the upward forces that counteract the downward loading. Figure 12.61 shows a beam loaded in more than one place. The calculations show how to work out the reaction at each end of the beam.

To calculate the Reaction A, think of taking the support at A away; the beam is then free and clockwise turning moments can be taken from A.

Reaction A plus Reaction B must be equal to the total loading on the beam for it to be in equilibrium.

Total loading = 300 kN
Length of the beam = 8 m

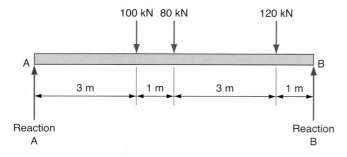

Fig 12.61 Loaded beam

Taking moments about A:

Reaction B x 8 = (100 x 3) + (80 x 4) + (120 x 7)
= 300 + 320 + 840
= 1460

Reaction B = 1460 / 8 = 182.5 kN

Reaction A = 300 – 182.5 = 117.5 kN

Stress

Stress (σ) is a measure of how much force is applied to a cross-sectional area using the formula:

stress = force / cross-sectional area

Figure 12.62 shows two pillars each carrying a force of 5 kN. The stress in each can be calculated using the formula.

The square pillar in Figure 12.62 has a cross-sectional area of:
10 x 10 = 100 mm^2

The stress will be 5000 / 100 = 50 N/mm

The round pillar in Figure 12.62 has a cross-sectional area of:
π x r^2 = 78.5 mm^2

The stress will be 5000 / 78.5 = 64 N/mm

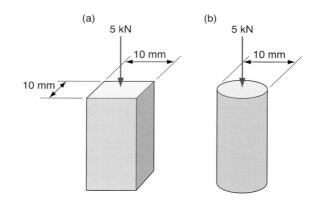

Fig 12.62 Stress (square column and round column)

Strain

Strain (ε) is a measure of the change in length of a component (compared to its original length) when either a compressive or tensile load is applied. Figure 12.63 shows a structural member being compressed by a load.

Strain = change in length / original length

= 5 / 1000 = 0.005 or 0.5%

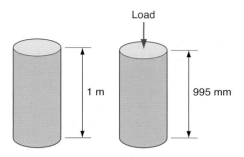

Fig 12.63 Strain (column with load applied)

Elasticity

Stress is proportional to strain up to a certain point, shown on the graph in Figure 12.64 as a straight line. This is the elastic stage and, providing point A is not passed, the material will return to its original length.

Along the elastic stage stress (σ) is proportional to strain (ε).

σ = constant x ε

The constant is known as Young's modulus (E). The steeper the gradient of the straight-line part of the graph, the stiffer the material.

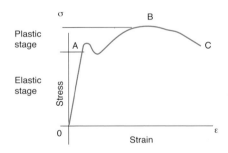

Fig 12.64 Elasticity graph

Safety factor

If a structure is designed to a maximum load that keeps it just below the ultimate stress point B, in theory it should be safe. In practice this does not take into account the unknown factors. If designing a child's swing there is no guarantee that an adult would not use it. Bridges could be subjected to unusual weather conditions and extra loading of future super-sized lorries.

In order to take into account these unknown factors the working stress value would be only part way along the straight-line part of the graph in Figure 12.64.

The ratio of ultimate stress is known as the factor of safety.

Working stress

The factor of safety can be any number, depending on such things as how catastrophic failure would be: obviously a bridge failing could kill people, but the collapse of a toy bridge with too many toy cars on it would not cause too much concern about the safety factor.

In general, designers of crucial structures use a factor of safety of 4. Aircraft structures are designed with a safety factor of 6+. This would mean that, for a structure that must support a load of 500 N, an ultimate figure of 2000 N must be used in the design calculations.

Shear

Garden shears have a shearing action that is put to good use in that they force blades of grass, or other garden growth, to be cut by making one part slide past another, creating two pieces.

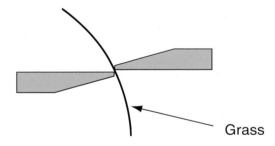

Fig 12.65 Cross-sectional view of shears

In other situations shear forces can cause problems and must be considered when designing structures.

As an example, if two struts are joined together and carry a load then the joining rivet is subject to a shear force. The rivet must be strong enough to withstand the force otherwise a failure will occur. Figure 12.66 shows before and after the loading of a joined strut where the rivet is not strong enough. The ultimate shear strength is the point where the maximum loading occurs before any failure starts.

A shear force failure occurs when there is a total collapse at the joint because the rivet or its equivalent has sheared.

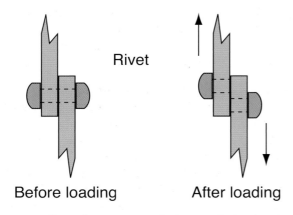

Rivet

Before loading   After loading

Fig 12.66 Shear force on rivet before and after loading

It is important to choose the correct shape and orientation of the members when designing a structure. There are various shapes of beams; some cross sections are shown in Figure 12.59a. The beam should be used with the orientation as shown for the greatest vertical strength.

Mechanical systems

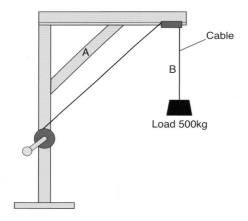

Cable

A

B

Load 500kg

Fig 12.67 Jib crane lifting a 500 kg load

1 Figure 12.67 shows a jib crane lifting a 500 kg load.
  a) i) State the force acting on beam A.
     ii) State the force acting on cable B.
  b) The testing of materials used in structures is done using destructive and non-destructive testing.
    i) State one industrial method of destructive testing of a material.
    ii) State one industrial method of non-destructive testing of material.
  c) When determining a safe working stress for a structure a factor of safety is used. State two areas that would be considered when determining a factor of safety for the structure shown in Figure 12.67.
  d) The cable in Figure 12.67 is 8 mm diameter. If the load of 500 kg causes a 5 m long cable to extend 2 mm:
    i) Calculate the stress in the cable.
    ii) Calculate the strain in the cable. Assume 1 kg = 10 N.

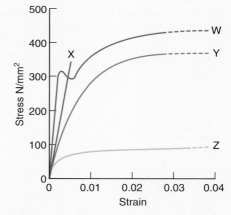

Fig 12.68 Graph of stress and strain

2 Figure 12.68 shows the stress versus strain curves for a number of common materials.
  i) State which of the four – W, X, Y or Z – is likely to be the most ductile.
  ii) Explain the difference between the elastic and plastic regions. Use sketches where appropriate.
3 Discuss the implications of selecting materials based on their mechanical properties.

Mechanisms

Mechanisms are mainly concerned with movement through a system that transforms one kind of motion or direction into another.

Gear systems

Gears are toothed wheels, fixed to the driver and driven shafts, that mesh together. When a number of gears are meshed together it is called a gear train.

Spur gears are fixed on parallel shafts and can vary in size. The example shown in Figure 12.69 has 16 teeth on the smaller driver gear and 40 teeth on the larger driven one. The difference in their speeds (velocity ratio) can be calculated from the number of teeth on each gear. The direction of rotation is shown by the arrows.

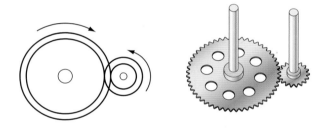

Fig 12.69 Spur gears

VR = No of teeth on driven gear / No of teeth on driver gear
= 40 / 16
= 2.5 (or 2.5:1)

If the driver shaft and driven shaft are required to rotate in the same direction, an idler gear can be fitted between them as shown in Figure 12.70.

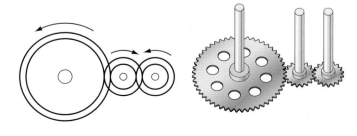

Fig 12.70 Spur gears with idler gear

Figure 12.71 shows a bevel gear. Bevel gears have teeth cut on a cone. They are used in pairs to transmit rotary motion where the bevel gear shafts are usually at right angles to each other.

Figure 12.72 shows a worm gear and wormwheel, which is an arrangement of gears that can be used to make large speed reductions. The worm, which looks similar to a screw thread, is fixed to the driver shaft, and the wormwheel is fixed to the driven shaft. The two shafts are at right angles to each other. The worm could be considered as a spur gear with one tooth when it comes to calculate VR.

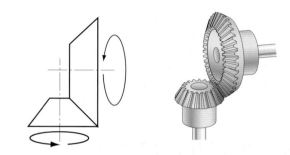

Fig 12.71 Bevel gear

VR = Driven / Driver
= 44 / 1 (or 44:1)

The mechanism locks if the wormwheel tries to turn the worm; this characteristic can be put to advantage when winding up a heavy weight and the winder is released accidentally.

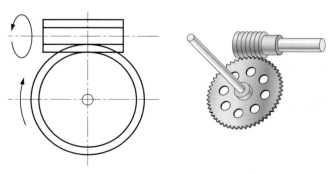

Fig 12.72 Worm gear and wormheel

Figure 12.73 shows a rack and pinion mechanism, which is used to transform linear into rotary motion, or rotary to linear. The pinion is the round spur gear; the rack has teeth in a straight line and meshes with the pinion. A good example of this is a pedestal drilling machine: as you rotate the handle the drill chuck moves up or down.

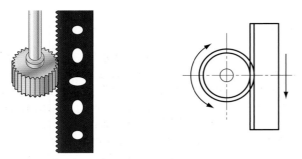

Fig 12.73 Rack and pinion mechanism

Transfer of drive

Figure 12.74a shows a belt and pulley system that transmits motion and force from the driver shaft to the driven shaft. The V-belt is one of the more common types in use; to change the speed a pulley block that has different diameter pulleys can be used (Figure 12.74b). A pulley belt could slip in a situation where the output shaft became jammed; a toothed belt is used where the pulleys must not slip. The driver and driven shafts rotate in the same direction.

VR = Driven pulley diameter / Driver pulley diameter
= 25 / 100
= 1 / 4 (or 1:4)

Driver
Ø 100mm

Driven
Ø 25mm

Fig 12.74

This shows that one turn of the driver shaft will turn the driven shaft four times. If the driver shaft is rotating at 1500 rpm and the VR is 1:4, then:

The output speed of the of the driven shaft = input speed / velocity ratio

= 1500 x 4 = 6000 rpm

The chain and sprocket system in Figure 12.75 is used to transmit rotary motion from the driver to the driven shaft with both rotating in the same direction. The sprockets are toothed wheels on which the chain runs. Unlike the belt and pulley system the chain and sprocket cannot slip, like the toothed belt.

The most common example of a chain and sprocket system is on a bicycle, and if it has a derailleur type then it also has a set of sprockets that allow the VR to be changed for different riding conditions. The derailleur makes use of a jockey wheel that keeps the chain in tension.

When working out the speed of the system the number of teeth on each wheel is used:

VR = Number of teeth on the driven sprocket / Number of teeth on the driver sprocket
= 25 / 75
= 1 / 3 (or 1:3)

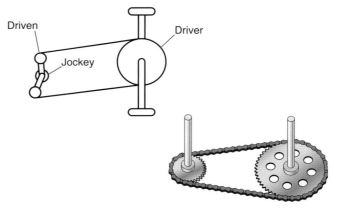

Driven

Driver

Jockey

Fig 12.75

The crank and slider mechanism shown in Figure 12.76 can convert rotary motion to reciprocating, or reciprocating to rotary. The slider reciprocates while the crank rotates. An example of reciprocating to rotary is the car engine, where the pistons are driven in a linear motion to rotate the crankshaft in a circular motion.

An example of rotary to reciprocating is a mechanical saw, where the saw is mounted on the slider and the crank is driven round by an electric motor.

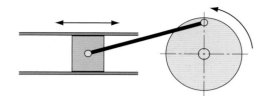

Fig 12.76

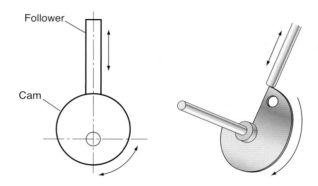

Fig 12.77

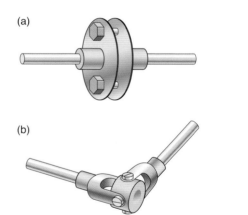

Fig 12.78

Figure 12.77b shows a snail cam, which produces a slow rise of the follower and, at the end of one revolution, a rapid return.

Couplings

Where two shafts are to be joined together and are aligned, and will be when they are rotating, then an arrangement as shown in Figure 12.78a can be used. This is known as a flanged coupling and an example could be the connection between the shaft holding the blades of a wind-powered generator and the generator itself.

Where the two shafts are not always aligned a flexible coupling is required. If the angle of misalignment is slight a rubber disc could be inserted between the flanges. If the shafts are likely to be more than a few degrees out of alignment then a universal joint is often used, as shown in Figure 12.78b. An example of this could be the shaft connecting the engine to the rear wheels of a car, where the rear wheels can move up and down.

Torque

Where the rotary motion of a turning force is transferred from the driver to the driven shaft, there is a turning force which is called the torque. Figure 12.79a shows the shaft before it is subject to a turning force and 12.79b after; the lines indicate what is happening in an exaggerated way. The far end of the shaft is fixed.

output torque = input torque x VR

The torque rating of an electric motor is often provided on the data sheet or specification plate.

Cams are available in various shapes and sizes. Figure 12.77a shows a circular cam with a follower. The rotary motion of the cam causes the reciprocating motion of the follower; this is the simplest type and the movement it produces is known as simple harmonic motion. Using imagination the cam can be designed to create a variety of follower movements. The example in Figure 12.77a could be an electric motor turning the cam to move the follower to operate a diaphragm of a small pump.

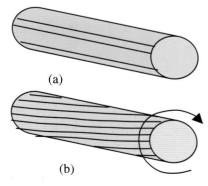

Fig 12.79

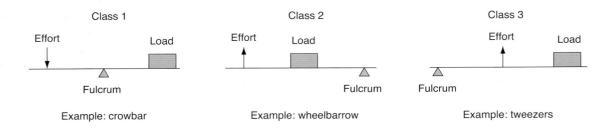

Fig 12.80

Levers

Levers are probably the oldest type of mechanism to be used. There are three classes depending on the relative position of the load, the effort and the fulcrum, as shown in Figure 12.80.

Levers in class 1 and 2 give you a mechanical advantage (MA), which is why they are also more common.

mechanical advantage = load / effort

So, in the class 1 example of a crowbar, if an effort of 100 N just lifts the load of 400 N then the mechanical advantage is:

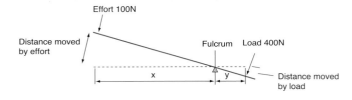

Fig 12.81

MA = 400 / 100 = 4

If the MA can be calculated by measuring the effort required to just lift the load, and the load is known, then the velocity ratio can be calculated:

VR = distance moved by effort / distance moved by load

When the crowbar balances the load horizontally, in equilibrium (as shown in Figure 12.81), then the product of the effort and distance from the fulcrum is called the moment of the force. When in equilibrium the anticlockwise moments about the fulcrum must be equal to the clockwise moments. Using the example in Figure 12.81:

ACTM (effort x distance) = CTM (load x distance)
100 x Nm = 400 y Nm

x / y = 400 / 100

Using similar triangles, or any other method, VR = 4 / 1 (or 4:1)

efficiency = (MA / VR) x 100

For the crowbar example, efficiency = (4 / 4) x 100 = 100%

The arrangement shown in Figure 12.82 shows one rope connecting four pulleys but, for calculations, it is equivalent to four ropes each supporting a quarter of the load. The MA will be 4 and the VR will be 4.

It may be that the MA is not quite 4 because there will be some friction in the system; it may only be 3.8.

In that case:

Efficiency = (3.8 / 4.0) x 100 = 95%

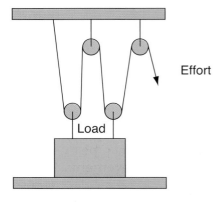

Fig 12.82

Pneumatics

Pneumatic input devices

Input devices connect the outside world to the circuit. A pneumatic circuit operates using valves that can be operated in various ways to cause an input. The basic internal structure of these valves is similar and, as a general rule, they are activated and cause a change in the direction of the air flow. Figure 12.83 shows the two states of the valve; it has three ports. These are known as a 3 port, 2 or 3/2 way, push button, spring return valves.

Valves can be operated in various ways; Figure 12.84 shows some of those alternatives.

Another valve in this family is the 5 port 2 way; one of the uses for this type is where there is a need for the air to be directed in one of two ways, as shown in Figure 12.85.

Pneumatic output devices

Cylinders are the main components used as output devices in pneumatics. A cylinder has a piston that is moved by air pressure entering either end. Figure 12.86 shows the commonly used cylinders.

In a single-acting cylinder air pressure forces the piston out and the internal spring returns it. In a double-acting cylinder air pressure forces the piston out and air pressure at the other end returns it. In a cushioned cylinder the final part of the movement is slowed down to prevent an abrupt stop.

Single-acting cylinders

Figure 12.87a shows a simple circuit diagram using a 3/2 valve to activate a single-acting cylinder. When the push button is pressed, main air sends the piston out against the spring, where it stays until the button is released. When the button is released the spring expands and the piston returns.

An example where this type of circuit could be used is as a date-stamping machine.

Double-acting cylinders

As a double-acting cylinder requires air pressure to both outstroke and instroke the piston, it is necessary to use a 5 port 2 way valve. A useful circuit can be built by combining the valves as shown in Figure 12.88 to produce an automatically reciprocating sawing machine.

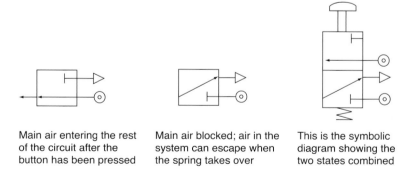

Main air entering the rest of the circuit after the button has been pressed

Main air blocked; air in the system can escape when the spring takes over

This is the symbolic diagram showing the two states combined

Fig 12.83

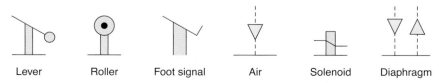

Lever   Roller   Foot signal   Air   Solenoid   Diaphragm

Fig 12.84

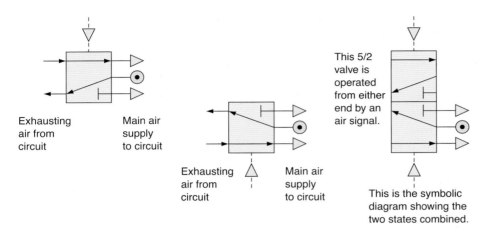

Exhausting
air from
circuit

Main air
supply
to circuit

Exhausting
air from
circuit

Main air
supply
to circuit

This 5/2
valve is
operated
from either
end by an
air signal.

This is the symbolic
diagram showing the
two states combined.

Fig 12.85

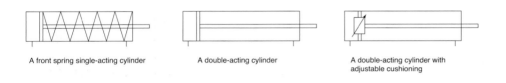

A front spring single-acting cylinder

A double-acting cylinder

A double-acting cylinder with
adjustable cushioning

Fig 12.86

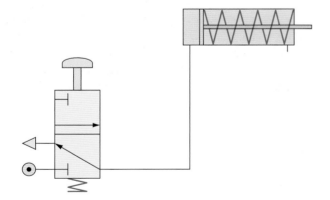

Fig 12.87

When valve A is operated by the end of the saw
it switches the 5 port 2 way valve over and main
air enters the opposite end of the double-acting
cylinder, causing it to outstroke and move the
saw across the log to cut it. When it gets to the
end of the stroke valve B is operated and the
cylinder instrokes, and the saw returns to start
again.

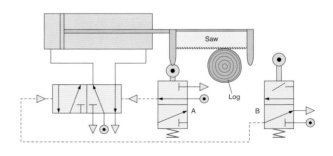

Fig 12.88

Cushioned cylinders

Sliding doors are often operated using
cushioned pneumatic cylinders because the
cushioning at the end of each stroke prevents
the door from slamming open and shut. The
cushioning of each stroke is bought about by
trapping a small amount of air – a cushion – at

the end of the cylinder and allowing it to exhaust much more slowly than the rest of the air (or, in the case of more sophisticated applications, a separation piston with a regulated bleed orifice).

Shuttle valves

Many sliding doors are operated from either side; the component that is used to allow valve A or B to open the door shown in Figure 12.89 from either direction is a shuttle valve. If valve A or B is operated, main air will pass to the sliding door circuit because the shuttle valve has a component that blocks one exit or the other, and prevents the air from exhausting through the opposite valve.

Restrictors

Another pneumatic component used in sliding doors and many other circuits is the restrictor or flow control valve. It comes in two varieties: bi-directional and uni-directional. Figure 12.90 shows the symbols. As the name implies, a bi-directional restrictor restricts the air flow in both directions while a uni-directional restrictor restricts flow in one direction and allows full flow in the other. As in the shuttle valve there is a component that blocks the flow of air in one direction.

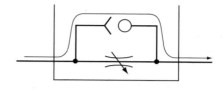

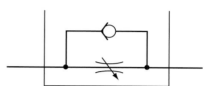

Fig 12.90 Two states of the uni-directional restrictor

Air bleed occlusion

A useful way of detecting movement or the covering of a small air hole is to use an air bleed occlusion circuit. The circuit makes use of low-pressure air venting through a small hole: if the hole becomes covered, the low pressure change is detected by a diaphragm valve. The diaphragm valve has a large surface area on which the low pressure acts, and the valve changes state in the usual way. Figure 12.91 shows an air bleed occlusion circuit and the small hole at 'a' being open. If the hole at 'a' is covered then air will be directed to the diaphragm operated 2 way 3 port valve, which will send the single-acting cylinder outstroking. An application for this could be on a conveyor belt where the date is automatically stamped on the product when it arrives in the correct place.

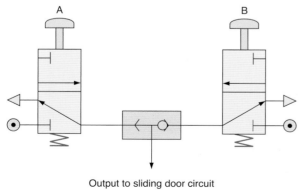

Output to sliding door circuit

Fig 12.89

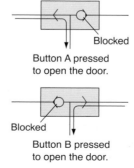

Button A pressed to open the door.

Button B pressed to open the door.

The two states of a shuttle valve

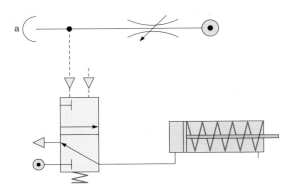

Fig 12.91

Valve port identification

The ports on the valves are identified by numbers as in Figure 12.92.

- No 1 is main air with the symbol 1
- 2 and 4 can be either input or return from the circuit
- 3 and 5 are exhausts with the symbol 3 or 5
- 12 and 14 are named 'one two' and 'one four' because they refer to the fact that when a signal is applied to port 12, main air (1) is connected to output (2); 14 connects 1 to 4
- Main air is a solid line, and signal and exhaust lines are broken

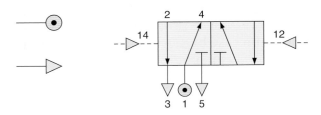

Fig 12.92

If a piston rod has extended out of the cylinder, it has outstroked or gone positive (+); if a piston rod has retracted back into the cylinder it has instroked or gone negative (-).

Sequential control of more than one cylinder

In many pneumatic circuits it is necessary to combine the operation of more than one cylinder. This is called a sequential control circuit. Figure 12.93 shows the movement of two doors, A and B, controlled by cylinders A and B in a sequence of:

- Cylinder A instroking (A-)
- Cylinder B outstroking (B+)
- Cylinder A outstroking (A+)
- Cylinder B instroking (B-)

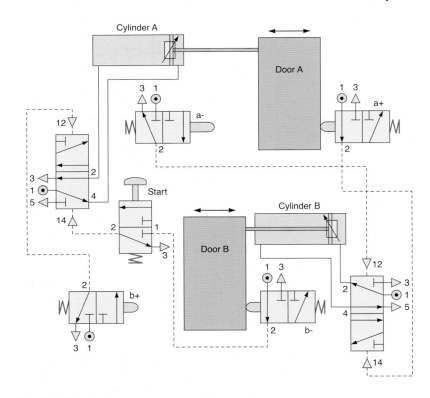

Fig 12.93

The circuit in Figure 12.93 will cause door A to move left, then door B to move left, then door A to move right and finally door B to move right. The circuit works as follows:

- The start button is pressed and released then a signal is sent to port 14 of the 5/2 valve controlling cylinder A
- Main air out of port 4 is sent to cylinder A, causing it to instroke and door A to move to the left
- Door A hits 3/2 valve a- and a signal is sent to port 12 of the 5/2 valve controlling cylinder B
- Main air out of port 2 is sent to cylinder B, causing it to outstroke and door B to move to the left

- Door B hits 3/2 valve b+ and a signal is sent to port 12 of the 5/2 valve controlling cylinder A
- Main air out of port 2 is sent to cylinder A, causing it to outstroke and door A to move to the right
- Door A hits 3/2 valve a+ and a signal is sent to port 14 of the 5/2 valve controlling cylinder B
- Main air out of port 4 is sent to cylinder B, causing it to instroke and door B to move to the right
- The circuit will now wait until the start button is pressed again and the sequence will start again

Section B	Programmable and control devices

Learning outcomes

By the end of this section you should have developed a knowledge and understanding of:

- The use of programmable devices to solve problems in system design
- The integration of programmable control devices with electrical/electronic, mechanical and pneumatic systems

Programmable devices in electronics

In the last few years this area of electronics has seen considerable growth. There are many advantages, particularly in the lower parts count and ease of development of a project. When designing a circuit using a microcontroller the main section of the circuit will be identical when solving many problems. The area that will change is the inputs and outputs.

A popular device used in schools is the PIC, manufactured by Microchip Inc. Although PIC stands for programmable interface controller the term PIC is often used generically to describe a programmable device.

A microcontroller will contain a processor, program and data memory alongside the pins for inputs and outputs. The first generation of this type of device required a 12 V supply to carry out the programming; this meant that a separate programming unit was needed in addition to a PC. Current versions can be programmed directly from a PC, leading to a cut in costs and easier program development.

Facilities offered by microcontrollers include digital input and output, analogue to digital conversion, and PWM output for servo motors and sound output. The microcontroller is an electrically erasable and programmable device, known as an EEPROM. Programs can be written on a PC and transferred to the controller using a wired connection. Once the program is installed the wired connection can be removed. Programs may be developed using a flow chart system, or by using the basic programming language.

External devices can be driven by a microcontroller as easily as by a circuit using discrete components. Outputs can be used to operate a Darlington transistor or MOSFET allowing transducers with a high current draw to be operated. The supply voltage for a microcontroller is normally quite critical and should be supplied from a low-voltage battery or from a voltage regulator such as the 7805.

Fig 12.94 Golf scorer

Electrical/electronic control of pneumatic circuits

Many manually operated valves and other components can be replaced by an electrical/electronic version. Figure 12.95a shows the circuit of a reed switch cylinder controlled by a solenoid-operated 3 port 2 way valve. Figure 12.95b shows the actual components

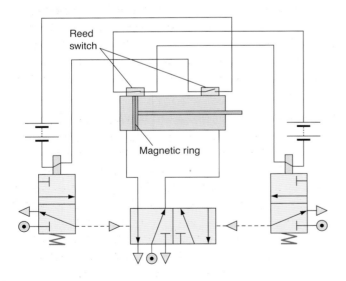

Fig 12.95

A reed switch cylinder differs from a normal cylinder. It has a magnetic ring built into the piston that causes the reed switches to close when it comes close; the switch opens when the piston moves away. The closing and opening of the switches can complete an electrical circuit,

which can then operate a solenoid 3 port 2 way valve.

The 3 port 2 way solenoid valve, as with all solenoid valves, has a coil inside that, when energised, causes the valve to switch states. When the solenoid valve changes state it reverses the direction of movement of the piston until the magnetic ring on the piston influences the other reed switch, which again reverses the direction of movement of the piston. This could be another way of operating a saw as shown in Figure 12.88.

Computer control of pneumatic circuits

With the availability of electrical/electronic pneumatic components it is possible to control circuits from a school computer or microcontrollers, such as programmable logic controllers (PLC) used in industry, a PIC or other programmable chips. A microcontroller is often described as a 'computer on a chip' because it contains memory, processing units and input/output circuitry, which can be programmed from an external device to run a control program.

The circuit in Figure 12.95 uses a reed switch cylinder to provide the movement, which is detected by the two reed switches being open or closed depending on where the piston is positioned.

These two switches can be the inputs to the computer system, and signal either a 0, meaning the piston is not at that reed switch, or a 1, meaning the piston is at that reed switch.

The circuit uses solenoid-operated 3/2 valves that switch when a voltage is applied to them; a computer has the facility to output a voltage to these valves when told to do so from a program.

Using these components the computer can run a program that monitors the reed switches: when a switch closes, and an input occurs, the program can output a voltage to the correct solenoid 3/2 valve and cause the piston to reverse direction. This is a simple program but, with some imagination, the computer can control a highly complicated industrial robotic assembly.

Hydraulics

Hydraulics are used where high precision of piston positional control and more force is required. There are many similarities between pneumatics and hydraulics: the obvious one is that pneumatics use air pressure and hydraulics use a pressurised liquid in place of air. Similar valves can be used but the main difference is that, in pneumatics, the exhaust air can be vented into the atmosphere but, in hydraulics, the liquid must be retained and sealed within the system for continuous use. This continuous use and the fact that it is a sealed system has the advantage that a hydraulic piston can be stopped anywhere along the stroke and held there, giving the advantage of positional control. The fact that the hydraulic oil will not compress gives another advantage over pneumatics, because air can be compressed.

Section C — Manufacturing processes

Learning outcomes

By the end of this section you should have developed a knowledge and understanding of:

- How to design and present information using standard symbols
- Production methods, comparing small-scale techniques used in school/college and large-scale techniques used in industry
- How to interface systems to provide working solutions

Designing and drawing electronic circuits

Although hand drawing is suitable for initial ideas, much of the circuit design carried out in schools is now done using circuit design software. In many cases this has the advantage of allowing the circuit operation to be simulated and developed before converting it to a PCB.

When carrying out the design stage it is good practice to treat the circuit as a system, splitting it into input (including the power supply), process and output sections. Each section can then be designed and tested to ensure that it meets the specification. Care is needed to ensure that process circuits can cope with the current draw requirements of the output stage.

To ensure that each stage of circuit development is recorded when using circuit design software it is good practice to save the file regularly using a different version number: this allows the complete development of the circuit to be printed out and annotated.

PCB production

The usual method of making a PCB in school is the photo-etch process. For this process artwork is produced to show tracks, pads and any text that will appear on the copper or track side of the board. PCB design software is usually used at this stage.

The board consists of a substrate or base layer that is laminated to a copper layer, which has a coating of UV-sensitive resist. When the board is exposed to UV light through the artwork the resist is softened where the UV light has exposed it. A developer will react with the soft resist and it can then be washed away leaving pads and tracks still coated in resist. Chemical removal of excess copper is carried out in a bubble etch tank using ferric chloride as the etchant.

At this stage the remaining resist is removed and the board can be tinned. Cold tinning deposits a thin layer of tin onto the copper, providing a surface that will solder easily. An alternative to this is to use a PCB flux-based soldering varnish. At each stage of the process the board is washed to remove traces of chemical.

The time taken for production of a single board is approximately 20 minutes from start to finish. To reduce the time for each board, where possible, artwork is blocked together on a single sheet to produce as many boards as possible in one go. Commercially produced boards have component information screen printed onto the component side of the board. In a school situation this process is not normally available so critical information is etched into the copper side of the board.

Fig 12.96 Flow or wave soldering

Flow or wave soldering

Flow soldering, sometimes known as wave soldering, is an automated process that can solder all joints on a board as it travels through the wave soldering machine.

The process starts with the boards being placed onto a conveyor that will carry them through the machine. As the board enters the machine a synthetic flux is applied by spraying through multiple jets. The next part of the process is preheating, which will both evaporate the flux solvent and preheat the board and components.

A resist strip can be stuck to the board to protect areas where solder is not required. The solder wave is the main part of the machine and can consist of two types of wave. With a *lambda wave* the solder flows backwards against the travel of the board. The board is in contact with the solder for three to five seconds. A *chip wave* is sometimes used before the lambda wave; this can improve the soldering of surface mount components. The chip wave contacts a smaller

length of the board, typically 10 to 15 mm against the 25 to 30 mm of the lambda wave.

Fig 12.97

After passing through the machine the boards are allowed to cool. Modern fluxes will require no further processing but with more active fluxes a wash process is used. Most wave soldering is carried out in air, though some manufacturers use a sealed unit that operates in a nitrogen environment, preventing oxide formation. Since July 2006 and the EU Restriction of Hazardous Substances (RoHS) directive, manufacturers have had to adapt machines to cope with lead-free solder that melts at a slightly higher temperature.

Surface mount technology (SMT)

The components used are extremely small in comparison to conventional through-hole components. Resistors and capacitors are encased in strips that can be fed into a pick-and-place machine, which will accurately place the component in the correct orientation.

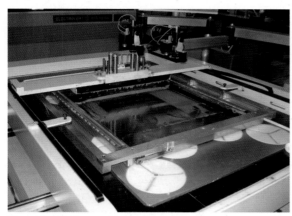

Fig 12.98 Spreading solder paste

Fig 12.99 Mask for solder paste

Larger components such as ICs are held either in tubes that feed into the machine or in blister reels for smaller ICs. The assembly process consists of applying a thin layer of solder paste or adhesive to the PCB: solder paste is applied directly to the pad positions, adhesive to the

gaps between pads. This is done by a similar process to screen printing. A metal foil with the pad cut-outs is the mask through which a roller or squeegee pushes the solder paste. The amount of solder applied is determined by thickness of the metal sheet. Following the application of solder the board is 'populated' in a CNC (computer numerical control) pick-and-place machine.

Fig 12.100 SMT components

Fig 12.101 Vacuum-operated head

The machine in Figure 12.101 uses a vacuum-operated head to pick up components from the reel or tube and place them on the PCB. When programming the machine component orientation can be specified for polarised items such as electrolytic capacitors and ICs. The PCB has to be placed accurately in the machine and, to do this, 'registration' or 'fiducial' marks that are about 1 mm in diameter are designed into the PCB. The marks are used by the pick-and-place machine for absolute reference when aligning the placing head.

Soldering surface mount boards

Populated boards can either be flow soldered or reflow soldered using one of the methods described below.

Reflow soldering – vapour phase

This process, which was patented in 1975, uses a liquid with a known boiling point to produce a vapour blanket 60 to 80 mm in height, into which the populated board is immersed. In the case of lead-free solders the vapour will boil at approximately 230°C. The heat of the PCB cannot exceed the boiling point of the liquid so overheating of the board and components can be avoided. Any liquid remaining on the board is evaporated before it leaves the machine. Heat transfer with this process is uniform and the inert atmosphere is particularly suited to lead-free solders. A typical soldering cycle using this process will last about five minutes.

Fig 12.102 Vapour phase soldering unit

Fig 12.103 Placing boards on the conveyor

Reflow soldering – convection

In this process several zones with differing heat can be used as the PCB passes through the oven. The temperature of the process zones and the rate at which the board passes through them can both be varied; convection is forced from outside the zones by fans or blowers. Heat transfer depends on the length of time taken to pass through the oven as well as the temperature difference between the oven and PCB. In a short oven the temperature of the oven has to be greater to achieve a similar board temperature to that achieved in a long oven. The atmosphere in a convection oven can be air or nitrogen.

Inspection of boards

Automated optical inspection can be carried out that compares a known 'good' board to the one being inspected. In this way missing components or incorrect component values can be quickly identified.

Using flow charts

Flow charts use simple drawings to show the flow of computer programs or subroutines. They are a design tool that can be used to plan the control program to input to the programmable control devices. These devices, which integrate the program to the outside world, could be a computer, a PIC, a PLC or a black box. In schools it is usually either a computer with inputs and outputs or PICs, but PICs still need to be written to via a keyboard. In industry PLCs are more likely to be found controlling complicated systems.

The flow charts can be written to control different systems, and these could be electrical/electronic, mechanical or pneumatic. These programmable control devices read the program that has been written to them and respond by making decisions, opening or closing switches, waiting for a given time or any other action written into the program.

The simple drawings that flow charts use are symbols that define a function, and there are only a small set of them needed as shown in Figure 12.104. The names used may vary from user to user, and there are other symbols, but at this level these are sufficient. The standard used is ISO 5807.

The **terminator**, usually found at the beginning and end of a flow chart, also known as the entry or exit symbol. Often used words are 'Start' and 'End', 'Start Prog' and 'End Prog' or 'Return'.

The **process** symbol is used to describe a single action or sequence. An example could be 'Send signal to instroke cylinder A' or 'Time = 10 seconds'.

The decision symbol is used where a branch may occur within a program. An example could be 'Has reed switch a- closed?' or 'Has 10 seconds past?'. There is an exit point that can continue the program if the decision has been satisfied, but there is also an exit point that can be fed back into the program before the question if the answer is 'No', and will continue being fed back until the answer is 'Yes'.

The data symbol is used to display an input or output when an interaction has taken place. An example could be 'Display the state of reed switch a-' or 'Display the countdown of the 10 seconds'.

The predefined process symbol is a subroutine that indicates a call to a procedure or function. An example could be a subroutine that causes a delay, 'Delay 10 seconds'. At school level this is not seen often but is used where there are a number of embedded subroutines.

Flowlines join the symbols together and indicate the direction and order of execution.

There are other symbols that are rarely used in schools but are available and can be found on many drawing programs.

To plan a computer-controlled pneumatic circuit a flow chart is used. This puts the process in logical order, which is needed to be able to write the computer-control program.

The sequence of A-/B+/A+/B- which was the one used on the circuit in Figure 12.93 is used again here to compare the mechanically controlled door circuit to this one, which is computer controlled. The flow chart in Figure 12.105 is a logical progression of the steps that need to be planned in the computer program.

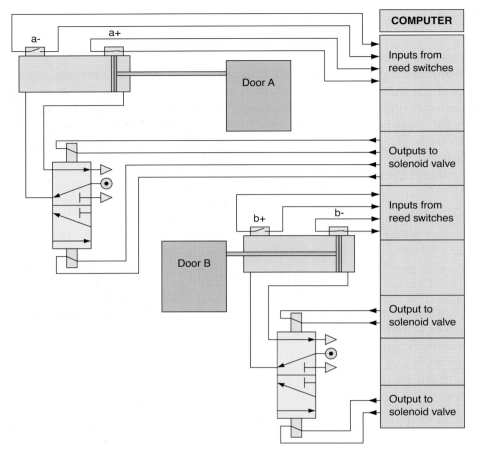

Fig 12.104

The comparison between the circuit in Figure 12.93, the mechanically controlled door operation, and the computer-controlled equivalent in Figure 12.104 shows that the mechanical one has nine pneumatic components against four for the computer control: this is significant as pneumatic components are expensive.

A computer or a microprocessor is needed for the computer-control system.

Computer control also gives more flexibility: it would be easy to program in a delay before a cylinder made any movement. As a safety precaution the program could make a warning sound before there was any door movement, or even speech synthesise: 'beware door closing'.

With a computer program running the sequence it is quick and easy to go back in and change the order of the cylinder operations, so that if the circuit is a prototype then alternatives can be

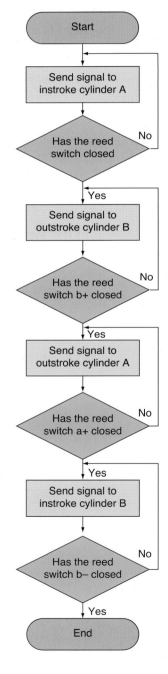

Fig 12.105

explored. Significantly more time would be needed to change a piped up mechanical sequence.

Pneumatics are often used in hazardous industrial situations: with computer control the operation can be monitored remotely from a safe distance.

The theory of force, pressure and area

Pneumatic cylinders are available in various piston diameters and stroke lengths, and the correct type must be chosen for the particular application. The supply pressure is another variable, as is the required force that the cylinder must produce. All these variables can be calculated using the equation:

force = pressure x area

$F = P \times A$

The area of the piston on which the air pressure is acting is measured in mm².
The force produced by the cylinder is measured in Newtons (N).
The air pressure acting on the piston can be expressed as N/mm². Figure 12.106 shows the cylinder.

If the pressure supplied by the compressor was 0.1 N/mm² and the diameter of the piston was 25 mm, then the force produced by the piston rod can be calculated using the formula:

$F = P \times A$
$F = 0.1 \times (\pi \times 12.5^2)$
$F = 49.08 \text{ N}$

This calculation is for an outstroking piston. If the calculation is for an instroking piston the area of the piston will differ because the piston rod takes up some of that area. Compare the answer with the previous example if we give the piston rod a diameter of 8 mm:

$F = P \times (\text{area of piston} - \text{area of piston rod})$
$F = 0.1 \times ((\pi \times 12.5^2 - (\pi \times 4^2))$
$F = 0.1 \times (490.87 - 50.26)$
$F = 44.06 \text{ N}$

If the piston is outstroking it can exert a force of 49.08 N but on the instroke it is only 44.06 N, so if the cylinder is doing the same work in both directions, the calculations must be based on the instroking piston.

Key Points

Safety precautions to be taken when working with pneumatics:
- Never allow compressed air to be directed on to the skin
- Always check for good connections and be ready to shut off the air if a leak appears

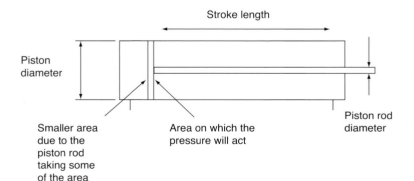

Piston diameter

Stroke length

Smaller area due to the piston rod taking some of the area

Area on which the pressure will act

Piston rod diameter

Fig 12.106

Pneumatics

1 Programmable control devices are used to control electro-pneumatic circuits.

   a) Name two programmable control devices that could be used for the operation of an electro-pneumatic circuit.

   b) State two benefits of using programmable control devices.

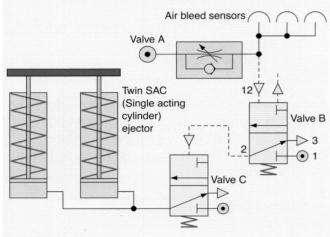

Fig 12.107

2 Figure 12.107 shows part of the pneumatic circuit for a conveyor system. When a box blocks all the air bleed sensors, an ejector pushes the box onto an output conveyor placed at 90 degrees to the input conveyor.

   a) Explain how the circuit operates once the air bleed sensors are blocked.

   b) The twin ejector cylinders must act in the same direction simultaneously. Explain how this can be ensured.

   c) A foot pedal-operated 3 port valve was piped in parallel with valve C so that the operator can now manually eject a box. Sketch the new circuit diagram, naming any additional components.

   d) Both ejector cylinders have a diameter of 40 mm and air is supplied at a pressure of 0.6 N/mm². Calculate the force exerted by the pair of cylinders if the efficiency of both is 95 per cent.

3 Discuss the implications for the manufacturer of using ICT in developing a prototype for volume production.

Electronics

1 A company is producing a voltage converter to allow a mobile phone to be powered from a car electrical system. Data for the two systems is given below:

Car electrical system: 11 V to 13.8 V (nominal 12 V)

Mobile phone: 6 V (at 750 mA when recharging)

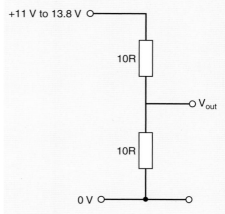

Fig 12.108

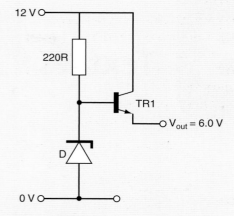

Fig 12.109

a) Figure 12.108 shows a prototype design based around a potential divider.
   i) State the maximum and minimum output voltages when the mobile phone is not connected. Ignore the resistor tolerance.
   ii) Explain one limitation of this circuit if used as a voltage converter for a mobile phone.

b) Figure 12.109 shows an improved circuit for the voltage converter.
   i) Explain the purpose of diode D.
   ii) State and justify, for a V_{BE} of 0.8 V, the most suitable voltage for diode D.
   iii) Calculate the maximum power dissipated in transistor TR1 when recharging the mobile phone.
   iv) Calculate the base current of transistor TR1 at full load if it has an h_{FE} of 150. Transistor gain = collector current / base current.
   v) Discuss the implications of the extensive use of mobile phones.

Section A — Performance characteristics

Fig 13.1 A polyamide tent

Learning outcomes

By the end of this section you should have developed a knowledge and understanding of the performance characteristics of fibres and fabrics, including:

- Tensile strength, elasticity, resilience, durability, weight, thermal conductivity/insulation.
- Chemical resistance, flammability, absorbency, water repellence.
- Reaction to light and micro-organisms, air and water permeability, and electrostatic charge.

Introduction

All fibres have different performance characteristics or properties that make them suitable for a range of end uses. The designer has to decide which properties and characteristics are essential to make the textile product perform and function well, but they must balance this against the cost and aesthetics of the product.

Aesthetic properties

Aesthetic properties are related to the look, touch and style of the finished item. For example, fashion garments may want a good drape, soft handle or trendy colours.

Comfort properties

Comfort properties are related to comfortable wear. These include thermal conductivity, insulation, absorbency and electrostatic charge. For example, a fleece should keep the wearer warm.

Functional properties

Functional properties are the most important as the product must function as the designer intended. For example, a waterproof jacket must protect the wearer from getting wet. These properties include tensile strength, elasticity, resilience, durability, weight, chemical resistance, flammability, water repellence, reaction to light, reaction to micro organisms, air and water permeability.

You must be able to make informed decisions when carrying out product analysis and when selecting fabrics for making your designs.

Fig 13.2 Children's playsuits

Fig 13.3 Trainers

Fig 13.4 Denim jeans

Select a product made from a range of different textiles and make a list of the performance characteristics of the fabrics used to produce it.

- Research the key terms that describe the performance characteristics of fibres. Record what each means in a chart and identify products where this property is essential when selecting suitable fabrics.
- Some key terms are listed in the glossary on the website. Print off a reference copy and add to it as you work through this chapter.

See www.hodderplus.co.uk/ocrd&t
Chart to identify performance characteristics

See www.hodderplus.co.uk/ocrd&t
Key terms describing performance characteristics of fibres

1 Figure 13.2 shows children's playsuits.
 (a) (i) Give two performance characteristics needed by the fabric used for the playsuits.
 (ii) Name two suitable fabrics for the playsuits.
 (b) Describe two standardised tests that would be carried out on the fabric as part of the development of the product. Use sketches and notes where appropriate.
 (c) Discuss the importance of fabric testing to both the manufacturer and the consumer.

Learning outcomes

By the end of this section you should have developed a knowledge and understanding of the performance characteristics, selection and use of natural and manufactured fibres:

- Natural fibres, such as cotton, flax, wool and silk
- Regenerated fibres, such as viscose
- Manufactured and synthetic fibres, such as polyamide, polyester, acrylic, elastomeric, flurofibres, chlorofibres, glass and metal

Introduction

The textile industry uses many different kinds of fibre for raw material. Some, such as wool and silk, have been used since the earliest time but the textile industry is continuously developing exciting new fibres to meet the technological demands of today's society.

Fibres are classified according to their origin: some are from plants and animals, for example cotton, flax, wool and silk; some are from plant material, for example viscose and lyocell; and some are made entirely by chemical reaction, for example polyester and polyamide.

Table 13.1 The main fibre groups

Natural	Vegetable	Cotton, linen (flax), kapok, jute, hemp, sisal
	Animal	Wool, silk, camel, cashmere, mohair, angora
Manufactured	Regenerated	Viscose, lyocell
	Synthetic	Polyester, polyamide (nylon), acrylic, elastomeric, flourofibres, chlorofibres
	Inorganic	Glass, metal

Fibres

A fibre is a fine and flexible raw material which could be short or very long depending on where it comes from and how it is manufactured. There are three main fibre types:

- **Staple fibres** – staple fibres are short, ranging from a few millimetres, such as in cotton, to around a metre as in flax.
- **Continuous filaments** – continuous filaments are an indefinite length. All synthetic fibres start off as continuous but they may be cut up to form staple yarns. Silk is the only natural continuous filament.
- **Microfibres** – microfibres are very fine fibres, around 60 to 100 times finer than a human hair!

- Examine a range of fibres under the microscope. Record the staple and filament fibres. Make sketches to show their appearance under the microscope. It would also help you understand the origins of all the fibres if you also carried out burning tests on them.
- Produce an illustrated A3 sheet to use as a classroom display for younger pupils to show the origins and use of all of the different fibres.

Natural vegetable fibres – cotton

Cotton is a natural vegetable fibre from the seed pod of the cotton plant. It is a staple fibre and is the most widely produced natural fibre. The fibres are formed from plant cellulose.

Key Points

Cotton production has considerable environmental implications as well as being a crop of major global importance. Cotton production accounts for 25 per cent of total global pesticide use, and there are concerns about health risks for farm workers and soil contamination.

Fig 13.5 A cotton plant

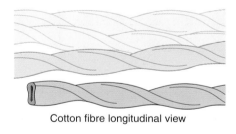

Cotton fibre longitudinal view

Cross section

Mature   Immature

Fig 13.6 Cotton fibres

Performance characteristics of cotton

- Because of the cavities in the fibres it can absorb up to 65 per cent of its own weight in moisture
- It always contains some moisture so it is non-static
- Fibres are strong when wet because they become swollen with water and this distributes stresses more evenly along the length of the fibre
- Soft handle, good drape and moisture absorption make it a comfortable choice for clothing
- Durability and abrasion resistance combined with strength make it a good choice for a range of products
- It is biodegradable and recyclable.
- Poor elasticity and a tendency to crease are disadvantages
- Cotton dries slowly

Finishes applied to cotton to improve its properties

- **Mercerising** – this is the treatment of cotton fibres with caustic soda, which causes the fibres to become more circular. It improves the strength and lustre.
- **Crease resist** – synthetic resins improve the elasticity of cotton fibres by cross linking the cellulose chains. This makes it easy care.
- **Anti-shrink finish** – household tumble-dryers can cause cotton to shrink so the fabric is deliberately shrunk before processing.
- **Stain resist** – Teflon® or silicone finishes make cotton repellent to water and stains.
- **Flame retardant** – a finish using the PROBAN® or PYROVATEX® process will improve cotton to make it flame proof and suitable for upholstery and nightwear.

Typical cotton fabrics include calico, chintz, corduroy, denim, drill, poplin, terry, velvet and lawn.

- Start a collection of fabric samples (called a swatch) and add to it as you work through this section.
- Find a variety of cotton samples, label them to show the different performance characteristics and uses of cotton fibres. There is an example on the website.

See www.hodderplus.co.uk/ocrd&t
Chart to identify performance characteristics

Fig 13.7 Cotton towelling

1 Explain why cotton is a suitable fibre to use in the manufacture of:
   (a) Towels
   (b) A tennis shirt
   (c) A pair of denim jeans
2 There has been an increase in organic cotton production.
   (a) Give reasons for this increase.
   (b) Explain the advantages of growing and using organically grown cotton.
   (c) Discuss how genetic modification could improve cotton plants and fibres.

Natural vegetable fibres – flax (linen)

The flax plant is the source of fibres for linen yarns and fabrics; it is an annual crop grown from seeds in many parts of Europe. The fibres are obtained from the inner fibrous parts of the stem. They are long fibres, which make it suitable for producing crisp, attractive, cool fabrics. It is produced as a staple fibre.

The flax fibres, like cotton fibres, are long-chain cellulose molecules. They are about 25 to 40 cm long and are cemented together by a mixture of ligins, pectins and hermicelluses. The properties of flax are influenced by this composition. Flax is stiffer than cotton because of the cement holding the fibres together. It also has a smoother surface and a darker colour than cotton.

Fig 13.8 A flax plant

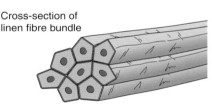

Linen fibre

Cross-section of linen fibre bundle

Linen fibre (flax) cross-section

Fig 13.9 Flax fibres – a fibre bundle and individual fibres

Performance characteristics of flax (linen)

- Strong, durable and long lasting. It is strong when wet, cool to wear and is fast drying because it is highly absorbent but releases water quickly
- Anti-static as it always contains some moisture
- Crisp, stiff handle due to the cements holding the fibres together
- Easy to wash and dry, and also shrink proof as it takes up water rapidly and releases it quickly
- Low elasticity, so it creases badly
- Poor insulation as the smooth fibres do not trap any air, making it cool to wear

- Linen fabric has a low lustre and does not soil easily
- The coarse fibres give linen a firm handle and affect its next-to-skin feel
- It resists micro-organisms
- It is biodegradable and recyclable

Finishes applied to linen to improve its properties

- Crease resist – a synthetic resin treatment is applied to give linen a crease-resistant finish
- Stain resist – a stain resistant finish can be obtained using Teflon® or silicone
- Enzyme treatments – these affect the surface of the fibre by smoothing it, making it softer to handle

Typical linen fabrics include crash, half linen (union), holland, interlining and mattress ticking. Linen is used for leisure and summer wear (suits, skirts, blouses and accessories), table linen and technical uses such as ropes and tarpaulins.

- **Add some samples of linen fabrics to your swatch.**
- **Explain why linen is chosen for garments for hot countries.**

Fig 13.10 A linen shirt

1 (a) (i) Give four reasons why linen is a good choice of fabric for the shirt.
   (ii) A non woven interfacing is used on the shirt. State two places on the shirt where interfacing is applied. Give two reasons for this.

(b) Describe, using annotated sketches, how to work the buttonholes and attach the buttons to the front of the shirt.
(c) Discuss the environmental implications of industrial textiles production.

Natural animal fibres – wool

Fig 13.11 A basket of unprocessed sheep wool

Natural animal fibres are formed from proteins. The most common animal fibre is wool, usually from sheep but also from goats, llamas and angora rabbits.

Wool fibres are very similar to human hair; they are made of protein molecules (keratin) formed into bundles and can vary in length. The best woollen yarns are made from staple fibres that are 50 to 120 mm in length, highly crimped and very fine. These are used to produce worsted cloth, which is used for fine suiting.

Coarser, longer fibres with less crimp are used to produce woollen cloth for heavier, more robust clothing. Carpets and upholstery are produced from the most coarse, long fibres.

Wool fibres have a very distinctive structure; when wool fibres absorb moisture and the temperature rises, the fibres swell and the protein bundles move apart. The bonds between the fibres are broken down. These bonds reform as the bundles dry and cool. This makes the fibres very elastic and gives wool its recovery and reshaping properties.

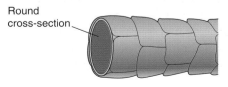

Round cross-section

Wool fibre cross-section

Fig 13.12 Wool fibre cross section

Performance characteristics of wool

- Three factors make wool a good insulator: the scales, the crimp and the length of the fibres.
- Wool is highly absorbent (hydrophilic): it can readily absorb 33 per cent of its weight in water without feeling wet. This also makes it anti-static.
- The fibres are hydrophobic because they are covered with a very thin skin – the epicuticle – which causes liquid to roll into droplets while allowing the passage of water vapour.
- The softness of wool gives it a comfortable feel next to the skin.
- Textiles made from wool are not particularly durable.
- Elasticity and springiness are excellent in wool so creases easily drop out.
- Wool can be formed into durable shapes because the molecular structure can be adjusted under heat.
- Laundering can be a problem as wool will stretch when it is wet due to the high amount of absorbed water, and felting can occur under the influence of mechanical action, heat and water.
- Wool is biodegradable and recyclable.
- High natural fire resistance and built-in UV protection.

Finishes applied to wool to improve its properties

- **Water repellency** – a silicon treatment for improved weatherproofing.
- **Moth proofing** – the fabric is treated with a chemical to make the fibres inedible to moths.
- **Machine washable** – chemical treatments that reduce the tendency of the fibres to felt.

Typical wool fabrics include flannel, herringbone, tartan, tweed and Viyella.

- **Collect samples of woollen, worsted and wool-blend fabrics and add them to your fabric swatch.**
- **Research recent developments that have improved the performance characteristics of wool.**
- **Research other luxury hair fibres such as mohair, angora and cashmere.**

1 Explain why woollen and worsted yarns have different properties.
2 Wool has been described as the 'original high-performance, all-weather fibre'. Discuss this statement.

Natural animal fibres – silk

Silk fibre is the only naturally produced continuous filament. It is produced by the caterpillar of the silk moth when it pupates. The long protein fibre comes from a spinneret below the mouth of the caterpillar which then forms the cocoon. Silk fibre is made from two long protein filaments glued together. The physical, chemical and comfort properties of silk fabric depend on the way the long chain protein molecules lie inside each filament of the fibre. The more closely they are layered together, the stronger and more resilient the fibre. The direction of the protein chains affects the ability to absorb moisture and the lustre of the fabric.

Cultivated filament silk is made into fine fabrics. Any broken fibres or short lengths are spun into spun silk, which is not as strong or lustrous as filament silk. Wild silk cocoons are also made into spun silk.

Performance characteristics of silk

- Silk is both cool and warm. Filament silk lies smoothly on the skin and gives a cooling effect, but it is also a good insulator because the layer of warm air between it and the skin cannot escape through the compact silk fabric.
- It can absorb and hold about 33 per cent of its weight in water vapour without feeling wet.
- The next-to-skin feel is excellent because of its fineness, softness and elegant drape.
- It is very strong, durable and light. Silk fibres can be up to 1 km long.
- Creases fall out because it has excellent resilience.
- It does not build up any electrostatic charge because the moisture contained in the fibres conducts any charge away.

Finishes applied to silk to improve its properties

- **Waterproofing** – the silk fabric is coated with polyurethane.
- **Reflective** – The fabric is layered with metallic foil.

Silk can be made heavier and firmer by the addition of metallic salts or other chemicals. Typical cultivated silk fabrics include chiffon, damask, organza, satin and taffeta. Wild silk fabrics include dupion and shantung.

Fig 13.13 Silk caterpillar cocoons

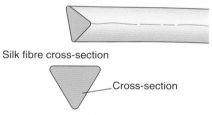

Silk fibre cross-section

Cross-section

Fig 13.14 Silk fibres cross section

Fig 13.15 Soaking silk cocoons in a factory in China

- Find samples of cultivated and wild silk fabrics and add them to your fabric swatch.
- Trend predictions for the summer season suggest that silk garments in natural colours will be in demand. You have been asked to produce garment designs for both men and women that:
  - Use silk and other natural fibres
  - Use a theme of nature
  - Use natural colours
  - Incorporate appropriate components
  - Can be batch produced

    Use notes and sketches to produce a range of initial ideas. Include details of all fabrics and components.

Regenerated vegetable fibres

Regenerated fibres are made by chemically changing natural materials that come from plants.

Viscose fibres are produced from eucalyptus, pine or beech wood that is dissolved in a solvent and extruded through a spinneret with fine holes. Viscose and acrylics are extruded into a bath containing chemicals that solidify the filaments. This is wet spinning. The holes in the spinneret can be circular or any other shape, which affects the lustre and handle of the fabric. The long filaments are smooth, fine and soft. Staple viscose fibres are produced by cutting short lengths as the fibres are extruded. It is cheap to produce.

Fig 13.16 Pine produced for regenerated fibres

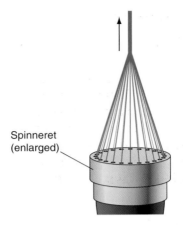

Spinneret (enlarged)

Fig 13.17 A spinneret

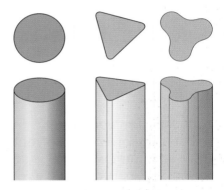

Spinneret shapes and fibre cross-section

Fig 13.18 Spinneret shapes and fibre cross sections

The chemical structure of viscose is similar to cotton but the molecules are shorter, which is the main reason for the lower strength of viscose.

Performance characteristics of viscose

■ Viscose filament yarns are made into smooth fabrics and therefore have poor thermal insulation. When cut into staple fibres the resulting fabric can trap air and improve insulation.

■ Viscose is more absorbent than cotton. The fibres can absorb 11 to 14 per cent of their weight in water vapour.

■ The fibres are fine and soft, so it has a good drape and can be very comfortable to wear.

■ Viscose has low strength, particularly when wet.

■ Elasticity is poor in all cellulose, man-made fibres, so it creases easily.

■ Electrostatic charge is very low because the fibres always contain some water.

■ It dyes easily and can be printed with bright colours, so it is a fabric of choice for high street fashion designers.

■ As it is made from cellulose it is biodegradable and recyclable.

■ It tends to shrink in water as the fibres absorb water and swell, but this can be reduced by treatment with synthetic resins.

Traditional methods of producing regenerated cellulose fibres involve the use of chemicals that are costly to recycle. Lyocell is the generic name for a solvent-spun regenerated cellulose fibre that is environmentally friendly.

> **Find samples of viscose fabrics and add them to your fabric swatch.**

Fig 13.19 A pair of children's pyjamas

> 1 The woven fabric used for the pyjamas is viscose.
> (a) (i) Give four reasons why viscose fibres are suitable for the fabric.
> (ii) The viscose fabric has been given a finish called 'brushing'. State one advantage and one disadvantage of applying this finish.
> (b) Using notes and sketches, describe how to apply the pocket onto the pyjamas.
> (c) Discuss the environmental implications of using regenerated fibres in the manufacture of textile products.

Cellulose man-made fibres can be classified according to the solvent used to convert the raw material into a spinable solution. Find out more about acetate and lyocell. Draw the cycle to show the production of lyocell.

Manufactured/synthetic fibres

Fig 13.20 An oil-drilling platform

Synthetic fibres are formed entirely by chemical synthesis from oil or coal sources. They were developed during the 1940s and 1950s and are now used to produce fibres that can be used in products as diverse as soldier's protective headwear to an elegant prom dress. Man-made, synthetic fibres can be designed and engineered to meet whatever functions and properties are required; they are the most innovative of all the fibres.

They are made from chemical units called polymers that are formed from single-molecule units called monomers, linked together like a bead necklace.

All synthetic fibres are produced from a non-renewable source and are not biodegradable.

Manufactured fibres – polyamide (nylon)

Polyamides are produced from chemical chips by melting and extruding into fibres. As they pass through the spinneret into a stream of cold air the filament fibres are stretched to give a fine, strong yarn. Staple fibres can be produced by chopping up the filament fibres. The shape of the spinneret can be varied to give textured fibres.

Fig 13.21 Polyamide sportswear for cyclists

Performance characteristics of polyamides

- Insulation depends on the type of fibre. Flat filament fibres do not trap air and have low insulation. Texturing increases the volume and gives better insulation. Staple yarns may be either fine and smooth or more bulky.
- Polyamides absorb very little water, which means that they are susceptible to electrostatic charge. It also means that they are easy to wash and dry.
- They are very strong, have excellent abrasion resistance, and are resilient and durable.
- Elasticity is good so products have good crease recovery.
- They are thermoplastic so can be textured and heat set.

- They are resistant to moulds and fungi, but will yellow and loose colour with long exposure to sunlight.
- Fineness ranges from microfibres (60 times finer than a human hair) to coarse fibres.
- Polyamides dye well.
- The handle of polyamides can vary. It can be soft and lightweight or coarse and firm.

Aramids

Aramids are a group of fibres in the polyamide category. They are highly flame retardant and able to withstand very high temperatures. Nomex® and Kevlar® are expensive fibres to produce and are used for fire fighters and racing drivers' clothing. Kevlar® also has a high strength to weight ratio – four times that of steel wire – which makes it suitable for bullet-proof vests, high-performance tyres and protective helmets.

Fig 13.22 Polyamide protective wear

- Find samples of polyamide fabrics and add them to your fabric swatch. Try to show the wide range of fabrics available.
- A local primary school is planning to celebrate a special event by providing all the pupils with a kite to fly. Design and make a themed kite from polyamide.

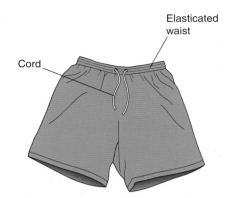

Fig 13.23 A range of shorts made from rip-stop nylon

1 (a) (i) Give four reasons why rip-stop nylon is a good choice of fabric for the shorts.
   (ii) Give three reasons why the elasticised waist is a useful design feature on the shorts.
   (iii) Give one reason why the designer has included a drawcord as well as the elasticised waist.
 (b) The shorts are to be batch produced. Describe, using a flow chart supported by annotated sketches, the order of manufacture of the shorts.
 (c) Discuss the implications of using modern materials in the production of leisurewear.

Manufactured fibres – polyester

Polyester is the most important synthetic fibre because it is very versatile and has a wide range of uses. The fibres, derived from petrochemicals, are macromolecules produced by a melt spinning process. The molten polyester is extruded into an air stream, which cools the melt and solidifies the filaments. It is inexpensive to manufacture and about 60 per cent of it is produced as staple fibres. It can be engineered to produce a wide variety of properties and characteristics.

Although polyester is made from a non-renewable source, the fibres can be produced from plastic bottles (it takes 25 bottles to make one Polartec® jumper).

Fig 13.24 Plastic bottles used for a Polartec® fleece

Performance characteristics of polyester

- Insulation depends on the type of fibre. Flat filament fibres do not trap air and have low insulation. Texturing increases the volume and gives better insulation. Staple yarns may be either fine and smooth or more bulky.
- Polyesters scarcely absorb water, which means that they are susceptible to electrostatic charge. It also means that they are easy to wash and dry.
- They are very strong, have excellent abrasion resistance, and are resilient and durable.
- Elasticity is very high so products have a good crease resistance.
- Polyester fibres are thermoplastic so can be textured and heat set. It is used for permanently pleated garments.
- They are resistant to most acids, alkalis and solvents, as well as moulds and fungi.
- Fineness ranges from microfibres to coarse fibres. They can be fine and soft or firm depending on the fibre fineness, fabric construction and finishing.
- Polyesters dye well.
- The handle of polyester can vary. It can be soft and lightweight or firm and stiff.

Staple fibres are used mainly as blends with other fibres, especially wool, cotton and viscose. They are used in a wide range of clothing and bedding. Filament yarns are usually textured and are used for dresses, blouses, ties, rainwear and linings.

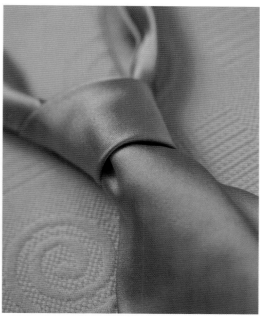

Fig 13.25 A polyester tie

- Find samples of polyester fabrics and add them to your fabric swatch. Try to show the wide range of uses from luxury to utility. Include some polyester blends.
- Design a range of themed hats and scarves to be made in Polartec® fleece for a major chain of supermarkets.

Manufactured fibres – acrylics

Acrylics are made from the petrochemicals propylene and ammonia and can be either wet or dry spun into staple fibres. They have a wool-like handle, low density and a good resistance to light and chemicals. Acrylic can be spun as microfibres. Acrylic fibres are made from a non-renewable source and are cheap to produce.

Performance characteristics of acrylics

- They are voluminous, very soft and warm and have a wool-like handle with good drape.
- Acrylics are wrinkle resistant, machine washable and dry quickly as they do not absorb moisture.
- Like all synthetic fibres they are thermoplastic

and can be easily deformed by heat; they shrink easily and must be ironed quickly!

Acrylics are used for blankets, furnishing fabrics, knitting yarns, fake fur and fleece.

Fig 13.26 A fancy dress costume made from acrylic

Manufactured fibres – elastomeric

Elastane is manufactured from segmented polyurethane. Its outstanding property is its elastic recovery. An elastomeric yarn can stretch by up to 500 per cent and recover its original length after the tension is released. Its molecular structure is composed of soft, flexible segments bonded with hard rigid segments, which allow for the extensibility of the yarn.

Elastane is produced as a filament yarn; it is very fine and is usually covered by another yarn. Bare elastane filaments are used in sheer hosiery and foundation garments. Covered yarns are used in fashion garments. They are unseen in the garments and never come into contact with the body, providing invisible elasticity and wrinkle resistance.

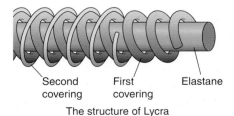

Second covering   First covering   Elastane

The structure of Lycra

Fig 13.27 The construction of an elastomeric yarn

Performance characteristics of elastomeric fibres

- Capacity to stretch up to five times its original length, making the fabric drape well.
- It adds comfort, softness and crease resistance to garments.
- Body shape is improved by wearing garments with elastane in them.
- It dyes well and is easy care.
- It can be manufactured to enhance the performance of other fibres, for example chlorine resistance and comfort in swimwear.

Manufactured fibres – flourofibres

Flourofibres (PTFE) are produced from a milky-white synthetic polymer that can be produced in films, staple fibres or filament yarns. It is produced in films as a microporous membrane about 0.02 mm (the thickness of domestic cling film) containing microscopic pores (holes). It is laminated onto to textiles or interleaved between two fabrics as in the production of GORE-TEX®, a breathable, water-repellent outdoor fabric.

Performance characteristics of flourofibres

- They are chemically resistant to water so are water repellent.
- Because they do not absorb any moisture they are practically undyeable.

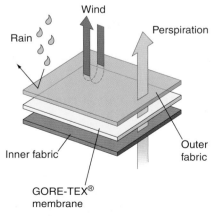

Fig 13.28 GORE-TEX®

- They are windproof and stain resistant.
- Flexible and durable.
- Breathable when use as a microporous film.

Flourofibres are used in the production of garments, upholstery, sportswear, work wear and shoes.

Manufactured fibres – chlorofibres

Polyvinylchloride (PVC) is manufactured as filament yarns and staple fibres but has only a limited use in clothing. Knitted thermal underwear is sometimes made from chlorofibres as they can be very warm to wear. They are also used in protective clothing as they have a high chemical resistance.

Performance characteristics of chlorofibres

- Strong, flexible and durable.
- Provides good insulation.
- Thermoplastic, so care must be taken when laundering.

Chlorofibres are used in weatherproof coatings, raincoats, shower curtains, underwear and active sportswear.

1 (a) Elastane fibres, such as LYCRA®, have had a major impact on sportswear design. Explain using examples the effect of LYCRA® on the performance of textile products.
   (b) Explain why elastane fibres are never used on their own to make fabrics.
   (c) Describe how elastane fibres are combined with other fibres.
   (d) Describe how to use and care for fabrics containing elastane.

Inorganic manufactured fibres – glass

Glass can be manufactured into filament yarns and staple fibres to produce long lasting, durable, woven-mesh glass fabrics. Glass textiles are used by the aerospace and military industries to produce flame and heat barriers, as well as light and ultraviolet filters. A glass textile was used for the roof of The O$_2$ Arena (Millennium Dome) in London.

Fig 13.29 The O$_2$ (Millennium Dome)

Performance characteristics of glass fibres

- They are heat and cold resistant
- Non-toxic and non-stick
- Resistant to chemicals, mildew and fungus
- They can bring reflective properties to garments

Other uses of glass textiles include architectural roof coverings, sterile wall coverings for hospitals and protective garments.

Inorganic manufactured fibres – metal

Aluminium, copper and steel are the metals most commonly used for textile construction. Silver and gold can also be used for more expensive applications. Traditionally used in industrial clothing and medical areas, metal fibres have more recently been used in decorative textiles, fibres for fashion and active sportswear garments.

Thin sheet metal can be slit into fine strips to form a delicate yarn, or the wire can be wound tightly round a more traditional yarn such as cotton. Lurex® is a metallised polyester yarn that is given a fine plastic coating to prevent tarnishing. Lamé is a fluid, sensuous and dramatic fabric made from flat metallic thread used for evening and stage wear.

Performance characteristics of metal fibres

- Strong, lightweight and abrasion resistant
- Protects against electromagnetic pollution, such as potentially harmful waves from mobile phones
- Conductive metal fibre within fabric is detectable by radar and heat-seeking devices:

this could provide protection to mountaineers or skiers in adverse weather conditions. Conductive properties also protect against static electricity
- Anti-microbial: it is used in medical and therapeutic applications

Add a selection of the synthetic fibres to your fabric swatch.

Section C   Yarns

Learning outcomes

By the end of this section you should have developed a knowledge and understanding of the performance characteristics, selection and use of staple and filament yarns:

- Yarns as single yarn
- Spun yarns
- Filament yarns
- Finishes applied to yarns and fabrics produced

Introduction

Yarn is a continuous length of fibres or filaments that have a substantial length, with or without a twist.

Spun yarns

Spun yarns are made by twisting together (spinning) staple fibres. Spinning is used for staple fibres from cotton, wool and flax; any short fibres from silk; and any broken or cut synthetic fibres.

Filament yarns

Filament yarns can be made from one or more filaments; they can be made from silk or synthetic fibres:

- Monofilament yarn consists of a single continuous filament
- Multifilament yarn is a yarn made from several filaments with or without a twist

Table 13.2 Yarn construction

Staple yarns	Filament yarns	
	Monofilament	**Multifilament**
Spun yarn	Monofilament	Flat filament
		Plied filament
From staple fibres such as cotton, flax, wool and silk	Microfilament consists of a single continuous yarn. Synthetic fibres.	Synthetic fibre
		Reeled silk

Twist

Twist is put into yarns during spinning to make them stronger so that they are suitable for weaving or knitting. Yarns made from twisted fibres vary according to the direction of the twist and the twist level. Yarns can be spun clockwise (Z twist) or anticlockwise (S twist). Light is reflected in opposite directions from Z and S yarns, so striped effects can be achieved in fabrics.

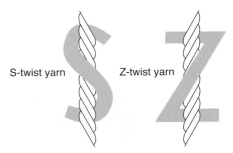

S-twist yarn   Z-twist yarn

Fig 13.31 S twist and Z twist

The twist level determines how the yarn can be used. Low-twist yarns are softer, weaker and more bulky, making them suitable for weft yarns and for knitting. High-twist yarns are strong and hard, making them suitable for warp yarns in weaving as they can stand the tension in the loom.

Core-spun yarns are multi-component yarns with one filament in the centre and staple fibre yarn spun around it. A good example of this is stretch yarns with elastane filament as the core, covered by non-elastic natural or man-made yarn such as cotton, polyamide or wool. Core-spun yarns enhance fabrics, making them more comfortable to wear. Think about your pairs of jeans and how comfortable they are when they have five per cent elastane added to the fabric.

Sewing threads are often made from a polyester core covered with a cotton yarn.

In the design of textile products yarns are first selected on the basis of the performance characteristics of the fibres they are made from. However yarns may also be selected for their appearance.

Fig 13.32 Sewing threads

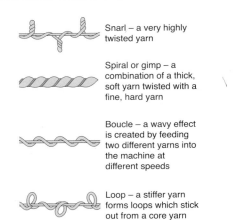

Snarl – a very highly twisted yarn

Spiral or gimp – a combination of a thick, soft yarn twisted with a fine, hard yarn

Boucle – a wavy effect is created by feeding two different yarns into the machine at different speeds

Loop – a stiffer yarn forms loops which stick out from a core yarn

Fig 13.33 Yarns

Key Point

The fibre and spinning method affect the final performance characteristics and appearance of the yarn.

Fancy yarns are produced by special spinning processes to give:

- Texture in the construction, for example slub, loop, chenille and bouclé
- Colour effects by mixing different coloured yarns
- Lustre effects, for example mixing with Lurex®
- Bulked effects from yarns that are a blend of different staple fibres, for example acrylic and cotton to give an inexpensive yarn that is warm, lightweight, easy care, and with a soft handle
- Textured yarn made from thermoplastic synthetic filament yarns that are heat processed to give crimps, coils or loops. This adds bulk and makes the yarn warmer, more elastic, absorbent and gives a softer feel

Examine a selection of yarns from some fabric samples. Pull off some warp and weft threads and pull them apart. Examine the yarns closely. Untwist the yarns and describe their structure (you could use a microscope). Try to identify a single fibre. You should be able to see if it is staple or filament, whether there are any fabric blends, and any bulking or texturing that has taken place. Record the details in your fabric fact file.

1 Describe how the insulation properties of fibres, yarns and fabrics could be improved.

Section D   Mixtures, blends and laminates

Learning outcomes

By the end of this section you should have developed a knowledge and understanding of the reasons for mixing and blending fibres and yarns, and the performance characteristics of the yarns and fabrics produced.

Introduction

Even in ancient times, fabrics were made from two types of fibres. It was common to weave fabrics with one type of fibre for the warp and then mix another fibre for the weft. This was called 'union fabric'. Union fabric gave the textile designer the opportunity to achieve striking effects. True blending is achieved by spinning yarns from a blend or mixture of two or more types of fibre, or by mingling two types of continuous filaments to produce a multifilament yarn.

A laminated fabric is produced by two or more layers of fabric joined together with resin, rubber, foam or adhesive to form one material. The laminate is applied in sheet form, for example a weatherproof membrane.

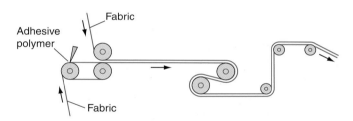

Fig 13.34 Laminating fabric using adhesive

The reasons for blending, mixing and laminating are:

- To combine the most desirable properties of different fibres into one fabric, for example polyester cotton
- Effects of texture and handle can be enhanced, for example the addition of wool increases warmth and fullness
- Novelty effects can be achieved by dyeing fabrics from blended yarns. One fibre component may remain undyed or dyed a different colour: this is known as cross dyeing
- The cost of producing fabric is always important and blending can be used to control the price, usually to reduce the cost.

Common blends are polyester and wool, polyester and viscose, polyester and cotton, wool and acrylic, and polyester and linen. If you examine a range of textile items you will see that the list is endless.

GORE-TEX® is a hydrophobic membrane that is laminated onto any type of synthetic or natural fabric.

Examine six textile items that are made from blended fibres. Photograph them or download images of similar products, then mount and annotate the pictures to explain why the specific fibre blends were chosen for that product.

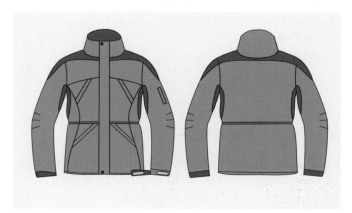

Fig 13.35 Waterproof, breathable jackets

1 (a) Give two performance characteristics the fabric for the jacket should have.

(b) Give two important design features of the jacket that make it suitable for outdoor pursuits.

(c) Describe, using annotated sketches, how a breathable/waterproof layer system works.

(d) Discuss the benefits to a designer of using a prototype to test the performance of an outdoor pursuits jacket.

Section E | Fabric construction

Learning outcomes

By the end of this section you should have developed a knowledge and understanding of the performance characteristics, selection and use of a variety of fabrics:

- Woven fabrics – plain, twill, sateen, loop and cut pile
- Knitted fabrics – warp knits, such as single and double jersey, and weft knits, such as tricot and lock knit
- Non-woven fabrics – felts and bonded fabrics

Introduction

Textile fabrics can be made by a variety of methods:

- From yarns, as in woven and knitted fabrics

- From fibres, as in the non-wovens, felt and bonded fabrics
- By a combination of both fibres and yarns, as in stitch bonded and laminated

Woven fabrics

Woven fabrics are made by interleaving two sets of yarns, warp and weft, at right angles on a loom.

The warp yarns are those that lie in the length direction of a fabric while it is being woven. These are the stronger yarns.

The weft yarns are those that are introduced between the warp yarns across the width direction of the fabric.

Examine a piece of woven fabric. Identify the selvedge, the warp and weft, the straight grain and the bias.

Plain weave

Fig 13.36 Plain weave

Plain weave is the simplest and tightest method of weaving. Each warp yarn is lifted over alternate weft yarns. Typical plain weave fabrics are calico, taffeta, poplin, muslin and rip-stop nylon.

Twill weave

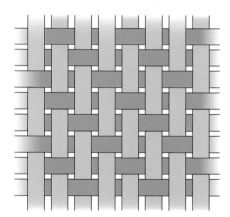

Fig 13.37 Twill weave

Twill weave makes a pattern of diagonal lines. Each warp yarn lifts over or remains under more than one weft. The twill order of interlacing causes diagonal lines to appear on the fabric. Weaving twills in opposite directions produces weave variations such as herringbone.

The most common twill weave fabric is denim, woven from strong, hardwearing cotton. Usually the warp is dyed blue and the weft white.

Satin weave

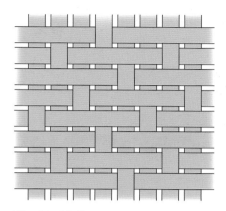

Fig 13.38 Satin weave

Satin weave has a lustrous appearance and good drape due to the fact that the warp floats over four or more wefts and remains under only one.

Satin fabric has a smooth, shiny finish; the weft only show on the back, which is dull. The pattern of the weave is random so that there is no twill line generated.

Examples of satin weave fabrics include satin, sateen and duchesse.

Loop and cut pile

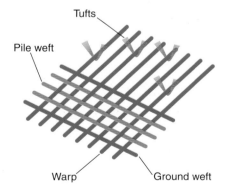

Fig 13.39 Velveteen

Loop and cut pile fabrics are made by creating a loop of yarn from an additional warp or weft. An additional warp is used in velvet or a weft yarn is used in velveteen to form a cut pile on the face of the fabric, which is called a nap. The third yarn forms the cut pile. For example, in velveteens the pile weft floats on the surface of the fabric; after cutting the tufts are held by the weft yarns. If the pile yarns are between the same warp yarns a corded effect is achieved.

Warp section

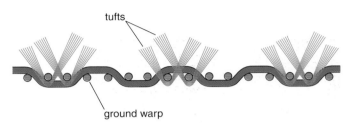

Fig 13.40 Corduroy

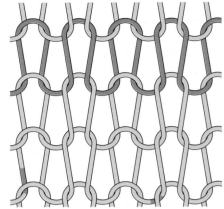

Fig 13.41 Weft knitting

Woven terry towelling is a plain woven cotton fabric made from two warps. The basic warp (ground warp) is highly tensioned and the pile warp is loose. The loose pile warps form loops on either one or both sides of the fabric as it is being woven.

The loop increases the absorbency of the cotton towelling by increasing the surface area.

- ■ Make your own samples of different weaves using paper, yarns or fine ribbons.
- ■ Research how velvet and towelling are woven.
- ■ Denim is a hard-wearing twill weave. Recycle a pair of old jeans into a new product. Look on the website for examples of students' work on this.

Knitted fabrics

Knitted fabrics are made from interlocking loops, formed from either a single yarn or many. They are classified into weft knitted and warp knitted.

Weft-knitted fabric

Weft-knitted fabric is made from a continuous length of yarn that is fed crosswise to the length of the fabric. It can be unravelled and may ladder. The knitting needles can work in sequence or all together. Single jersey is made with one set of needles. The two sides of the fabric have a different appearance. It produces a soft, comfortable fabric with variable elasticity,

but the fabric does curl at the edges. It is used for T-shirts, underwear and sweaters.

Double jersey is made on two sets of needles opposite each other. Double jersey fabrics are more compact, durable and retain their shape, but they are not very elastic. It is used for T-shirts, polo shirts, sportswear, skirts and leggings.

Warp-knitted fabric

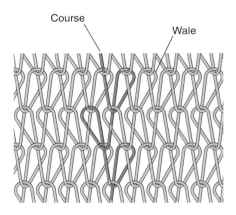

Fig 13.42 Warp knitting

Warp-knitted fabrics are made on straight or circular knitting machines and use more fabric than warp yarn. A separate warp yarn is fed to each needle producing a loop that interlocks with the loop below along the entire length of the fabric.

The main category of warp knitting is tricot, producing 50 m of fabric of 4 m wide in one hour. It uses mainly synthetic yarns to produce fabric for lightweight furnishings and underwear. Locknit is a combination of tricot and plain knit stitches – it produces a lightweight, smooth, non-laddering warp knit used for lingerie and linings.

Warp knits cannot be unravelled and do not usually ladder, but their use is limited mainly to swimwear linings, laces, ribbons and trimmings. Geotextiles used in the construction industry are often warp knitted.

> Obtain some knitting needles and ask someone to show you how to knit.

Non-woven fabrics

Non-woven fabrics are made directly from fibres that are initially entangled to produce a web. There are two types of non-woven: felts and bonded. In felts the web is strengthened by the entanglement of fibres, while in bonded the web is strengthened by adhesives. Felt has no strength, drape or elasticity, but it does not fray and it is warm and resilient. It is cheap to produce and, as it has no grain, pattern pieces can be placed in any direction, making it very economical to use.

Felted wool

Wool and other animal fibres are entangled to produce a fibre web using a combination of alkaline chemicals, water, heat, pressure and repeated mechanical action. It is used for hats, slippers and toys.

Needle felt

Needle felt can be made from any type of fibre but mostly synthetic fibres. The bulky fibre web is repeatedly punched with hot barbed needles that drag the fibres to the lower side of the web. This is used for waddings, upholstery, mattress covers and filters.

Bonded fabric

Bonded fabrics are made from a web of fibres bonded with adhesives, solvents or by heat. This kind of fabric is used mainly for interlinings, disposable cleaning cloths or disposable hospital clothing.

Fig 13.43 A non-woven bag made in Nepal

Collect samples of different woven, knitted and non-woven fabrics. Explain the properties of the fabric and what textile products it would be suitable for.

1 (a) Describe, using annotated sketches, how to weave a checked fabric.
  (b) Describe, using annotated notes and sketches, how to weave denim.
  (c) Explain why felt is commonly used in the manufacture of soft toys.

Learning outcomes

By the end of this section you should have developed a knowledge and understanding of new forms of fibres and fabrics and their applications in product design.

Introduction

There are exciting developments in the field of textile technology as the industry responds to consumer demands and changes in our society. You will probably have already researched some of the new fabrics available and it will be impossible to give you details of them in the space available here!

Textile designers have to keep themselves up to date and you need to do the same. Newspapers, magazines and the internet are good resources to use.

Modern materials

This includes fibres and fabrics with special properties that have been developed recently, such as the cellulose fibre lyocell and Tencel®, made from regenerated cellulose. Kevlar® and Nomex® are recent developments in polyamides that provide technical performance. Flourofibres are used in the production of GORE-TEX®.

Fig 13.44 A fire officer in protective wear

Smart fabrics

Fabrics that not only look and feel good, but which also respond to the environment, providing added properties such as being waterproof, windproof and breathable. They respond to the needs of the user and may even incorporate microelectronics. Simpatex® allows water molecules of perspiration to pass through but does not allow even the smallest drop of water to penetrate. Thermochromic dyes respond to heat; photochromic threads respond to light.

Performance textiles

These relate to a product's performance in a specific end use and include products used for sport and outdoor activities. Fastskin®, developed for use by Olympic swimmers combines fibres, knit and garment design to replicate the skin of a shark, enabling movement through water with less resistance. Stomax® fabric takes its inspiration from the way that leaves breathe and regulate temperature: it keeps the wearer dry and comfortable during exercise or exertion.

Industrial textiles

These are manufactured to meet technical specifications. Clothing that is resistant to cutting, tearing and abrasion for use by the emergency and military services is now used in upholstery and in clothing for skate borders and climbers. A polyethylene material called

Dyneema® is ten times stronger than steel: it is used for bullet-proof vests as well as bike helmets and building materials.

The average car contains 13 to 14 kg of textiles. This includes filters, tyres, pipes, hoses and belts, as well as upholstery. Geotextiles are widely used in landscaping.

- Keep up to date with developments in the textile industry. Collect articles from newspapers, magazines and the internet.
- Research textiles that are reflective, give antibacterial protection, have anti-allergenic properties, moisture management, chemical protection, have thermal insulation properties, ultraviolet protection, fire and spark resistance. You may be able to add more to the list yourself. Show how these exciting materials can be used in creative, innovative ways.

1 Explain how new fibres and fabrics have been developed to aid the performance of textile products.

Fig 13.45 Medical teams make use of innovative materials

Fig 13.46 Mountaineer in high-tech outdoor wear

Section G · Surface finishes

Learning outcomes

By the end of this section you should have developed a knowledge and understanding of:

- The nature and suitability of surface finishes and coatings across a range of products
- The functional purpose of surface finishes, such as resistance to water, stains, decay and wear; to improve insulation properties; crease resistance; anti-shrink; easy care; anti-static; flame resistance; anti-pilling; laminating; brushing; moth and rot proofing; and hygienic sanitising
- Finishes that enhance appearance, such as calendering, stone and sand washing
- Printing methods, such as screen, transfer, roller and digital printing
- Dyeing methods, such as batch and resist

Introduction

Dyeing and finishing are carried out to improve a product's appearance, properties and quality. They cover many different processes: some mechanical, some biological and some chemical. Finishing is carried out most efficiently on fabrics but, if the fabric is to be colour woven, dyeing must take place at the fibre or yarn stage.

Mechanical finishing processes

Mechanical finishing uses heat, pressure and rollers to improve the appearance of the fabric.

- Brushing cotton or polyamide fabrics makes them fluffy and warm, with a soft handle. The fabrics pass through rollers with wire brushes that lift the fibres to form a nap
- Calendering is the industrial equivalent of ironing: it smoothes the fabric and improves its lustre
- Engraved calendar rollers are used to emboss relief patterns on the fabric surface
- Heat-setting is used for thermoplastic fabrics (polyester and polyamide): the fabrics are set in permanent shapes or pleats

If you have a heat press in your school you should be able to experiment with pleating and creasing synthetic fabrics to create exciting effects.

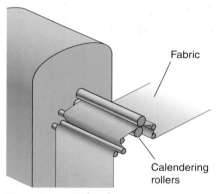

Fig 13.47 Calendering

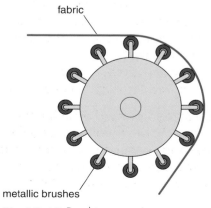

Fig 13.48 Brushing

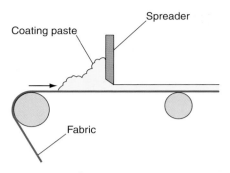

Fig 13.49 Coating

Chemical finishing processes

Chemical finishing involves the application of chemical solutions or resins to improve the appearance, handle or performance of a fabric:

- Stain resistant – silicones are applied to resist water-based stains, synthetic resins are applied to fabrics to resist oil-based stains
- Water repellent – spraying with water-repellent chemicals, for example silicones; can be temporary or permanent
- Shrink resistant – wool can be given a shrink-resist finish using silicone or Teflon®; this results in soft, smooth, lustrous yarns and fabrics that are machine-washable
- Crease resistant – cotton and viscose fabrics are given a crease-resistant finish using resin. This makes them easy care: fast drying and requiring little ironing
- Flame resistant – PROBAN® and PYROVATEX® are two chemicals used to make fabrics difficult to ignite. Children's nightwear and cotton/viscose furnishings must, by law, be given a flame-resistant finish
- Hygienic – anti-bacterial chemicals that hinder the growth of bacteria are applied to hospital fabrics and wall coverings; they also control odours in sports shoes and reduce infection in medical products

Coating

Coating involves applying a layer of polymer to the surface of a fabric. Teflon® coating makes fabrics stain resistant, water repellent and breathable. PVC-coated cotton is a good example of this.

Biological finishes

Biological finishes use natural enzymes to change a fabric's appearance. Bio-stoning gives a stone-washed finish to denim fabrics.

Dyeing

Batch dyeing

Before fabrics can be dyed they must be prepared, cleaned, faults rectified, contaminants removed and then bleached. Cotton or linen fabrics are also mercerized: treated with caustic soda under tension to make the fabric stronger, dye well and improve the lustre.

In batch dyeing a specific weight of fabric is treated in a machine containing a specific amount of dyestuff. In the jigger process the smoothly spread fabric is led backwards and forwards through the dye bath giving uniform colour distribution.

Resist dyeing

Methods include tie dyeing and batik. These are traditional Eastern methods of hand dyeing, where a method is used to stop the fabric taking up the dye. Tie dyeing uses tightly bound string to resist the dye; in batik, wax is drawn or painted onto the fabric.

You will probably have done some tie dyeing in previous art or textile lessons. Produce a piece of tie-dyed fabric and then experiment with other decorative techniques.

EXAMINER'S TIPS

Make sure that you can clearly explain, using annotated sketches, the stages in the production of a tie-dyed or batik fabric.

The stages in batik dyeing

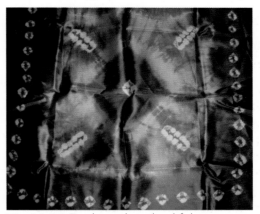

Fig 13.50 Traditional tie-dyed fabric

Fig 13.51 Traditional batik fabric

Printing

Screen printing

This is the most popular method of putting defined areas of colour onto a fabric.

The design is formed onto a screen by blocking off those parts of the screen where no printing is to occur. The screen is painted with a light-sensitive polymer and then selectively exposed through a stencil; unexposed areas are washed away. A modern alternative is to coat the screen with an insoluble polymer that is selectively etched away by a computer-driven laser. A separate screen is required for each colour.

Most fabric is rotary screen printed, which allows continuous production. Some is still done by flat screen printing, lowering successive screens onto the fabric, pressing the dye through the screen with a squeegee blade then moving the fabric onto the next screen.

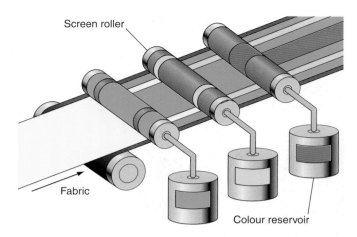

Fig 13.52 Rotary screen printing

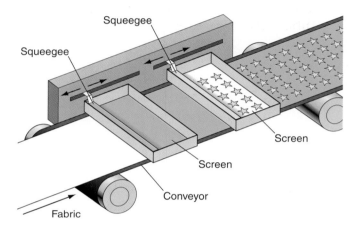

Fig 13.53 Flat screen printing

Transfer printing

The design is first printed onto a special type of paper with dyestuff. The pattern is then transferred to the fabric with the aid of a pressurised, heated calendar. The temperature is very high and causes the dye to pass into the vapour stage (sublimation). The vapour diffuses into the fabric.

You may have used transfer printing paper using an ink-jet printer and transfer paper in your workshop.

Roller printing

This is the oldest mechanised method of printing. The designed is engraved onto copper rollers, one for each colour. The fabric travels round a main cylinder and the dye paste is in a trough. The dye is transferred from the trough onto the fabric via the engraved roller.

Digital printing

This uses ink-jet printers to generate computer-aided designs directly on to lengths of fabric using special printing inks. It is an ideal method to use to produce samples or one-off designs.

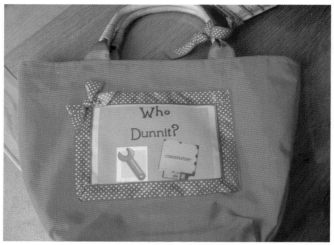

Fig 13.54 A bag made using transfer paper printing

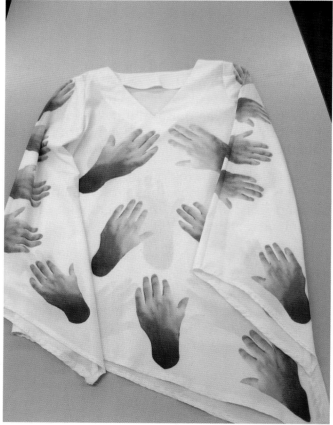

Fig 13.55 A kaftan printed using photographs of hands.

EXAMINER'S TIPS

You can make your own products exciting by printing your own fabric design.

- Experiment with fabric design:
  - Use drawing software (CorelDraw, MS Paint or similar), create a fabric design for a new range of interior design products.
  - Repeat the pattern.
  - Experiment with different colourways.
  - Use either ink-jet transfer paper or a sublimation printer and heat press to print your own fabric.
  - Produce a design sheet for a prospective customer with some product designs using your fabric.
  - Design and make a bag using your fabric design.
- Research the advantages and disadvantages of the different methods of printing and dyeing.

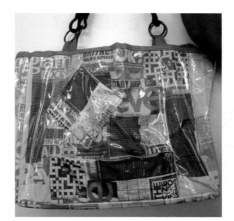

Fig 13.56 A bag produced by a student

1 Describe, using annotated sketches, how fabric is screen printed.

Section H Manufacturing processes

Learning outcomes

By the end of this section you should have developed a knowledge and understanding of:

- The production and use of pattern pieces to cut fabrics to accurate sizes and shapes, including plotters/cutters, laser cutters, water jet and die cutters
- Seams – plain/open, double-stitched and French seams
- Seam finishes – hems, facings, bias binding and overlocking
- Disposal of fullness – gathering, darts, pleating, tucks and casings
- Fastenings – zips, buttons and buttonholes or loops, velcro, poppers, parachute clips, hook and eyes, laces and ties, buckles and toggles
- Decorative techniques – appliqué, quilting, free machine embroidery and computer-controlled stitching
- Reinforcing materials to improve strength/durability – laminating, quilting, interlinings, reinforced stitching and taping of seams

Introduction

A pattern is a diagrammatic representation of the way that a garment is constructed. Flat pattern pieces are joined to make a three-dimensional shape.

The production and use of pattern pieces

Basic block patterns

Basic block patterns based on a British Standard using anthropometric data are used in the industry as the basis for designs. From the basic block, a designer will develop a working pattern to trial the garment. The final pattern will include all the seams, hems, grain lines and information needed to make the product.

Pattern development

Pattern cutting is now carried out using computer software that stores the basic blocks and allows new designs to be digitised and developed on screen. Once a design has been developed it will need to be altered to fit a range of sizes. This is called grading. It means the increase or decrease of a master pattern to create larger or smaller sizes without changing the design, shape or appearance. Computer-aided grading systems do this calculation very quickly.

Fig 13.58 A student adapting a pattern

Lay planning

In the next stage, the pattern pieces of all the individual components of the garment have to be laid as economically as possible onto the fabric. Account must be taken of the directional properties of the fabric such as pattern direction, nap or pile or to match stripes, checks or designs. The final lay plan is called the cutting marker. In a fully automated system the lay plan will be calculated by a computer programme and transmitted to the laying and cutting system. Plotters draw full-scale patterns for manual cutting.

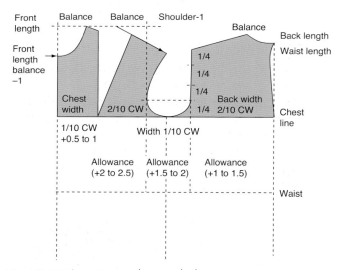

Fig 13.57 A pattern to be graded

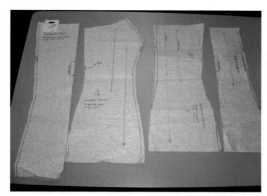

Fig 13.59 A student calculates their lay plan

Take apart a simple garment such a baby vest. Draw your own pattern pieces for a new one. Use newsprint or similar paper to draw and calculate the most economical lay plan for ten vests on 150 cm-wide fabric.

Spreading

This is the term given to how the fabric is laid out for cutting. Manual spreading is used for short lays (small numbers) and where changes of fabric are required. Automatic spreading is used for long lays (a large production run) and the fabric is spread in multiple plies (layers) for high-volume production.

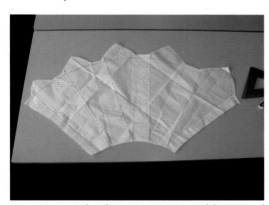

Fig 13.60 A bodice pattern spread by a student to include large folds and tucks

Cutting

Cutting means to cut out a garment from lays of fabric with the help of cutting templates. It is often carried out in two stages. Rough cutting which separates out the individual pieces and final cutting which accurately cuts the precise pieces.

Fabrics can be cut by:

- Manually controlled cutting blades using band knives, straight knives or circular cutters are the traditional way of cutting.

Fig 13.61 Cutting with straight knives

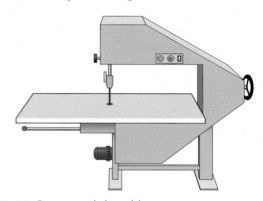

Fig 13.62 Cutting with band knives

- Die cutters which are mainly used for coated and laminated fabrics where the same patterns are used over a long period of time, eg work wear. The fabric pieces are stamped out on a base plate
- Fully automated computer controlled systems using special vertical knives, laser beams or water jets to cut the fabric.

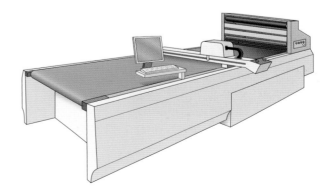

Fig 13.63 Automatic cutter

Disposal of fullness

Textile products need to have shape; this is achieved not only by the cutting but by techniques that dispose of fullness.

> **EXAMINER'S TIPS**
>
> You need to be able to describe all of the different techniques shown on the following pages with the aid of annotated drawings.

Gathers

Gathering can be done by hand or machine. Two parallel lines of even stitches are pulled up to evenly draw in the fullness to the required amount.

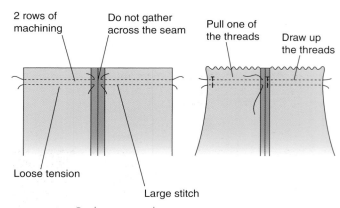

Fig 13.64 Gathering stitches

Fig 13.65 Student's work showing a gathered frill on a duvet cover

Darts

Darts are folds of fabric that end in a point, giving a smooth shape. They are used extensively in garment design. There are three main types:

- Single pointed, as used in the waist of skirts or trouser
- Double pointed, generally used on dresses where there is no waist seam
- Dart tuck, which is an opened dart used as a feature on blouse shoulders or waists.

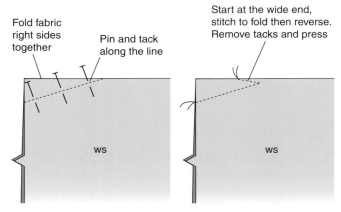

Fig 13.66 Darts

Pleats

Pleats are used to give shape with fullness in a garment by forming folds in the fabric. This can be done in synthetic thermoplastic fabrics by applying heat to make the pleats permanent where a creased line is firmly pressed in place. Unpressed pleats give a soft, folded look to a garment. Types of pleat include knife pleats, box pleats, inverted pleats and kick pleats.

(a)

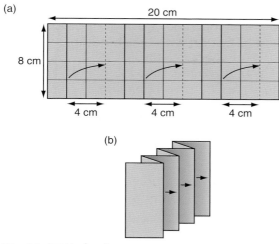

(b)

Fig 13.67 Knife pleating

- Experiment with different types of pleats using paper. Make box, inverted and kick pleats.
- Find examples of pleats in a range of textile products.

Tucks

Tucks are narrow folds held in place by a line of machining on the tuck. Very narrow tucks are referred to as pin tucks and are often used as a decorative finish as well as to dispose of fullness.

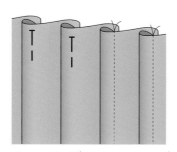

Fig 13.68 Experimenting with tucks

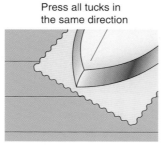

Press all tucks in the same direction

Fig 13.69 The stages in making tucks

Elasticated casing

This is an alternative to gathering where increased flexibility is needed. It is often used in the waist of children's garments where no other fastening is required. A casing is made either by folding the fabric over twice and machining or by machining a binding in place. This also neatens the top edge. Elastic is threaded through the casing.

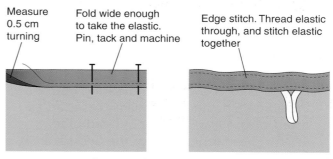

Measure 0.5 cm turning

Fold wide enough to take the elastic. Pin, tack and machine

Edge stitch. Thread elastic through, and stitch elastic together

Fig 13.70 An elasticated casing

Use old lampshade frames to model ways of incorporating fullness and interest. Develop new lampshade designs.

Fig 13.71 Modelling using lampshades

Pillow case with decorative design

Duvet cover

Main section

Pleated valance sheet

Fig 13.72 A coordinating bed set

1 Figure 13.67 shows a coordinating bed set including a duvet cover, pillow cases and a fitted, pleated valance sheet dyed blue.

(a) (i) The fabric used for the bed set is 50 per cent cotton and 50 per cent polyester fibres. Give two reasons why this is a good combination of fibres for the product.

(ii) Name two suitable fastenings for the duvet cover and state a benefit of each.

(b) The lower section of the valance is pleated. Describe, using annotated sketches, how to pleat the lower section and attach it to the main section of the sheet.

(c) Describe, using annotated sketches, the industrial process used to dye the fabric.

(d) Discuss the health and safety implications associated with industrial methods of dyeing.

Seams

Seams are used to permanently join two pieces of fabric together. Temporary joining methods include pinning and tacking (basting). This is then removed after the seam has been machine stitched permanently. Tacking is a row of alternate long and short stitches, usually done by hand.

Plain seams

Plain seams are used to give a flat finish. They are used on all thick fabrics and need neatening.

They can be neatened by turning a very narrow hem under and machining, using a zigzag stitch, overlocked or bound with bias binding.

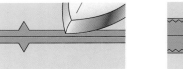

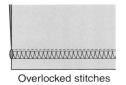

Stitch along the seam line and press open

Zig zag neatened

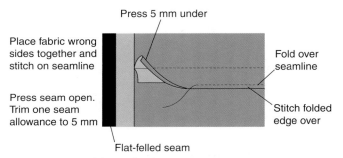

Overlocked stitches

Fig 13.73 Plain seams

Double-stitched seams

Double-stitched seams are used where strength is needed: the most common use is for jeans.

Press 5 mm under

Place fabric wrong sides together and stitch on seamline

Fold over seamline

Press seam open. Trim one seam allowance to 5 mm

Stitch folded edge over

Flat-felled seam

Fig 13.74 Double-stitched seams

French seams

French seams are particularly good on sheer or very fine fabrics. Seams are very neat as no unfinished edges of fabric show.

Stitch plain seam 1 cm from edge. Trim seam allowance to 3 mm

Turn to right side. Press flat. Stitch exactly on the seam line 4 mm away

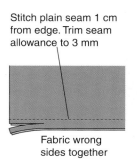

Fabric wrong sides together

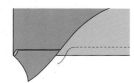

Fig 13.75 French seams

> Practice your seams by designing and making an organiser for a child's bedroom. Incorporate one or more decorative techniques.

Fastenings

Fastenings are mainly functional to close an opening but are sometimes used as a decoration. They are a temporary method of joining fabrics.

Zips

Zip fasteners are a neat, firm way of giving an edge-to-edge fastening. For lightweight and fine fabrics a plastic zipper should be used. For trousers the tab of the slider has a locking device. Metal zippers for casual and sports wear are heavier, broad and firm. Zips can be open ended, such as those used to fasten jackets, or closed for inserting in seams.

> Use recycled zips to practice inserting a zip into different places. Try at the waist of a skirt, a fly front, and in a seam in the back of a cushion.

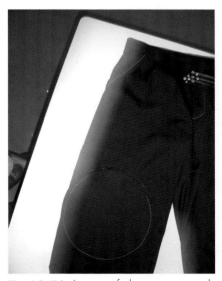

Fig 13.76 A pair of disco trousers designed and made by a student: they have secret zipped pockets in the knee and turn ups to conceal a mobile phone, keys and money

Buttons, buttonholes and loops

Buttons are used for decoration as well as for fastenings. Interesting effects can be achieved by grouping buttons or choosing contrasting colours. Modern buttonholes are worked by machine using built-in buttonhole programmers and specially designed foot attachments.

> - **Use your machine handbook to practise working buttonholes.**
> - **Describe, using annotated sketches, how to work buttons and buttonholes.**

Velcro

Velcro is a fastener made from two layers of polyamide. One layer has a furry surface and the other lots of tiny hooks. By lightly pressing the two together they will stick and give a strong bond. It makes an easy fastening for children's garments and shoes.

Fig 13.77 Trainers with a velcro fastener

Other fasteners

- **Poppers** used on the front of jackets are both functional and decorative
- **Press studs** are used for concealed extra fastenings; they are not very strong on their own
- **Parachute clips, hooks and eyes** are used in positions of strain; a good example is at the top of a zip
- **Laces and ties** give a decorative way of fastening; **rouleau loops** are fastened with laces or used with buttons

- **Buckles** used on belts
- **Toggles** used on sports and outdoor pursuits clothing and equipment

Fig 13.78 A selection of fasteners for a school bag

Design and make a child's educational toy. Experiment with different types of fasteners.

1 (a) Give four justified design requirements for the components used on a school bag.
  (b) Describe, using annotated sketches and notes, how ergonomics would be considered when selecting components for a school bag.
  (c) Discuss the implications of a manufacturer purchasing components from different parts of the world.

Decorative techniques

Appliqué

Appliqué means applying one fabric on top of another. Appliqué was traditionally carried out by hand but manufacturers use industrial machines. The stages in appliqué are:

1 Using patterns or markings to cut out the appliqué design in fabric.
2 Applying interfacing or Stitch'n Tear to the underneath of the background fabric to strengthen it.
3 Positioning the fabric design by pinning and tacking, or by ironing Bondaweb onto the right side of the background fabric (tip: the heat melts the Bondaweb and holds the pieces in place).
4 Machine stitching in place using a zigzag stitch or, if hand sewing, using a blanket stitch or herringbone.
5 Trim and press.

Fig 13.79 Examples of students' work using appliqué

Quilting

Quilting involves the sewing together of two fabrics with a thick layer of wadding in between. Quilts are usually filled with polyester wadding, which is warm, washable, lightweight and easy to work with. Quilting is essentially functional, to add insulation to the product, but it adds decoration as well. The process of quilting is:

1 Selection of a suitable fabric for the backing, usually the same or a cheaper, lighter fabric than the top.
2 Assembling the three layers in the correct order (lining/wadding/top layer).
3 Tacking the three layers together, working from the centre and ensuring the fabric is flat.
4 Using a special quilting attachment on the machine to keep rows parallel or to your own design.
5 Keeping the fabric very flat as it is machined; a frame can be used to do this.
6 Stitching lines can be marked on using tailors pencil or something similar.

Make a wrap and roll to store a small item. See the examples in Figure 13.80.

Free machine embroidery

Free machine embroidery has developed from the traditional hand processes and is carried out as follows:

1 The presser foot is removed from the machine and the feed teeth are lowered.
2 The fabric is stretched tightly in an embroidery ring, using it upside-down to keep the fabric flat on the bed of the machine.
3 The fabric is positioned under the needle and the presser foot is lowered.
4 The fabric is moved around under the needle to fill in the design.
5 Straight or small zigzag stitches are used for small areas and wide zigzag for larger areas.

Computer-controlled stitching

Industrial embroidery machines are usually controlled by computer software and are used to embroider on a large scale. The process is:

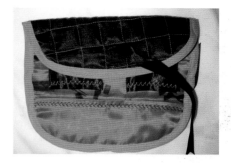

Fig 13.80 Quilted holdalls

1 The design is produced using appropriate software.
2 The design is transferred to the machine.
3 Fabric with an interfaced backing is inserted into a frame.
4 The frame is attached to the machine.
5 The machine stitches the design automatically, either selecting colours or prompting the operator to change the colour.

Design a logo for a new children's range of playwear. If you have access to an embroidery machine use CAD/CAM to produce an example. If not, try to produce it using free machine embroidery.

Reinforcing materials to improve strength/durability

Laminating

Laminating combines two or more layers of different materials that are bonded together by adhesive or by the thermoplastic quality of one or both of the materials. The fabrics are permanently fused together by melting an adhesive resin that is applied to one of the fabrics. Examples of laminating include:

- Iron-on interfacing, which is bonded to another fabric to stiffen and strengthen it
- GORE-TEX® is laminated to another fabric to provide a waterproof membrane for outdoor wear
- Foam is laminated to upholstery fabrics for added strength and softness

Research outdoor jackets and produce a consumer report on four jackets for a walking magazine.

Fusing/taping of seams

Fusing by hand with an iron can be used to stick two layers together, for example applying interfacing or using Bondaweb for a hem. Fusing by heat press is used in commercial manufacture.

Heat sealing is applied to fabrics made from thermoplastic fibres such as polyester and polyamide, which can be melted by heat. Heat-sealed seams are often used with tape to help waterproofing on all-weather gear and outdoor equipment.

Quilting

Quilting, as described earlier, will strengthen and make a fabric more durable. It also provides insulation.

Fig 13.81 A quilted bag and oven glove

Find out about tog rating on duvets and produce a guide for consumers.

Interlining

Interlining is a woven, non-woven or warp-kitted fabric attached to the back of another fabric either by fusing or stitching. Its primary purpose is to ensure that the garment maintains its shape. A large range of interfacings are available to suit a wide range of performance characteristics, such as stiffening, shape retention and extensibility.

Disassemble a tie. Examine the interlinings and also the way that the fabric has been cut. Make pattern pieces. Cut a new tie from fabric and interlining. Remember to cut the fabric on the bias and to join on the bias (see Section E Fabric construction).

See www.hodderplus.co.uk/ocrd&t
How to join bias fabrics

Reinforced stitching

Reinforced stitching may be needed in parts of a textile product to enhance its performance. Extra machine stitches are used to strengthen weak areas. For example the top of a jacket pocket is usually finished with a triangle of stitches to prevent the pocket from pulling away from the garment. Figure 13.82 shows how a bag handle is securely attached by reinforced stitching.

Examine a pair of jeans and look at all the places where it has been reinforced to enhance its performance. Draw the design for the jeans and identify the strengths and weaknesses of the product.

Fig 13.82 Reinforced stitching on a bag

Glossary

Advertising – bringing a product (or service) to the attention of potential and current customers. This is done using a range of media including signs, brochures, commercials, direct mailings or e-mail messages. The Advertising Standards Authority (ASA) is an independent regulator for advertisements and promotions in the UK.

Aesthetics – in its widest interpretation this is involved with our senses: vision, hearing, taste, touch, smell and our emotional responses to objects and things. Often associated with an appreciation of beauty.

Annealing – the process using heat to relieve the internal stresses set up in a metal caused by working that material.

Anodising – an electrochemical process used to create a hard surface finish on aluminium.

Anthropometrics – concerned with the measurement of the physical dimensions and shapes of humans. There are two types of measurement: static (structural, for example head width, needed for the design of goggles) and dynamic (when the body is in motion, needed for the clearance heights of door frames).

Automation – a method of production that uses control systems to operate mechanical and/or electronic technologies to do the work of humans.

Batch production – involves the production of a specified quantity of a product. Batches can be repeated as many times as required. This type of production is flexible and can be used to produce batches of similar products with only a small change to the tooling.

Bought-in components – some manufacturers purchase components or sub-assemblies from other companies to speed up production and keep costs low, for example most manufacturers of televisions would buy in loudspeakers.

Brand – the recognised identity of a company that provides a service or manufactures products. The identity is created through the use of a name, slogan, colour scheme or logo. Consumers may develop brand loyalty, where they will continue to purchase goods from a company that is perceived to produce good-quality products, is reliable and offers good value for money.

Brittleness – the opposite of toughness; the likelihood of the material to fracture, crack or break when force is applied.

BSI – British Standards Institution; an independent, non-profit making organisation that provides services including:

- Certification of management systems and products
- Product testing
- Development of private, national and international standards

Standards are codes of best practice that will improve safety and efficiency and will reassure consumers and users of products/services.

CAD/CAM – computer-aided design (CAD) and computer-aided manufacturing (CAM). CAD drawing software is used to describe geometries, which are used by CAM software to define a tool path. This will direct the motion of a computer numerically controlled (CNC) machine tool to create the exact shape that was drawn.

Cell production – a manufacturing system that uses a number of production cells that are grouped together to manufacture a component or sub-assembly of a larger product.

Characteristics – see 'sensory characteristics' and 'performance characteristics'.

CIE – computer-integrated engineering; a system that involves the analysis and simulation of design ideas prior to production.

CIM – computer-integrated manufacture; a manufacturing system in which the whole process from product definition to product manufacture is organised using computers in an integrated and efficient way.

Commercial practice – the range of business processes carried out by a company including marketing, assessing consumer needs, product development, pricing, promotion and distribution. It also includes advertising and legal issues such as design rights and patents.

Continuous flow production – (also called in-line production) a production system where components are processed and moved directly onto the next process one at a time. The components undergo each stage of production sequentially. Often involves uninterrupted, 24/7 production of a basic commodity such as steel, chemicals, oil or basic food products. This type of production is continuous because it is expensive to shut down and then restart.

Copyright – a set of exclusive rights or protection given to creators of original ideas, information or other intellectual works. It is often seen as 'the right to copy' an original creation. Most copyrights are of a limited duration. The symbol for copyright is ©.

COSHH – Control of Substances Hazardous to Health; many workplaces, including schools and colleges, use chemicals or other hazardous substances. COSHH is a law that requires employers to control the exposure to hazardous substances to prevent ill health. It includes:

- Clearly identifying substances
- Carrying out risk assessments on the use and handling of substances
- Correct and secure storage of substances
- Correct disposal of substances

Creativity – the generation and creation of new ideas and concepts. It is associated with imagination and results in novel and appropriate proposals.

Critical path analysis – the identification and prioritisation of the key stages in a manufacturing process.

Cultural influences – an understanding of cultural influences is very important for a product designer, especially in a global marketplace. Designers must consider social trends, diversity, race, ethnicity and religion and the socio-historical context.

Design rights – involve the rights of the originator of a design or designs, unless a third party commissions the work. Unregistered design rights protect the configuration or shape of a marketable (or potentially marketable) product, and are used to prevent copying of an original design without permission. Design rights, as with copyright, can be bought, sold or licensed.

Digital technology – a broad term covering a range of systems that employ digital technologies such as computing, electronics, audio and video through the use of cameras (still and movie), and the internet. Seen as a wider and more accurate version of the term ICT (information and communication technology).

Ductility – the ability of a material to be drawn into a length of material with a certain cross section.

Economies of scale – the savings that are possible when large quantities of products are manufactured. For example, materials may be purchased in bulk at discounted rates, and the set-up costs of equipment and machinery is spread over many more products, lowering the unit cost.

Elasticity – the ability of a material to return to its original shape and length after a force is applied. The elastic limit is the point beyond which permanent deformation takes place.

Electroplating – the use of electrolysis to apply a coating or surface finish to a metal. For example bright zinc-plated screws and fasteners.

Electrostatic spraying – an industrial process where an electrostatic charge is used to ensure an evenly sprayed paint finish. It is used when spraying complex items such as bicycle frames.

Ergonomics – often referred to as human factors, ergonomics is concerned with the design of products, systems and environments to suit people. It uses scientific data from the fields of anthropometry, physiology and psychology to help designers produce efficient and easy-to-use products and comfortable work environments.

Ethical influences – ethics is concerned with moral dilemmas and decision making. Designers have to consider ethical influences and have to make important decisions in their work such as 'will the product create a better world?' 'Am I making best use of resources?' 'Is it sustainable?'

Fashion – can be defined as the latest or current style. Fashions, by definition, can change very quickly. Although usually associated with clothing design, fashion is common to many fields such as hair styling, interior and landscape design, and architecture.

Fusibility – the ability of a material to be welded or to join itself to another material.

Gantt charts – used in project planning where tasks are plotted in sequence against time.

Hardness – the ability of a material to resist scratching, wear or indentation by other materials.

Hardening – hardening of metal (usually high-carbon steel) is the use of heat to increase the hardness of the material. It is usually followed by tempering to reduce brittleness in the material produced during the hardening process.

Health and Safety at Work Act (HASAWA) 1974 – the principal law covering the safety of people affected by work activities. It defines the role of employers to ensure that the following are protected:

- Employees
- Sub-contractors
- Others including visitors, passers-by and members of the public

The agency that enforces this law is the Health and Safety Executive (HSE).

Heat treatment – the use of heat to alter the structure and properties of a metal (see hardening and tempering).

High-volume production – often referred to as 'mass production'. It involves the production of high-demand products such as plastic cups, bottles and pens, which require minimal assembly, in large quantities. It requires expensive, often fully automated machinery.

Inclusive design – helps to ensure that goods, services and environments are accessible to a wider range of people. The very young, very old and disabled members of society are often disadvantaged by non-inclusive design. A good example of inclusive design is the 'Good Grips' range of kitchen products that can be used by a wider range of consumers.

Intellectual property – a broad term to describe the creative outcomes from the mind such as design ideas, written material, artistic and musical composition. There are four main types of intellectual property that can be protected by law: copyright, designs, trademarks and patents.

Innovation – occurs when existing ideas, concepts or inventions are used in a new and different way.

Invention – the first occurrence of an idea or concept for a new process or product. It can be based on a collaboration of existing ideas or concepts but involves a unique quality.

Job bag – a collection of research and resource materials required to help to carry out or solve a task. It is commonly used by product designers working on specific commissions; several software packages are available that include tools and information to carry out specific tasks. A job bag is required in Unit F521 Advanced Innovation Challenge.

Just-in-time (JIT) manufacture – the philosophy of JIT manufacturing is to meet consumer orders with a quality product with minimal delay and effective use of resources. The JIT system is sometimes referred to as lean manufacturing as it focuses on giving customers value for money by reducing wastage and minimising stock of components and finished products.

LCA – life-cycle assessment (sometimes referred to as life-cycle analysis) is a method of assessing the entire environmental impact, energy and resource usage of a material or product. It is sometimes known as a 'cradle-to-grave' analysis and encompasses the entire lifetime of a product or material, from extraction to end-of-life disposal.

Malleability – the ability of a material to be formed into shape by bending, pressing, rolling or hammering, without breaking or fracturing.

Marketing – carried out by companies to ensure that they meet consumer needs. It involves carrying out market research to identify target markets, (a specific cohort of consumers) and looking at what is available from competitors. The pricing, promotion and placement of the product are considered to maximise sales.

Modelling – modelling is used to test and/or communicate design ideas. Modelling can be carried out by hand using appropriate available materials or can be generated by computer. Models can be 2D or 3D.

Modular production – (often referred to as cell production) a manufacturing system that uses use a number of production modules that are grouped together to manufacture a component or sub-assembly of a larger product.

Obsolescence – products become out of date or are replaced by new products. This is when the product becomes obsolete. 'Built-in' obsolescence is often used to give a product a specific life-span beyond which it may be unsafe to use. For example, disposable syringes are designed to be used for one injection only.

One-off production – (often referred to as job production) a manufacturing system used when only one specialist item is required. An example of this could be an individually made piece of jewellery, a specific building project or a unique wedding dress. This type of production is very labour intensive, requires high skill levels and is therefore expensive.

Patent – a patent gives a designer protection against copying of the technical and functional aspects of their invention without permission. It covers details such as how they work, how they are made and what they are made of.

Performance characteristics – relates to the features or characteristics of products, materials or systems and how they perform under certain conditions or as a result of tests.

Primary research – the personal collection of research and information. It is carried out through methods such as visits and observations, interviews, testing and surveys.

Product life cycle – a product passes through a sequence of stages from its introduction as a concept to its final decline and disposal. The stages include conception and introduction, growth, maturity and decline.

Prototype – a working model built to test the function and feel of the new design before production commences. Very often the construction of a full-scale working prototype is the final check for design faults and allows last-minute changes to be made.

Purchasing logistics – the need for companies to be well organised and efficient when researching and negotiating deals with suppliers to ensure best quality, reliability and value for money.

Quality – the quality of a product refers to its collective properties such as: suitability for its stated purpose, value for money, safety, reliability, ease of maintenance and disposal.

Quality assurance – involves looking at quality procedures in all aspects of a manufacturing company to give consumers confidence in their products and/or services offered.

Quality control – carried out by means of a series of checks, tests or inspections to ensure that components or products meet specified standards or tolerances.

Rapid prototyping – a process used to quickly create a solid scale model of a component using 3D CAD data. Systems used include stereo-lithography (SLA), laminated object manufacturing (LOM), selective laser sintering (SLS), fused deposition modelling (FDM) and 3D printing.

Risk assessment – a careful examination of a process or activity that could cause harm to people, to identify hazards and the precautions required. Risks are assessed and control measures are proposed.

Sale of Goods Act 1979 – this Act has been amended three times since 1979, the most recent amendment being in 1995. The Act is designed to protect the fundamental rights of purchasers by ensuring that goods must be:

- Sold as described
- Of satisfactory quality considering factors such as price
- Fit for intended purpose

Scale of production – the numbers of products made; this will depend on the demand for the product. The most common manufacturing systems are one-off or job production, batch production and high-volume production.

Secondary research – the collection, collation and editing of readily available information. Such research would come from sources such as published details, company literature and existing test data.

Sensory characteristics – relates to human perception, through the use of the senses, of the characteristics of products or ingredients. For example, when referring to food products, sensory characteristics such as the taste, aroma, texture, appearance, temperature and sound are considered.

Smart and modern materials – smart materials respond to the environment, for example differences in temperature or light, and change in some way. They are referred to as smart because they sense prevailing conditions and respond. Examples include thermochromic inks. Some smart materials have a 'memory' as they revert back to their original state when conditions change, for example shape memory alloy. Modern materials are developed through the invention of new or improved processes. They are altered to perform a particular function, for example Kevlar®.

Standardised components – those components in regular demand that can be used in a range of products. Examples include machine screws, zips and window frames. Standardised components have guaranteed dimensions and will have passed appropriate quality control standards.

Stock control – careful planning and organisation to ensure that a business has sufficient stock of the right quality available at the right time. It involves controlling:

- The raw materials and supplies required
- Current work in progress
- Completed products ready for despatch

Sustainable design – designs that comply with basic principles of economic, social and ecological sustainability. These principles aim to reduce the use of non-renewable resources, minimise the energy required to manufacture a product and ensure minimal environmental impact when disposing of products.

Target market – the identified market segment expected to purchase a particular product. It is defined by factors such as gender, age and socio-economic grouping.

Tempering – a process using heat to reduce the brittleness produced in a metal during the hardening process.

Tolerance – an allowable variation from a given dimension. The diameter of a shaft may be Ø20 mm to fit into a 20 mm hole. To ensure that they fit, the shaft would have a tolerance of 20 minus 0.05 mm, and the hole would have a tolerance of 20 plus 0.05 mm.

TQM – total quality management; systems employed throughout a company to embed an awareness of quality in every aspect of its operation. Everyone involved in the company has responsibility for quality.

Toughness – the opposite of brittleness; the ability of a material to resist impact forces without breaking.

Trade Descriptions Act 1968 – this Act was introduced to prevent the misleading or false description of goods including:

- Selling goods incorrectly described by manufacturers
- False impressions given by incorrect images
- Incorrect details of sizes, method of manufacture and so on

USP – unique selling proposition; the specific qualities and features of a product that give it an advantage in the marketplace over similar products.

Index

Note: Page numbers in **bold** refer to key terms.